Advanced Chemistry Calculations

Edited by

Alec Thompson MA
Advisory Teacher, ILEA

Lambros Atteshlis BSc
Formerly Advisory Teacher
and member of the Science
Support Team, ILEA

John Murra

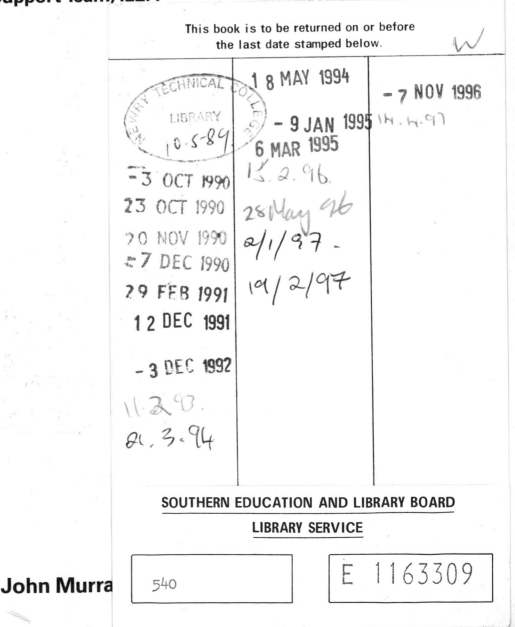

First published 1986
by John Murray (Publishers) Ltd
50 Albemarle Street
London W1X 4BD

Printed in Great Britain by
Martin's of Berwick

British Library Cataloguing in Publication Data

Advanced chemistry calculations.
1. Chemistry   Mathematics
I. Thompson, Alec   II. Atteshlis, Lambros
540'.1'51   QD39.3.M3

ISBN 0-7195-4189-1

PREFACE

This book, which covers the numerical aspects of all the major A-level Chemistry syllabuses, has been compiled largely from material developed as part of the Independent Learning Project for Advanced Chemistry (ILPAC) produced by an Inner London Education Authority team. All the ILPAC materials have been tested in schools and colleges inside and outside London but some of the background theoretical sections have been modified and a few new exercises have been added.

An important feature of the complete ILPAC scheme is that it enables students to work more effectively on their own and at their own pace. This feature has been retained, as far as is possible in a single volume, by developing concepts progressively with simple exercises leading on to more complex problems.

To aid the student's understanding, detailed answers to exercises are given, including full methods of working and some descriptive aspects. At the end of each chapter, a representative collection of problems is presented to test the student's knowledge of a broad topic and, for these questions, only the numerical answers are given.

LONDON 1986

ACKNOWLEDGEMENTS

Thanks are due to the following examination boards for permission to reproduce questions from past A-level papers in the exercises listed below:

Southern Universities Joint Board; 24(1977) 54(1975) 170(1976) 193(1979) 195(1975) 230(1980) 281(1975)

The Associated Examining Board; 167(1975) 190(1974) 212(1976) 227(1979) 314(1980) 320(1980) 321(1978) 326(1980) 332(1980) 337(1979) 342(1979) 344(1981) 382(1981)

University of Cambridge Local Examinations Syndicate; 38(1979) 65(1977) 243(1979) 363(1983)

University of London School Examinations Board; 16(1977) 35(1978) 36(1982) 53(1977) 59(1980) 60(1980) 61(1976) 62(1984) 77(1981) 87(1979) 101(1982) 102(1975) 103(1974) 107(1981) 108(1974) 109(1981) 110(1977) 111(1980) 171(1977) 172(1978) 189(1977) 219(1975) 231(1979) 232(1978) 233(1979) 234(1978) 240(1980) 242(1982) 279(1979) 285(1973) 286(1975) 304(1973) 312(1981) 313(1973) 340(1978) 357(1983) 360(1979) 361(1972) 362(1975) 368(1982) 369(1977) 377(1975) 378(1977) 379(1978) 380(1976) 381(1976) 383(1980) 385(1982)

University of Oxford Delegacy of Local Examinations; 105(1977) 224(1979) 241(1981) 256(1978) 359(1977) 384(1980)

Welsh Joint Education Committee; 93(1977) 97(1976) 98(1978) 317(1979)

Exercises 211, 213, 214 and 225 are reproduced, by permission, from Nuffield Advanced Science: Chemistry; Students' Book I, edited by B.J. Stokes and published by Longman Group Limited for the Nuffield Foundation.

Photograph, page 2 - Tony Langham

Graphics - Vanda Kiernan

Layout - Peter Faldon

Typing - Stella Jefferies

CONTENTS

Chapter 5. Gases 68

Chapter 6. Equilibrium I - principles 89

CHAPTER 7. Equilibrium II - acids and bases 120

INTRODUCTION

This book is more than just a collection of numerical problems. All the relevant theoretical background for solving the problems is given, either as reminders of the basic work you will have already done before you use the book, or in the course of a logical development of the topic.

You will find that the exercises at the beginning of the main sections of each chapter are easy in order to encourage you to work on but, as you do so, they tend to become more difficult. Many A-level questions, either complete or in part, are included but, since syllabuses have changed considerably in recent years, you should check with your teacher whether they, and the accompanying theory, are suitable for you. The same applies to the other exercises but they have been chosen to reflect recent trends in examining.

To help you work effectively on your own we have included worked examples for many different types of question and all the exercises have answers in a separate section at the back of the book (pages 209-282). Most of these answers are very detailed and include a full account of the method of working. Use them wisely - if you have difficulty with a problem, and you cannot relate it to a worked example, look at the first part of our answer just to help you get started and then try again.

You will need a data book for many of the exercises. Do not be surprised, however, if you find that your book gives values for some physical quantities which differ slightly from those we have used in our answers. Refinement of techniques means that data is constantly being revised and universal agreement is, therefore, not possible.

At the end of each chapter you will find a selection of questions which we feel are representative of the whole topic. You can use these as a test of your overall knowledge of the topic - we give only the numerical answers for those exercises.

The symbols below, together with the contents list on pages iv, v and vi, should help you to find your way around the book.

ACTIVITY SYMBOLS

W	Worked Example	A	A-level Question
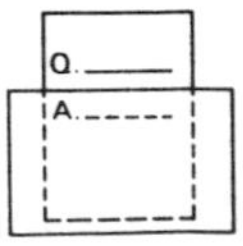	Revealing Exercise	A (part)	A-level Question (part only)
	Written Exercise	A (S)	A-level Question (Special paper)

AMOUNT OF SUBSTANCE – THE MOLE.

1

INTRODUCTION AND PRE-KNOWLEDGE

Everyday objects are often counted in unit amounts; for example, we buy a pair of socks, a dozen eggs and a ream (500 sheets) of paper. For convenience, the smaller the object, the larger the number in a unit amount. In counting atoms we also need a convenient unit, large enough to be seen and handled. Since atoms are so small, there are a great many of them in a convenient unit.

Fig.1.

The counting unit for atoms, molecules and ions is the mole (abbreviation: mol). It is defined as the amount of substance that contains as many elementary particles as there are atoms in 0.012 kg (12 g) of carbon-12. You must learn this definition.

The mass of an atom of carbon-12 is 1.99252×10^{-23} g. So the number of atoms in 12 g of carbon-12 is given by

$$\frac{12 \text{ g}}{1.99252 \times 10^{-23} \text{ g}} = 6.02252 \times 10^{23}$$

Note that, in this context, 12 is taken to be an integer and it does not, therefore, limit the significant figures in the answer to two. (See page 206.)

The Avogadro constant (symbol: L) relates the number of particles to the amount. It is represented as $L = 6.02252 \times 10^{23}\ \text{mol}^{-1}$ or, to three significant figures,

$$L = 6.02 \times 10^{23}\ \text{mol}^{-1}$$

The following examples try to convey the magnitude of the Avogadro constant:

If you had 6.02×10^{23} tiny grains of pollen they would cover the City of London to a depth of 1 mile.

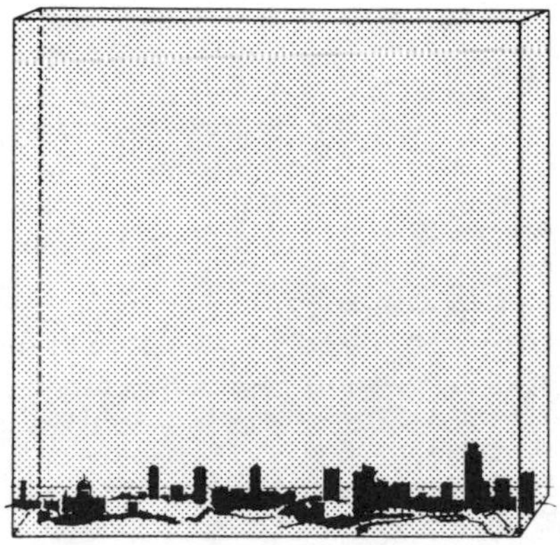

Fig.2

6.02×10^{23} marshmallows spread over the United States of America would yield a blanket more than 600 miles deep!

Computers can count about 10 million times per second. At this rate 6.02×10^{23} counts would require almost 2 billion years!

A similar example is included in the next exercise.

Exercise 1 A 5 cm^3 spoon can hold 1.67×10^{23} molecules of water.

(a) How long would it take to remove them one at a time at a rate of one molecule per second?

(b) How many years is this?

Molar mass

The mass per unit amount of substance is called its molar mass (symbol: M) and is the mass per mole of that substance. We usually use the unit: g mol^{-1}.

The molar mass of an element is the mass per mole. It follows from the definition of the mole that the molar mass of carbon is 12.0 g mol^{-1}. Similarly, since $M_r(\text{U}) = 238$, $M(\text{U}) = 238$ g mol^{-1}.

The term molar mass applies not only to elements in the atomic state but also to all chemical species - atoms, molecules, ions, etc.

For ethane, the molar mass is calculated from its formula, C_2H_6, which indicates that one molecule contains two atoms of carbon and six atoms of hydrogen. The relative atomic masses are: C = 12.0, H = 1.0. The relative mass of a molecule on the same scale is, therefore, given by:

$$C_2H_6$$

$$(2 \times 12.0) + (6 \times 1.0) = 30$$

Thus, the relative molecular mass of ethane is 30.0 and its molar mass, *M*, is 30.0 g mol^{-1}.

There are two important points which you must bear in mind when dealing with amounts of substances:

1. You must specify exactly what entity the amount refers to . The phrase '1 mol of chlorine', for instance, has two possible meanings because it does not specify whether it refers to atoms or molecules. To avoid confusion, you must always specify the entity, either by formula or in words:

   1.0 mol of Cl <u>or</u> one mole of chlorine atoms

   1.0 mol of Cl_2 <u>or</u> one mole of chlorine molecules

2. By weighing out the same number of grams as the relative atomic mass or the relative molecular mass (whether atoms, molecules or ions) you have measured out one mole, i.e. 6.02×10^{23} atoms, molecules or ions.

The following two exercises test your understanding of these concepts. You will need to refer to the table of relative atomic masses at the back of this book.

<u>Exercise 2</u> Calculate the molar masses of the following substances:

(a) ammonia, NH_3

(b) calcium bromide, $CaBr_2$

(c) phosphoric(V) acid, H_3PO_4

(d) sodium sulphate-10-water, $Na_2SO_4 \cdot 10H_2O$

<u>Exercise 3</u> Calculate the mass of 1.00 mol of:

(a) chlorine atoms

(b) chlorine, Cl_2

(c) phosphorus, P

(d) phosphorus, P_4

(e) iodide ions, I^-

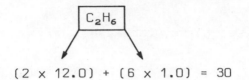

AMOUNT CALCULATIONS

The amount of substance is defined as the mass of the sample divided by the mass per mole (i.e. the molar mass):

$$\text{amount of substance} = \frac{\text{mass}}{\text{molar mass}}$$

The symbol for amount is n and, therefore, in symbols:

$$n = \frac{m}{M}$$

The following Worked Example shows you how to use this expression in mole calculations.

Reminder: the unit for amount is mole, for mass is gram, and for molar mass is grams per mole (abbreviation: g mol^{-1}).

Worked Example A sample of carbon weighs 180 g. What amount of carbon is present?

W

Solution

Calculate the mount by substituting into the key expression:

$$n = \frac{m}{M}$$

where m = 180 g and M = 12.0 g mol^{-1}

$$\therefore\ n = \frac{m}{M} = \frac{180\ \text{g}}{12.0\ \text{g mol}^{-1}} = \boxed{15.0\ \text{mol}}$$

At this point you should make sure that you are not using formulae in a mechanical way but understand what you are doing. So, here is an alternative method which relies on first principles. YOU SHOULD NEVER SUBSTITUTE INTO AN EXPRESSION WITHOUT FIRST UNDERSTANDING HOW IT WAS DERIVED OR DEFINED.

Alternative solution

1. Write down the mass of 1.00 mol of the substance in the form of a sentence:

   12.0 g is the mass of 1.00 mol of carbon.

2. Scale down to the amount of carbon in 1.00 g by dividing by 12.0 throughout

   $\frac{12.0}{12.0}$ g is the mass of $\frac{1.00}{12.0}$ mol of carbon

3. Scale up to the amount of carbon in 180 g by multiplying by 180 throughout

   $180 \times \frac{12.0}{12.0}$ g is the mass of $180 \times \frac{1.00}{12.0}$ mol of carbon

   i.e. 180 g is the mass of 15.0 mol of carbon

   $\therefore$ the amount of carbon in 180 g = $\boxed{15.0\ \text{mol}}$

Now attempt the following exercises. (We use the method of substituting into the expression in our answers, largely because it takes up less space. You can try the 'sentence method' if you get stuck.)

Exercise 4 Calculate the amount in each of the following:

(a) 30.0 g of oxygen molecules, O_2

(b) 31.0 g of phosphorus molecules, P_4

(c) 50.0 g of calcium carbonate, $CaCO_3$

You must also be prepared for questions which ask you to calculate the mass of substance in a given amount. To do this in the following exercises, use the expression:

$$n = \frac{m}{M} \quad \text{in the form} \quad m = nM$$

Exercise 5 Calculate the mass of each of the following:

(a) 1.00 mol of hydrogen, H_2

(b) 0.500 mol of sodium chloride, NaCl

(c) 0.250 mol of carbon dioxide, CO_2

Exercise 6 A sample of ammonia, NH_3, weighs 1.00 g.

(a) What amount of ammonia is contained in this sample?

(b) What mass of sulphur dioxide, SO_2, contains the same number of molecules as there are in 1.00 g of ammonia?

We now consider another type of amount calculation.

Calculating the number of particles in a given amount

Sometimes you are required to calculate the number of particles in a given amount of substance. This is easy because you know the number of particles in 1.00 mol (6.02×10^{23}). Use the expression:

$$\boxed{N = nL}$$

where N = the number of particles, n = the amount, L = the Avogadro constant.

Exercise 7 Calculate the number of atoms in:

(a) 18.0 g of carbon, C

(b) 18.0 g of copper, Cu

(c) 7.20 g of sulphur, S_8

Exercise 8 Calculate the number of molecules in:

(a) 1.00 g of ammonia, NH_3

(b) 3.28 g of sulphur dioxide, SO_2

(c) 7.20 g of sulphur, S_8

Exercise 9 Calculate the number of ions present in:

(a) 0.500 mol of sodium chloride, NaCl (Na^+, Cl^-)

(b) 14.6 g of sodium chloride, NaCl

(c) 18.5 g of calcium chloride, $CaCl_2$

STOICHIOMETRY

We now consider how we can use the mole concept to calculate how much of one substance will react with a given amount of another - this is called stoichiometry and is one of the main reasons why the mole concept is so important in the study of chemistry.

Stoichiometry (pronounced stoy-key-om-i-tree) in its broadest sense includes all the quantitative relationships in chemical reactions. It has to do with how much of one substance will react with a specified amount of another. A chemical equation such as

$$N_2(g) + 3H_2(g) \rightarrow 2NH_3(g)$$

is a kind of chemical balance sheet; it states that one mole of nitrogen reacts with three moles of hydrogen to yield two moles of ammonia. (It does not tell us about the rate of the reaction or the conditions necessary to bring it about.) The numbers 1, 3 and 2 are called the stoichiometric coefficients. Such an equation is an essential starting point for many experiments and calculations; it tells us the proportions in which the substances react and in which the products are formed.

We start with a Worked Example.

Worked Example What mass of iodine will react completely with 10.0 g of aluminium?

W

This problem is a little more complicated than those you have done previously, because it involves several steps. Each step is very simple, but you may not immediately see where to start. Before we present a detailed solution, let us look at a way of approaching multi-step problems. Even if you find this problem easy, the approach may be useful to you in more difficult problems. We suggest that you ask yourself three questions:

1. What do I know? In this case, the answer should be:

   (a) the equation for the reaction;

   (b) the mass of aluminium.

In some problems you may be given the equation; in this one, you are expected to write it down from your general chemical knowledge. In nearly every problem about a reaction, the equation provides vital information.

2. What can I get from what I know?

   (a) From the equation, I can find the ratio of reacting amounts.

   (b) From the mass of aluminium, I can calculate the amount, provided I look up the molar mass.

3. Can I now see how to get the final answer?

   In most cases the answer will be 'Yes', but you may have to ask the second question again, now that you know more than you did at the start.

   (a) From the amount of aluminium and the ratio of reacting amounts, I can calculate the amount of iodine.

   (b) From the amount of iodine, I can get the mass, using the molar mass.

Instead of writing answers to these questions, you can summarise your thinking in a flow diagram.

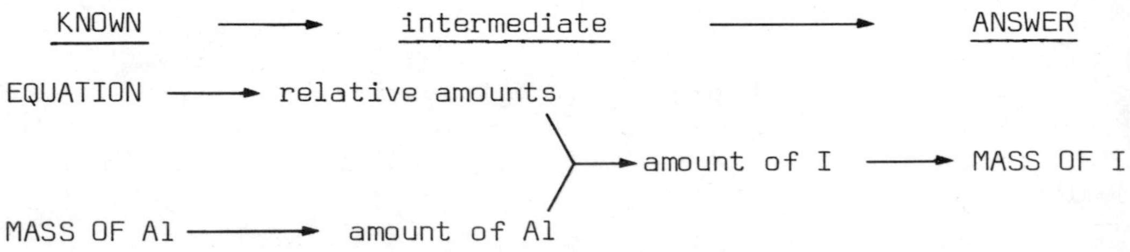

An alternative strategy is to work backwards from the answer towards the given information, or you may use a combination of both strategies, meeting up in the middle. In any case, the steps you use will probably be the same, although the order in which you take them may be different.

Now we go through each step, using the first strategy, and doing the necessary calculations.

Solution

1. Write the balanced equation for the reaction:

$$2Al(s) + 3I_2(s) \rightarrow 2AlI_3(s)$$

   This equation tells us that 2 mol of Al reacts with 3 mol of I_2; so we write the ratio:

$$\frac{\text{amount of Al}}{\text{amount of } I_2} = \frac{2}{3}$$

2. Calculate the amount of aluminium using the expression:

$$n = \frac{m}{M}$$

$$\therefore\ n = \frac{10.0\ \text{g}}{27.0\ \text{g mol}^{-1}} = 0.370\ \text{mol}$$

3. Calculate the amount of iodine which reacts with this amount of aluminium by substituting into the expression based on the equation:

$$\frac{\text{amount of Al}}{\text{amount of } I_2} = \frac{2}{3}$$

$$\therefore \quad \text{amount of } I_2 = \frac{3}{2} \times \text{amount of Al}$$

$$= \frac{3}{2} \times 0.370 \text{ mol} = 0.555 \text{ mol}$$

4. Calculate the mass of iodine from the amount using the expression:

$$n = \frac{m}{M} \quad \text{in the form} \quad m = nM$$

$$\therefore \quad m = nM = 0.555 \text{ mol} \times 254 \text{ g mol}^{-1} = \boxed{141 \text{ g}}$$

Now try some similar problems for yourself.

<u>Exercise 10</u> (a) What mass of magnesium would react completely with 16.0 g of sulphur?

$$Mg(s) + S(s) \rightarrow MgS(s)$$

(b) What mass of oxygen would be produced by completely decomposing 4.25 g of sodium nitrate, $NaNO_3$?

$$2NaNO_3(s) \rightarrow 2NaNO_2(s) + O_2(g)$$

<u>Exercise 11</u> What mass of phosphorus(V) oxide, P_2O_5, would be formed by complete oxidation of 4.00 g of phosphorus?

$$4P(s) + 5O_2(g) \rightarrow 2P_2O_5(s)$$

<u>Exercise 12</u> When 0.27 g of aluminium is added to excess copper(II) sulphate solution, 0.96 g of copper is precipitated. Deduce the equation for the reaction which takes place.

In the next exercise you have to decide first which one of two reactants is present in excess.

<u>Exercise 13</u> A mixture containing 2.80 g of iron and 2.00 g of sulphur is heated together. What mass of iron(II) sulphide, FeS, is produced?

$$Fe(s) + S(s) \rightarrow FeS(s)$$

Another important application of the mole concept is in the calculation of the empirical formulae of substances.

CALCULATING EMPIRICAL FORMULAE AND MOLECULAR FORMULAE

The empirical formula of a compound is the simplest form of the ratio of the atoms of different elements in it. The molecular formula tells the actual number of each kind of atom in a molecule of the substance. For example, the molecular formula of phosphorus(V) oxide (in the gas phase) is P_4O_{10}, whereas its empirical formula is P_2O_5.

Calculating empirical formulae from the masses of constituents

To determine the empirical formula of a compound, we must first calculate the amount of each substance present in a sample and then calculate the simplest whole-number ratio of the amounts.

It is convenient to set the results out in tabular form and we suggest that you use this method. However, in the following Worked Example, we will go through the procedure step by step, establishing the table as we go.

W

Worked Example An 18.3 g sample of a hydrated compound contained 4.0 g of calcium, 7.1 g of chlorine and 7.2 g of water only. Calculate its empirical formula.

Solution

1. List the mass of each component and its molar mass. Although water is a molecule, in the calculation we treat it in the same way as we do atoms.

	Ca	Cl	H_2O
Mass/g	4.0	7.1	7.2
Molar mass/g mol^{-1}	40.0	35.5	18.0

2. From this information calculate the amount of each substance present using the expression:

$$n = \frac{m}{M}$$

	Ca	Cl	H_2O
Amount/mol	$\frac{4.0}{40.0} = 0.10$	$\frac{7.1}{35.5} = 0.20$	$\frac{7.2}{18.0} = 0.40$

(For simplicity, we have omitted the units from the calculation - they cancel anyway.)

This result means that in the given sample there is 0.10 mol of calcium, 0.20 mol of chlorine and 0.40 mol of water.

3. Calculate the <u>relative</u> amount of each substance by dividing each amount by the smallest amount.

	Ca	Cl	H_2O
Amount/smallest amount = relative amount	$\frac{0.10}{0.10} = 1.0$	$\frac{0.20}{0.10} = 2.0$	$\frac{0.40}{0.10} = 4.0$

The relative amounts are in the simple ratio 1:2:4.

From this result, you can see that the empirical formula is $CaCl_2{\cdot}4H_2O$

To see if you understand this procedure, try Exercise 14.

<u>Exercise 14</u> A sample of a hydrated compound was analysed and found to contain 2.10 g of cobalt, 1.14 g of sulphur, 2.28 g of oxygen and 4.50 g of water. Calculate its empirical formula.

A modification of this type of problem is to determine the ratio of the amount of water to the amount of anhydrous compound. You have practice in this type of problem in the next exercise.

<u>Exercise 15</u> 10.00 g of hydrated barium chloride are heated until all the water is driven off. The mass of anhydrous compound is 8.53 g. Determine the value of x in $BaCl_2{\cdot}xH_2O$.

You should be prepared for variations to this type of problem. The following part of an A-level question illustrates such a variation.

<u>Exercise 16</u> When 585 mg of the salt $UO(C_2O_4){\cdot}6H_2O$ was left in a vacuum desiccator for forty-eight hours, the mass changed to 535 mg. What formula would you predict for the resulting substance?

Now look at another way of calculating empirical formulae.

Calculating empirical formulae from percentage composition by mass

The result of the analysis of a compound may also be given in terms of the percentage composition by mass. Study the following Worked Example which deals with this type of problem.

<u>Worked Example</u> An organic compound was analysed and was found to have the following percentage composition by mass: 48.8% carbon, 13.5% hydrogen and 37.7% nitrogen. Calculate the empirical formula of the compound.

Solution

If we assume the mass of the sample is 100.0 g, we can write immediately the mass of each substance: 48.8 g of carbon, 13.5 g of hydrogen and 37.7 g of nitrogen. Then we set up a table as before. The instructions between each step are omitted this time, but you should check our calculations.

	C	H	N
Mass/g	48.8	13.5	37.7
Molar mass/g mol^{-1}	12.0	1.00	14.0
Amount/mol	4.07	13.5	2.69
$\frac{\text{Amount}}{\text{Smallest amount}}$	$\frac{4.07}{2.69} = 1.51$	$\frac{13.5}{2.69} = 5.02$	$\frac{2.69}{2.69} = 1.00$
Simplest ratio of relative amounts	3	10	2

Empirical formula = $C_3H_{10}N_2$

In the preceding exercises you 'rounded off' values close to whole numbers in order to get a simple ratio. This is justified because small differences from whole numbers are probably due to experimental errors. Here, however, we cannot justify rounding off 1.51 to 1 or 2, but we can obtain a simple ratio by multiplying the relative amounts by two.

Now attempt the following exercises where you must decide whether to round off or multiply by a factor.

Exercise 17 A compound of carbon, hydrogen and oxygen contains 40.0% carbon, 6.6% hydrogen and 53.4% oxygen. Calculate its empirical formula.

Exercise 18 Determine the formula of a mineral with the following mass composition: Na = 12.1%, Al = 14.2%, Si = 22.1%, O = 42.1%, H_2O = 9.48%.

Exercise 19 A 10.00 g sample of a compound contains 3.91 g of carbon, 0.87 g of hydrogen and the remainder is oxygen. Calculate the empirical formula of the compound.

Empirical formulae from quantitative analysis

When an organic compound is burned in an excess of oxygen, the carbon is converted to carbon dioxide and the hydrogen to water. Other methods must be used to determine nitrogen, sulphur and halogen content. When oxygen is present, the amount must be determined by difference. The entire process is called 'quantitative analysis' or 'elemental analysis'.

In the following Worked Example we show you how to determine an empirical formula from the results of such an analysis.

Worked Example A 1.00 g sample of a compound was burnt in an excess of oxygen. The water and carbon dioxide produced were absorbed in previously weighed tubes containing anhydrous calcium chloride and soda-lime respectively. The tubes were found to have gained 2.52 g of carbon dioxide and 0.443 g of water. Determine the empirical formula of the compound, assuming that it contains carbon, hydrogen and oxygen only.

W

Solution

1. Calculate the amount of carbon in the sample from the mass of carbon dioxide absorbed.

$$\text{Amount of } CO_2 = \frac{\text{mass}}{\text{molar mass}} = \frac{2.52 \text{ g}}{44.0 \text{ g mol}^{-1}} = 0.0573 \text{ mol}$$

$\therefore$ amount of C = 0.0573 mol

2. Calculate the amount of hydrogen in the sample from the mass of water absorbed.

$$\text{Amount of } H_2O = \frac{\text{mass}}{\text{molar mass}} = \frac{0.443 \text{ g}}{18.0 \text{ g mol}^{-1}} = 0.0246 \text{ mol}$$

$\therefore$ amount of H = 2 x 0.0246 mol = 0.0492 mol

3. Now calculate the mass of carbon and mass of hydrogen.

Mass of C = amount x molar mass = 0.0573 mol x 12.0 g mol^{-1} = 0.688 g

Mass of H = amount x molar mass = 0.0492 mol x 1.00 g mol^{-1} = 0.0492 g

4. The mass of oxygen = mass of sample - (mass of C + mass of H)

= 1.00 g - (0.688 g + 0.0492 g) = 0.263 g

5. Now construct a table as in the previous worked examples.

	C	H	O
Mass/g			0.263
Molar mass/g mol^{-1}			16.0
Amount/mol	0.0573	0.0492	0.0164
$\frac{\text{Amount}}{\text{Smallest amount}}$	$\frac{0.0573}{0.0164} = 3.49$	$\frac{0.0492}{0.0164} = 3.00$	$\frac{0.0164}{0.0164} = 1.00$
Simplest ratio of relative amounts	7	6	2

Empirical formula = $C_7H_6O_2$

Use similar methods in the next two exercises.

Exercise 20 Complete combustion of a hydrocarbon, Z, gives 0.66 g of carbon dioxide and 0.36 g of water. (Relative atomic masses: H = 1.0, C = 12, O = 16.)

What is the empirical formula of Z?

Exercise 21 The combustion of 0.146 g of compound B gave 0.374 g of carbon dioxide and 0.154 g of water. Assuming that B contains carbon, hydrogen and oxygen only, determine its empirical formula.

As you will have found out from your reading, one of the ways of determining the halogen content of a compound is to make from it a precipitate of silver halide, which is filtered off, washed, dried and weighed. The next exercise includes information which will enable you to determine the empirical formula of a halogeno-compound.

Exercise 22 0.2243 g of compound C gave 0.3771 g of carbon dioxide, 0.0643 g of water and 0.2685 g of silver bromide. Determine the empirical formula of the compound.

Having established the empirical formula of a compound, the next step is to determine its molecular formula.

Molecular formulae from empirical formulae

In order to determine the molecular formula of a compound we need to know not only its empirical formula but also its molar mass. This can be determined for volatile substances from measurements of mass and volume, using the method described in Chapter 5 (Gases). You may wish to refer to this method before attempting the exercises.

To show you how to use empirical formula and molar mass to determine molecular formula, we give you a simple Worked Example.

Worked Example The empirical formula of compound X is found from quantitative analysis to be C_2H_5. Vapour density measurements show that the relative molecular mass is 58. What is the molecular formula of X?

Solution

1. Determine the relative molecular mass corresponding to the empirical formula, C_2H_5.

$$M_r(C_2H_5) = (2 \times 12) + (5 \times 1) = 29$$

2. Substitute into the expression: molecular formula = (empirical formula)$_n$

$$\text{where } n = \frac{M_r(\text{compound X})}{M_r(C_2H_5)} = \frac{58}{29} = 2$$

Molecular formula = $(C_2H_5)_2$ = $\boxed{C_4H_{10}}$

You should now be able to do the next exercise.

Exercise 23 The empirical formula of a compound is CH_2. At 100 °C, 0.24 g of this compound occupied 134 cm^3 at a pressure of 0.98 atm. (R = 0.0821 atm dm^3 K^{-1} mol^{-1}.) Calculate:

(a) the molar mass of the compound;

(b) its molecular formula.

You are not always required to calculate the molar mass of the compound - it may be given to you as it is in the next exercise.

Exercise 24 The organic compound X, which contains carbon, hydrogen and oxygen only, was found to have a relative molecular mass of about 85. When 0.43 g of X is burned in excess oxygen, 1.10 g of carbon dioxide and 0.45 g of water are formed. (H = 1, C = 12, O = 16.)

(a) What is the empirical formula of X?

(b) What is the molecular formula of X?

(c) Write an equation for the complete combustion of X.

We now show you how to apply the mole concept to reactions in solution.

AMOUNTS IN SOLUTION

So far, we have shown how to calculate the amount of substance from the mass of a substance and its molar mass. However, most chemical reactions take place in solution. If we want to know the amount of substance in solution, then we must know the concentration of the solution and its volume.

Concentration of solution

We express concentration of a solution as the amount of solute dissolved in a given volume of solution; i.e.

$$\text{concentration} = \frac{\text{amount}}{\text{volume}} \quad \text{or, in symbols,} \quad c = \frac{n}{V}$$

Normally, we measure amount in mol, and volume in dm^3, so the usual unit of concentration is mol dm^{-3}.

Standard solutions

A standard solution is one of known concentration. We can 'know' the concentration either by preparing the solution according to a given recipe or by analysing it.

Let us suppose we dissolve 0.15 mol (27.0 g) of glucose, $C_6H_{12}O_6$, in enough water to make 1.00 dm^3 of solution. Then its concentration is given as c = 0.15 mol dm^{-3}.

The letter M is sometimes used as an abbreviation for mol dm^{-3}, but you should <u>only</u> use it in conjunction with a formula. For example, 0.0100 M NaOH means a solution of sodium hydroxide, NaOH, having a concentration of 0.0100 mol dm^{-3}.

In this section, we go through the main types of problems you are likely to meet which involve solutions. For each one, read the Worked Example, then try the exercises which follow it.

Calculating concentration from volume and amount

<u>Worked Example</u> Calculate the concentration of a solution which is made by dissolving 0.500 mol of sodium hydroxide, NaOH, in 200 cm^3 of solution.

<u>Solution</u>

Calculate the concentration by substituting into the key expression:

$$c = \frac{n}{V}$$

where n = 0.500 mol and $V = \left(\frac{200}{1000}\right) dm^3$

$$\text{i.e.} \quad c = \frac{n}{V} = \frac{0.500\ \text{mol}}{0.200\ \text{dm}^3} = \boxed{2.50\ \text{mol dm}^{-3}}$$

Now attempt the following exercise.

<u>Exercise 25</u> Assume that 0.100 mol of $CuSO_4 \cdot 5H_2O$ is placed in each of the volumetric flasks shown (Fig. 3) and is properly diluted to the volumes shown. Calculate the concentration of each solution.

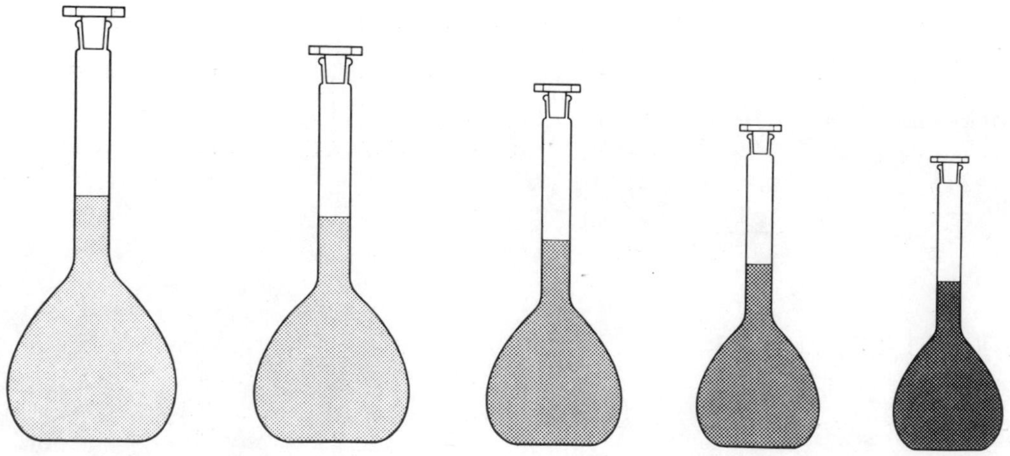

Fig.3 (a) V = 2000 cm^3 (b) V = 1000 cm^3 (c) V = 500 cm^3 (d) V = 250 cm^3 (e) V = 100 cm^3

Calculating concentration from mass of solute and volume

Now we take this calculation a step further - be prepared for problems which give the mass of a substance, not the amount. These need an extra step at the start, i.e. dividing mass by molar mass to get amount using

$$n = \frac{m}{M}$$

Now attempt the following exercise.

Exercise 26 The table below indicates the masses of various compounds that were used to prepare the solutions of the stated volumes. Calculate the concentration of each solution.

Compound	Mass/g	Volume/cm^3
(a) $AgNO_3$	8.50	1000
(b) KIO_3	10.7	250
(c) $Pb(NO_3)_2$	11.2	50.0
(d) $K_2Cr_2O_7$	14.3	250
(e) $CuSO_4 \cdot 5H_2O$	11.9	500

Calculating the amount of substance in a solution

Often you want to know the amount of substance contained in a given volume of solution of known concentration.

To do this, you substitute values for c and V into the expression:

$$c = \frac{n}{V} \text{ in the form } n = cV$$

You use this in the next exercise.

Exercise 27 Calculate the amount of solute in each of the following solutions:

(a) 4.00 dm^3 of 5.00 M NaOH

(b) 1.00 dm^3 of 2.50 M HCl

(c) 20.0 cm^3 of 0.439 M HNO_3

Now we take this calculation a step further, calculating the mass of solute contained in a given volume of solution of known concentration. This needs an extra step at the finish, i.e. multiplying amount by molar mass to get mass $m = nM$.

Now try the next exercise.

Exercise 28 Calculate the mass of solute in the following solutions:

(a) 1.00 dm^3 of 0.100 M NaCl

(b) 500 cm^3 of 1.00 M $CaCl_2$

(c) 250 cm^3 of 0.200 M $KMnO_4$

(d) 200 cm^3 of 0.117 M NaOH

END-OF-CHAPTER QUESTIONS

Exercise 29 What mass of material is there in each of the following?

(a) 2.00 mol of SO_3

(b) 0.0300 mol of Cl

(c) 9.00 mol of SO_4^{2-}

(d) 0.150 mol of $MgSO_4 \cdot 7H_2O$

Exercise 30 What amount of each substance is contained in the following?

(a) 31.0 g of P_4

(b) 1.00 x 10^{22} atoms of Cu

(c) 70.0 g of Fe^{2+}

(d) 9.00 x 10^{24} molecules of C_2H_5OH

Exercise 31 The mass of one molecule of a compound is 2.19 x 10^{-22} g. What is the molar mass of the compound?

Exercise 32 What mass of aluminium, Al, is required to produce 1000 g of iron, Fe, according to the equation:

$$3Fe_3O_4(s) + 8Al(s) \rightarrow 4Al_2O_3(s) + 9Fe(s) \ ?$$

Exercise 33 A solution is made containing 2.38 g of magnesium chloride, $MgCl_2$, in 500 cm^3 of solution.

(a) What is the concentration of magnesium chloride, $MgCl_2$, in this solution?

(b) What is the concentration of chloride ions in this solution?

Exercise 34 A solution is made by dissolving 8.50 g of sodium nitrate, $NaNO_3$, and 16.40 g of calcium nitrate, $Ca(NO_3)_2$, in enough water to give 2000 cm^3 of solution.

(a) What is the concentration of sodium ions in the solution?

(b) What is the concentration of nitrate ions in the solution?

Exercise 35 A hydrated aluminium sulphate, $Al_2(SO_4)_3 \cdot xH_2O$, contains 8.20% of aluminium by mass. Find the value of x.

A part

Exercise 36 A compound, P, contains carbon 59.4%, hydrogen 10.9%, nitrogen 13.9% and oxygen 15.8%, by mass. Calculate its empirical formula. (Relative atomic masses: H = 1, C = 12, N = 14, O = 16.)

A

Exercise 37 An organic compound A, has a relative molecular mass of 178 and contains 74.2% C, 7.9% H and 17.9% O. Determine:

A

(a) the empirical formula of A;

(b) the molecular formula of A.

Exercise 38 Two substances, Q and R, each have the elemental composition C = 60.0%; H = 13.3%; O = 26.7% (by mass).

A part

0.60 g of each substance occupied a volume of 0.336 litre (dm^3) at 137 °C and standard pressure.
(R = 0.0821 atm dm^3 K^{-1} mol^{-1}.)

(a) Calculate the empirical formula and the relative molecular mass of Q and R.

(b) What is their molecular formula?

Exercise 39 Compound A is a liquid hydrocarbon of molecular weight (relative molecular mass) 78 and contains 92.3% carbon. What is its molecular formula?

VOLUMETRIC ANALYSIS (TITRIMETRY)

2

INTRODUCTION AND PRE-KNOWLEDGE

A titration is a laboratory procedure in which a measured volume of one solution is added to a known volume of another reagent until the reaction is complete. This operation is an example of volumetric (titrimetric) analysis. The stoichiometric point (equivalence point) is usually shown by the colour change of an indicator, and is then known as the end-point.

Volumetric analysis is a powerful technique which is used in a variety of ways by chemists in many different fields.

We consider three types of titration in this chapter:

(a) an acid-base titration,

(b) a redox titration,

(c) a precipitation titration.

ACID-BASE TITRATIONS

We start with a Worked Example to illustrate titrimetric calculations. For this, we use an acid-base titration, but the method is applicable to all titrations.

Worked Example 20.0 cm^3 of a solution of barium hydroxide, $Ba(OH)_2$, of unknown concentration is placed in a conical flask and titrated with a solution of hydrochloric acid, HCl, which has a concentration of 0.0600 mol dm^{-3}. The volume of acid required is 25.0 cm^3. Calculate the concentration of the barium hydroxide solution.

W

$$Ba(OH)_2(aq) + 2HCl(aq) \rightarrow BaCl_2(aq) + 2H_2O(l)$$

Solution

This is a multi-step calculation. You may find it helpful to look again at the advice we gave on such calculations on pages 7 and 8. A summarising 'flow-chart' for the solution to this problem is:

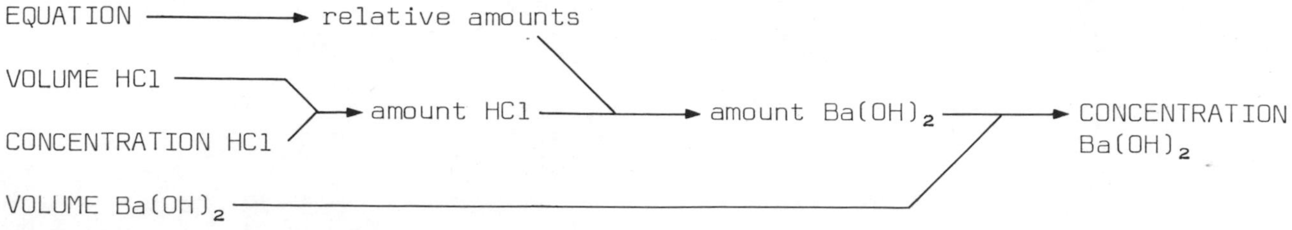

Now we look at the calculations, step by step.

1. Calculate the amount of HCl delivered (the solution of known concentration) by substitution into the expression

$$c = \frac{n}{V} \quad \text{in the form} \quad n = cV$$

where c = 0.0600 mol dm^{-3} and V = (25.0/1000)dm^3 = 0.0250 dm^3

$\therefore n = cV = 0.0600 \text{ mol dm}^{-3} \times 0.0250 \text{ dm}^3 = 1.50 \times 10^{-3}$ mol

2. Calculate the amount of $Ba(OH)_2$ (the solution of unknown concentration) which reacts with this amount of HCl by substituting into the expression derived from the equation:

$$\frac{\text{amount of } Ba(OH)_2}{\text{amount of HCl}} = \frac{1}{2}$$

$$\begin{aligned} \therefore \quad \text{amount of } Ba(OH)_2 &= \tfrac{1}{2} \text{ amount of HCl*} \\ &= \tfrac{1}{2}(1.50 \times 10^{-3} \text{ mol}) \\ &= 7.50 \times 10^{-4} \text{ mol} \end{aligned}$$

*Always check this step carefully - it is easy to put the '½' in the wrong place.

3. Calculate the concentration of $Ba(OH)_2$ by substitution into the expression

$$c = \frac{n}{V}$$

where n = 7.50 x 10^{-4} mol and V = 0.0200 dm^3

$$\therefore c = \frac{n}{V} = \frac{7.50 \times 10^{-4} \text{ mol}}{0.0200 \text{ dm}^3} = \boxed{0.0375 \text{ mol dm}^{-3}}$$

It is possible to derive a general expression to solve this problem and many like it. Remember, however, you should not use an expression unless you understand its derivation.

A general expression

Consider the general equation:

$$a\text{ A} + b\text{ B} \rightarrow \text{Products}$$

We can derive an expression relating concentration of A (c_A), concentration of B (c_B), volume of A (V_A), volume of B (V_B) and the stoichiometric coefficients a and b.

We start with a relation obtained directly from the chemical equation. It is:

$$\frac{\text{amount of A}}{\text{amount of B}} = \frac{a}{b}$$

We know that: amount of A = $c_A V_A$

and amount of B = $c_B V_B$

So, substituting for amount of A and amount of B in the first expression we get:

$$\boxed{\frac{c_A V_A}{c_B V_B} = \frac{a}{b}}$$

To help you use this expression correctly, remember that both A and a are on the top and both B and b are on the bottom.

This expression is very useful for solving titrimetric problems and we use it in our answers to such problems wherever appropriate. However, it may not always provide the best way to tackle a particular problem.

Now we give you some practice in doing volumetric calculations.

Exercise 40 (a) Solve the Worked Example on page 20 using the expression we derived above.

(b) Why is it not necessary to convert volumes in cm^3 to dm^3 when using the expression?

Exercise 41 What volume of sodium hydroxide solution, 0.500 M NaOH, is needed to neutralize each of the following solutions?

(a) 50.0 cm^3 of nitric acid, 0.100 M HNO_3

$$NaOH(aq) + HNO_3(aq) \rightarrow NaNO_3(aq) + H_2O(l)$$

(b) 22.5 cm^3 of sulphuric acid, 0.262 M H_2SO_4

$$2NaOH(aq) + H_2SO_4(aq) \rightarrow Na_2SO_4(aq) + 2H_2O(l)$$

Although you most often start your calculations knowing the equation for the reaction, you may sometimes have to derive the equation, as in the next exercise.

Exercise 42 As a result of a titration, it was found that 25.0 cm^3 of silver nitrate solution, 0.50 M $AgNO_3$, reacted with 31.3 cm^3 of barium chloride solution, 0.20 M $BaCl_2$. Use these results to determine the stoichiometric coefficients, a and b, in the equation:

$$a\ AgNO_3(aq) + b\ BaCl_2 \rightarrow \text{Products}$$

REDOX TITRATIONS AND OTHER REDOX REACTIONS

Reactions which involve both REDuction and OXidation are called REDOX reactions. Before we deal with calculations concerning redox titrations, we show you how to separate the equation for a redox reaction into two half-equations, one for oxidation and one for reduction.

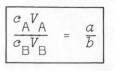

Redox half-equations

Every redox reaction involves the transfer of electrons. For example, consider the following reaction:

$$Zn(s) + Cu^{2+}(aq) \rightarrow Zn^{2+}(aq) + Cu(s)$$

The result is that two electrons from each zinc atom are transferred to a copper ion. We can split up the overall reaction into an oxidation reaction and a reduction reaction:

$$Zn(s) \rightarrow Zn^{2+}(aq) + 2e^- \qquad \text{OXIDATION}$$

$$Cu^{2+}(aq) + 2e^- \rightarrow Cu(s) \qquad \text{REDUCTION}$$

These are the two half-reactions for this reaction: one which shows electron loss and the other electron gain. Any redox reaction can be written as two half-reactions.

If you experience difficulty in remembering the difference between oxidation and reduction, the following mnemonic may help you.

O xidation
I s
L oss
R eduction
I s
G ain

Fig. 4.

Sometimes, it is difficult to 'pin down' electrons, and to say where they are or to which atoms they 'belong'. But, if we are not sure where the electrons are, how can we say anything useful about electron transfer? The concept of oxidation number has been developed, using ideas of electronegativity, to overcome this difficulty and enable us to identify redox reactions more readily.

Oxidation number

The oxidation number of an atom in a simple ion is, therefore, equal to the charge on the ion. Use this definition in the next exercise.

<u>Exercise 43</u> What are the oxidation numbers of the following elements?

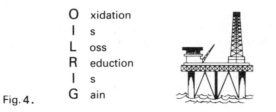

(a) Na in Na^+ (d) H in H^+

(b) O in O^{2-} (d) H in H^-

(c) Al in Al^{3+}

The *s*-block metals, with the exception of beryllium, always form ionic compounds, and the charge on the positive ion is always the same as the group number. This fact enables you to do the next exercise.

<u>Exercise 44</u> What are the oxidation numbers of the elements in the following compounds?

(a) KCl (d) Na_3P

(b) Na_2O (e) Mg_3N_2

(c) BaF_2 (f) CsH

In the last exercise you made use of the fact that the sum of the ion charges in an ionic compound is zero. Since ion charge is, by definition, the same as oxidation number, it follows that the sum of the oxidation numbers for each atom in a compound is zero. This rule also applies to covalent compounds, as we now show in a Worked Example.

<u>Worked Example</u> What are the oxidation numbers of the elements in these compounds?

W

(a) HCl (b) H_2S

<u>Solution to (a)</u>

1. Consider the relative electronegativities of hydrogen and chlorine.

   Electronegativity of H = 2.1 Electronegativity of Cl = 3.0

   (You should <u>know</u> that chlorine is more electronegative than hydrogen, without having to look up values.)

2. Imagine that the more electronegative atom gains both the bonding electrons so that ions are formed.

   $\overset{\delta+}{H}$—$\overset{\delta-}{Cl}$ imagined as H^+ Cl^-

2. Write down oxidation numbers from the ion charges

   $Ox(H) = +1$ $Ox(Cl) = -1$

<u>Solution to (b)</u>

1. Consider the relative electronegativities of hydrogen and sulphur.

   Electronegativity of H = 2.1 Electronegativity of S = 2.5

2. Imagine the partial separation of charge to be completed so that ions are formed.

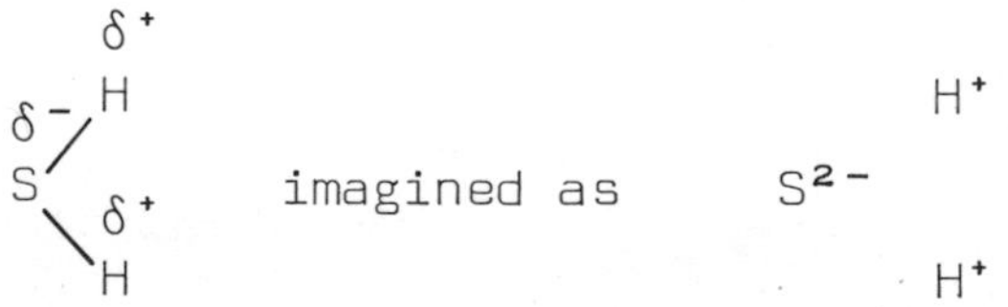

3. Write down oxidation numbers from the ion charges.

   $Ox(H) = +1$ $Ox(S) = -2$

   Note that the <u>sum</u> of the oxidation numbers for <u>each atom</u> is zero, because there are two hydrogen atoms:

   $2 \times (+1) + (-2) = 0$

Now do the following exercise in the same way.

Exercise 45 What are the oxidation numbers of the elements in these compounds?

(a) NH_3 (b) Cl_2O

Repeated application of the method used above has shown that certain elements in compounds always (or nearly always) have the same oxidation number.

Group I metals	-	always +1
Group II metals	-	always +2
Fluorine	-	always -1
Hydrogen	-	always +1, except in ionic hydrides where it is -1
Oxygen	-	always -2, except in peroxides (which contain an O—O linkage), and when combined with fluorine.

In the next exercise, you consider some cases where oxygen does not have an oxidation number of -2.

Exercise 46 What is the oxidation number of oxygen in the following molecules?

(a) H_2O_2 (H—O—O—H) (b) F_2O

The knowledge that some elements have fixed oxidation numbers enables us to take a short cut in finding the oxidation numbers of other elements. We show this in a Worked Example.

Worked Example What is the oxidation number of As in H_3AsO_3?

Solution

Write $Ox(\text{As}) + Ox(\text{all other atoms}) = 0$

$$\therefore\ Ox(\text{As}) = -\,Ox(\text{all other atoms})$$

$$= -\,[3\,Ox(\text{H}) + 3\,Ox(\text{O})]$$

$$= -\,[3(+1) + 3(-2)] = -3 + 6 = \boxed{+3}$$

Use the same method in the following exercise.

Exercise 47 What are the oxidation numbers of the following atoms?

(a) As in H_3AsO_4 (b) Cr in $HCr_2O_7^-$

You have now covered all the simple rules which help you to assign oxidation numbers. Apply them, as appropriate, in the next exercise.

Exercise 48 Calculate the oxidation numbers of the underlined elements in the following compounds:

(a)	$K_2\underline{Cr}_2O_7$	(d)	$Na_2\underline{S}_2O_3$	(g)	$\underline{N}H_4\underline{N}O_3$
(b)	$Pb\underline{S}O_4$	(e)	$Ca\underline{C}O_3$	(h)	$Na_2\underline{S}_4O_6$
(c)	$H\underline{N}O_3$	(f)	$Na_8\underline{Ta}_6O_{19}$	(i)	$\underline{Fe}_3O_4$

We mentioned earlier that oxidation numbers are useful in identifying redox reactions, and we now consider the point briefly.

Oxidation number and redox reactions

We bring together two statements which should, by now, be familiar to you:

Oxidation is a loss of electrons.

Oxidation number is the number of electrons 'lost' by an atom in going from its uncombined state to a particular combined state.

The combination of these two statements gives an alternative definition of oxidation:

> Oxidation involves an increase in oxidation number

Similarly,

> Reduction involves a decrease in oxidation number

You can apply these alternative definitions in the next exercise.

Exercise 49 By assigning oxidation numbers to the elements in both reactants and products, decide whether or not the following are redox reactions.

(a) $Fe(s) + CuSO_4(aq) \rightarrow FeSO_4(aq) + Cu(s)$

(b) $2KBr(aq) + Cl_2(aq) \rightarrow 2KCl(aq) + Br_2(aq)$

(c) $CuSO_4(aq) + Pb(NO_3)_2(aq) \rightarrow Cu(NO_3)_2(aq) + PbSO_4(s)$

(d) $C_2H_4(g) + H_2(g) \rightarrow C_2H_6(g)$

(e) $MnO_4^-(aq) + 5Fe^{2+}(aq) + 8H^+(aq) \rightarrow$

$Mn^{2+}(aq) + 5Fe^{3+}(aq) + 4H_2O(l)$

You learned from the Worked Example on page 20 that, in order to solve a titrimetric calculation, you need to know the balanced equation for the reaction. Equations for redox reactions may be difficult to balance by trial and error, so we now show you how a knowledge of oxidation numbers and electron transfer can be used to balance equations.

Balancing redox equations

The following Worked Example shows you how to balance a fairly cumbersome redox equation by two different methods.

W

Worked Example Balance the following redox equation:

$$ClO_3^-(aq) + I^-(aq) + H^+(aq) \rightarrow I_2(aq) + Cl^-(aq) + H_2O(l)$$

Solution

1. Identify which species is oxidized and which is reduced. This can be done by looking at oxidation numbers.

   (+5 above Cl in ClO_3^-; −1 above I^-; 0 above I_2; −1 above Cl^-)

   $$\overset{+5}{Cl}O_3^-(aq) + \overset{-1}{I^-}(aq) + H^+(aq) \rightarrow \overset{0}{I_2}(aq) + \overset{-1}{Cl^-}(aq) + H_2O(l)$$

   ∴ the species oxidized is I^-, [Ox(I) $-1 \rightarrow 0$]

   and the species reduced is ClO_3^- [Ox(Cl) $+5 \rightarrow -1$]

2. Construct two balanced half-equations, one representing oxidation and one representing reduction.

   (a) Balance the elements in each half-equation: if necessary, include H^+, OH^- and H_2O, as appropriate.

   Oxidation $2I^-(aq) \rightarrow I_2(aq)$

   Reduction $ClO_3^-(aq) \rightarrow Cl^-(aq)$

   The three oxygen atoms become incorporated in water molecules, i.e.

   $ClO_3^-(aq) \rightarrow Cl^-(aq) + 3H_2O(l)$

   Since the reaction occurs in an acid medium, include H^+ ions to balance the hydrogen in the water molecules, i.e.

   $ClO_3^-(aq) + 6H^+(aq) \rightarrow Cl^-(aq) + 3H_2O(l)$

   (b) Balance the charge in each half-equation. This can be done either by looking at the change in oxidation number or by balancing the charge on each ion.

Oxidation number method

$2I^-(aq) \rightarrow I_2(aq)$

Ox(I) changes from -1 to 0, i.e. each I atom loses one electron. Since there are two iodide ions, two electrons must be released in forming each iodine molecule.

Balancing charge method

$2I^-(aq) \rightarrow I_2(aq)$

Count the charges on each side of the half-equation:

2 negative charges → zero charge

Therefore, two electrons must be released from two iodide ions in order to form an iodine molecule.

$$2I^-(aq) \rightarrow I_2(aq) + 2e^-$$

Similarly, for

$ClO_3^-(aq) + 6H^+(aq)$

$\rightarrow Cl^-(aq) + 3H_2O(l)$

Ox(Cl) changes from +5 to −1, i.e. 6 units. Therefore, for each ClO_3^- ion reduced, 6 electrons are required.

Similarly, for

$ClO_3^-(aq) + 6H^+(aq)$

$\rightarrow Cl^-(aq) + 3H_2O(l)$

5 positive charges → 1 negative (6 − 1)
Therefore, 6 electrons are needed for each ClO_3^- ion reduced.

$$ClO_3^-(aq) + 6H^+(aq) + 6e^- \rightarrow Cl^-(aq) + 3H_2O(l)$$

3. Scale the half-equations up or down, so that the number of electrons required by one half-equation is the same as the number provided by the other.

$ClO_3^-(aq) + 6H^+(aq) + 6e^- \rightarrow Cl^-(aq) + 3H_2O(l)$

$2I^-(aq) \rightarrow I_2(aq) + 2e^-$ (multiply this equation by 3)

$6I^-(aq) \rightarrow 3I_2(aq) + 6e^-$

4. Combine the two half-equations by adding the left sides together and the right sides together.

$ClO_3^-(aq) + 6H^+(aq) + 6I^-(aq) \rightarrow 3I_2(aq) + Cl^-(aq) + 3H_2O$

Decide which of the two methods you prefer and attempt the next exerçise. We use the balancing charge method in the answers, but you should be aware of both methods.

Exercise 50 Balance the following equations.

(a) $IO_3^-(aq) + I^-(aq) + H^+(aq) \rightarrow I_2(aq) + H_2O(l)$

(b) $S_2O_3^{2-}(aq) + I_2(aq) \rightarrow I^-(aq) + S_4O_6^{2-}(aq)$

(c) $MnO_4^-(aq) + H^+(aq) + Fe^{2+}(aq) \rightarrow$
$Mn^{2+}(aq) + Fe^{3+}(aq) + H_2O(l)$

(d) $MnO_4^-(aq) + H^+(aq) + C_2O_4^{2-}(aq) \rightarrow$
$Mn^{2+}(aq) + CO_2(g) + H_2O(l)$

(e) $MnO_4^-(aq) + H^+(aq) + H_2O_2(aq) \rightarrow$
$Mn^{2+}(aq) + O_2(g) + H_2O(l)$

(f) $Cr_2O_7^{2-}(aq) + H^+(aq) + Fe^{2+}(aq) \rightarrow$
$Cr^{3+}(aq) + Fe^{3+}(aq) + H_2O(l)$

Now attempt the following titrimetric calculations which are based on redox reactions. Remember that the procedure for solving these problems is the same as for acid-base volumetric calculations.

Further calculations

Exercise 51 9.80 g of pure ammonium iron(II) sulphate, $FeSO_4 \cdot (NH_4)_2SO_4 \cdot 6H_2O$, was made up to 250 cm^3 with dilute sulphuric acid. 25.0 cm^3 of the solution reacted completely with 24.6 cm^3 of potassium manganate(VII) solution. Calculate the concentration of potassium manganate(VII) solution.

Exercise 52 25.0 cm^3 of a solution containing ethanedioic acid, $H_2C_2O_4$, and sodium ethanedioate, $Na_2C_2O_4$, required 14.8 cm^3 of 0.100 M sodium hydroxide solution for neutralization. A further 25.0 cm^3 of the $H_2C_2O_4$/$Na_2C_2O_4$ solution required 29.5 cm^3 of 0.0200 M potassium manganate(VII) solution for oxidation in acidic conditions at about 60 °C. Calculate the mass of each anhydrous constituent per dm^3 of the solution.

You may sometimes have to derive the equation, as in the following two exercises.

Exercise 53

A

(a) What mass of potassium iodate(V) (KIO_3) would be required to make 250 cm^3 of a solution containing one-sixtieth of a mole per dm^3?

(b) When 25 cm^3 of the solution of potassium iodate(V) of the concentration in (a) was added to excess of acidified potassium iodide solution, the iodine liberated reacted with 20 cm^3 of a solution of sodium thiosulphate. Calculate the concentration of the thiosulphate solution in moles per dm^3.

$$IO_3^- + 5I^- + 6H^+ \rightarrow 3I_2 + 3H_2O$$

$$I_2 + 2S_2O_3^{2-} \rightarrow 2I^- + S_4O_6^{2-}$$

(c) State how you would use the iodate(V)/iodide reaction and a standard solution of sodium thiosulphate to find the concentration of a solution of hydrochloric acid.

(d) 50 cm^3 of a solution containing 0.10 mol dm^{-3} of bromine (Br_2) was added to 10 cm^3 of a 0.10 mol dm^{-3} sodium thiosulphate solution. Excess potassium iodide was then added to the solution; the bromine which was left over from the first reaction liberated enough iodine to react with exactly 20 cm^3 of the same thiosulphate solution.

(i) What is the apparent oxidation number of S in the thiosulphate ion ($S_2O_3^{2-}$)?

(ii) From the figures given, calculate the number of moles of bromine which reacted directly with one mole of thiosulphate.

(iii) To what oxidation number of the S did the bromine oxidize the thiosulphate?

(iv) Derive the equation for the reaction between bromine molecules and thiosulphate ions.

Exercise 54 Under certain concentration and temperature conditions, 0.8 g of iron is found to react with 0.9 g of pure nitric acid, evolving an oxide of nitrogen (N_2O, or NO, or N_2O_4).

(a) Calculate the molar ratio of iron and nitric acid which react.

(b) Complete:

gain of ... electrons

$$\ldots Fe(s) + \ldots HNO_3(aq) \rightarrow \ldots Fe^{3+}(aq) + \begin{bmatrix} N_2O \\ \text{or } NO \\ \text{or } N_2O_4 \end{bmatrix}$$

loss of ... electrons

(c) From the number of electrons gained by the nitric acid, deduce the change in oxidation number of nitrogen and decide which of the three oxides of nitrogen is formed.

(d) Complete:

$$\ldots Fe(s) - \ldots e^- \rightarrow \ldots Fe^{3+}(aq)$$

$$\ldots H^+ + 1HNO_3(aq) + \ldots e^- \rightarrow \begin{bmatrix} \tfrac{1}{2}N_2O \\ \text{or } 1NO \\ \text{or } \tfrac{1}{2}N_2O_4 \end{bmatrix} + \ldots H_2O$$

and so, by addition, balance the equation for the reaction between iron and nitric acid under these conditions.

(H = 1; N = 14; O = 16; Fe = 56)

PRECIPITATION TITRATIONS

The most common precipitation titrations involve the reaction between silver ions and chloride ions in solution, according to the equation:

$$Ag^+(aq) + Cl^-(aq) \rightarrow AgCl(s)$$

The necessary data for the calculations are the results of titrations in which the indicator is usually potassium chromate(VI). The end-point is the first trace of a precipitate of silver chromate, Ag_2CrO_4, which appears as soon as precipitation of silver chloride is complete. Alternatively, an adsorption indicator such as fluorescein may be used.

Exercise 55 1.420 g of a chloride of sulphur was dissolved in water and the solution made up to a volume of 250 cm^3. 25.0 cm^3 of this solution, after neutralization, needed 21.0 cm^3 of 0.100 M $AgNO_3$ for complete precipitation of the chloride.

(a) Calculate the number of moles of chloride combined with one mole of sulphur.

(b) If the relative molar mass of the chloride is 135.2, what is its molecular formula?

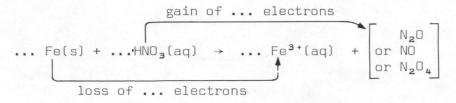

Exercise 56 0.767 g of a chloride of phosphorus was dissolved in water and made up to 250 cm^3 of solution. After neutralization, 25.0 cm^3 of this solution required 18.4 cm^3 of 0.100 M $AgNO_3$ for complete precipitation of the chloride.

(a) Calculate the amount of chloride ion in 25.0 cm^3 of solution.

(b) Calculate the amount and hence the mass of chloride ion in the whole sample.

(c) Calculate the mass and hence the amount of phosphorus in the whole sample.

(d) Calculate the amount of chlorine combined with one mole of phosphorus.

(e) Hence, state the simplest formula of the chloride.

END-OF-CHAPTER QUESTIONS

Exercise 57 100 cm^3 of concentrated hydrochloric acid was diluted accurately to 1000 cm^3 with distilled water. The diluted acid was titrated against 25.0 cm^3 of 0.500 M sodium carbonate, which required 27.2 cm^3 of acid for neutralization, using methyl orange indicator. What was the concentration of the original acid?

Exercise 58 50.0 cm^3 of a 0.400 M hydrochloric acid solution was added to 25.0 cm^3 of a solution of sodium hydroxide and the resulting solution required 21.6 cm^3 of 0.200 M sodium hydroxide solution for neutralization. Calculate the concentration of the original sodium hydroxide solution.

Exercise 59 Iodine dissolves in hot concentrated solutions of sodium hydroxide according to the equation

$$3I_2(s) + 6NaOH(aq) \rightarrow NaIO_3(aq) + 5NaI(aq) + 3H_2O(l)$$

In one experiment, 3.81 g of iodine was dissolved in 4 M sodium hydroxide solution. (Relative atomic masses: H = 1, O = 16, Na = 23, I = 127.)

(a) How many moles of iodine were used?

(b) What volume of 4 M sodium hydroxide solution would be just sufficient to react with the iodine?

Exercise 60 When 0.203 g of hydrated magnesium chloride, $MgCl_m \cdot nH_2O$, was dissolved in water and titrated with 0.100 M silver nitrate ($AgNO_3$) solution, 20.0 cm^3 of the latter were required. A sample of the hydrated chloride lost 53.2% of its mass when heated in a stream of hydrogen chloride, leaving a residue of anhydrous magnesium chloride. From these figures, calculate the values of m and n.

Exercise 61 White phosphorus reacts with dilute aqueous solutions of copper(II) sulphate to deposit metallic copper and produce a strongly acidic solution.

A part

In an experiment to investigate this reaction, 0.31 g of white phosphorus reacted in excess aqueous copper(II) sulphate giving 1.60 g of metallic copper. (P = 31, Cu = 64.)

(a) (i) Calculate the number of moles of phosphorus atoms used.

(ii) Calculate the number of moles of copper produced.

(iii) Hence, calculate the number of moles of copper deposited by one mole of phosphorus atoms.

(b) (i) State the change in oxidation number of the copper in this reaction.

(ii) Calculate the new oxidation number of the phosphorus after the reaction.

(iii) In the reaction, the phosphorus forms an acid, HPO_n. What is the value of n?

(c) Now write a balanced equation showing the action of white phosphorus on copper(II) sulphate in the presence of water.

Exercise 62 This question concerns a hydrated double salt *A*, $Cu_w(NH_4)_x(SO_4)_y \cdot zH_2O$, whose formula may be determined from the following experimental data.

A

(a) 2 g of salt *A* was boiled with excess sodium hydroxide and the ammonia expelled collected by absorption in 40 cm^3 of 0.5 M hydrochloric acid in a cooled flask. Subsequently, this solution required 20 cm^3 of 0.5 M sodium hydroxide for neutralization.

(i) Calculate the number of moles of ammonium ion in 2 g of salt *A*.

(ii) Calculate the mass of ammonium ion present in 2 g of salt *A*.

(b) A second sample of 2 g of salt *A* was dissolved in water and treated with an excess of barium chloride solution. The mass of the precipitate formed, after drying, was found to be 2.33 g.

(i) Calculate the number of moles of sulphate ion in 2 g of salt *A*.

(ii) Calculate the mass of sulphate ion in 2 g of salt *A*.

(c) A third sample of 10 g of salt *A* was dissolved to give 250 cm^3 of solution. 25 cm^3 of this solution, treated with an excess of potassium iodide, gave iodine equivalent to 25 cm^3 of 0.1 M sodium thiosulphate solution.

$$2Cu^{2+} + 4I^- \rightarrow 2CuI + I_2$$

$$I_2 + 2S_2O_3^{2-} \rightarrow 2I^- + S_4O_6^{2-}$$

(i) Calculate the number of moles of copper(II) ion in 2 g of salt *A*.

(ii) Calculate the mass of copper(II) ion in 2 g of salt *A*.

(d) (i) Calculate the mass of water of crystallization in 2 g of salt *A*.

(ii) Calculate the number of moles of water of crystallization in 2 g of salt *A*.

(e) Determine the formula of the hydrated salt *A*.

Exercise 63 A solution made from pure iron(II) sulphate became partially oxidized on standing. 25.0 cm^3 of this solution gave a precipitate of mass 0.829 g when treated with excess barium chloride solution. A further 25.0 cm^3 was acidified and titrated with 0.0200 M potassium manganate(VII) solution, of which 26.2 cm^3 was required to give a permanent pink colour. Calculate the ratio of Fe^{2+} to Fe^{3+} ions in the partially oxidized solution.

MASS SPECTROMETRY AND NUCLEAR REACTIONS

3

INTRODUCTION AND PRE-KNOWLEDGE

We assume that you have read about the mass spectrometer, and know how the instrument produces beams of ions which can be detected separately according to their mass/charge ratios. Measurement of the relative intensities of the beams enables us to determine the relative atomic masses of elements and, more important, the relative molecular masses of compounds, particularly organic ones.

When a beam of ions strikes the detector in a mass spectrometer, it produces an electrical impulse, which is amplified and fed into a recorder. The mass/charge ratio and the relative abundance of the particular type of ion making up the beam are then shown by a peak on a chart. A set of such peaks is a mass spectrum (or mass spectrometer trace, or mass spectrogram).

Fig. 5 shows a mass spectrum for rubidium. The horizontal axis shows the mass/charge ratio of the ions entering the detector. If it is assumed that all the ions carry a single positive charge, the horizontal axis can also be labelled 'mass number', 'isotopic mass' or 'relative atomic mass'. The vertical axis shows the abundance of the ions. It can be labelled 'detector current', 'relative abundance' or 'ion intensity'.

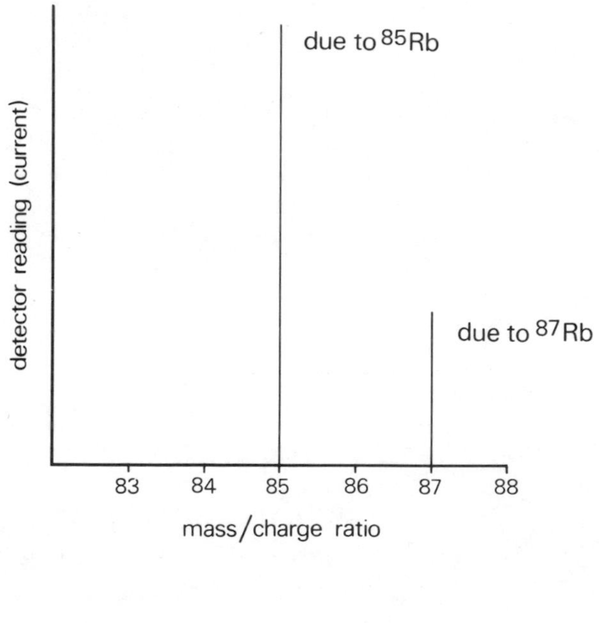

Fig.5.

We use the information from mass spectra to calculate the relative atomic mass of an element.

CALCULATING THE RELATIVE ATOMIC MASS OF AN ELEMENT

We start with a Worked Example, using the mass spectrum shown in Fig. 5.

W

Worked Example Use Fig. 5 to calculate the relative atomic mass of rubidium.

Solution

1. Measure the height of each peak. The height is proportional to the amount of each isotope present.

   height of rubidium-85 peak = 5.82 cm

   height of rubidium-87 peak = 2.25 cm

   Therefore, the relative amounts of ^{85}Rb and ^{87}Rb are 5.82 and 2.25 respectively.

2. Express each relative amount as a percentage of the total amount. This gives the percentage abundance.

$$\% \text{ abundance} = \frac{\text{amount of isotope}}{\text{total amount of all isotopes}} \times 100$$

$$\% \text{ abundance of } ^{85}\text{Rb} = \frac{5.82}{5.82 + 2.25} \times 100 = 72.1\%$$

$$\% \text{ abundance of } ^{87}\text{Rb} = \frac{2.25}{5.82 + 2.25} \times 100 = 27.9\%$$

   The percentage abundance figures mean that for every 1000 atoms, 721 are the ^{85}Rb isotope and 279 are the ^{87}Rb isotope.

3. Find the total mass of a sample of a hundred atoms.

$$\text{Total mass} = [(85 \times 72.1) + (87 \times 27.9)] \text{ amu*}$$

$$= (6129 + 2427) \text{ amu} = 8556 \text{ amu}$$

$$= 8.56 \times 10^3 \text{ amu (3 sig. figs.)}$$

4. Find the average mass. This gives the relative atomic mass of rubidium.

$$\text{Average mass} = \frac{\text{total mass}}{\text{number of atoms}} = \frac{8.56 \times 10^3 \text{ amu}}{100} = 85.6 \text{ amu}$$

   ∴ relative atomic mass = **85.6**

*amu stands for atomic mass unit. 1 amu = 1.660×10^{-27} kg. This value was chosen because it is exactly one-twelfth the mass of an atom of the carbon-12 isotope, which comprises 98.89% of natural carbon.

Now try a similar calculation for yourself by doing the next exercise.

Exercise 64 Use the mass spectrum shown in Fig. 6 to calculate:

(a) the percentage of each isotope present in a sample of naturally-occurring lithium;

(b) the relative atomic mass of lithium.

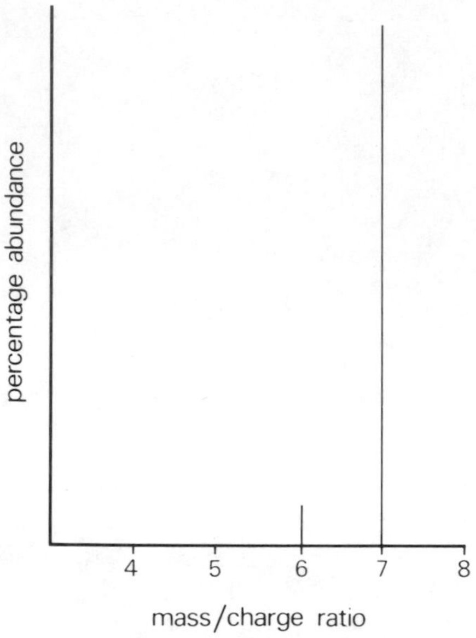

Fig.6.

The next exercise is similar, but you are given data rather than a diagram of a mass spectrum.

Exercise 65 The mass spectrum of neon consists of three lines corresponding to mass/charge ratios of 20, 21 and 22 with relative intensities of 0.91, 0.0026 and 0.088 respectively. Explain the significance of these data and hence calculate the relative atomic mass of neon.

A part

You could also be asked to do the reverse of this type of problem - sketch a spectrum from percentage abundance data, as in the next exercise.

Exercise 66 (a) The percentage abundances of the stable isotopes of chromium are:

$^{50}_{24}Cr$ - 4.31%; $^{52}_{24}Cr$ - 83.76%;

$^{53}_{24}Cr$ - 9.55%; $^{54}_{24}Cr$ - 2.38%

(b) Sketch the mass spectrum that would be obtained from naturally-occurring chromium. (Let 10.0 cm represent 100% on the vertical scale.)

(c) Calculate the relative atomic mass of chromium, correct to three significant figures.

(d) Label each peak on the mass spectrum using isotopic symbols.

You may be asked to calculate relative atomic masses directly from percentage abundance data, as you have just done in part (c) of Exercise 66. For more practice at this type of problem, see the End-of-Chapter Problems.

The next exercise concerns a molecular element.

Exercise 67 The element chlorine has isotopes of mass number 35 and 37 in the approximate proportion 3:1. Interpret the mass spectrum of gaseous chlorine shown in Fig. 7, indicating the formula (including mass number) and charge of the ion responsible for each peak.

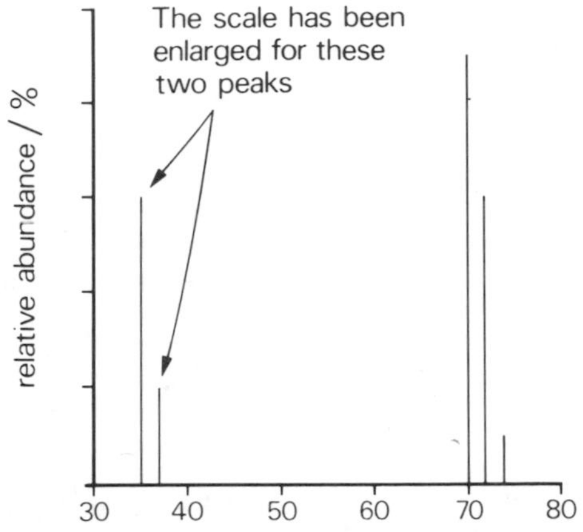

Fig.7.

You may be asked to interpret the more complex mass spectra produced by organic compounds, but this is not likely to involve calculations.

The second half of this chapter concerns nuclear reactions and the simple calculations required in order to write nuclear equations.

NUCLEAR REACTIONS

Whereas chemical reactions leave the nucleus untouched and involve only the electrons surrounding it, nuclear reactions rearrange the particles within the nucleus. Also, different elements may be formed, in a process known as transmutation, which does not, of course, occur in chemical reactions.

The basic nuclear reactions are alpha and beta particle emission, often called α and β decay. We consider alpha decay in detail in a Worked Example.

Worked Example When thorium-228 decays, each atom emits one alpha particle. Write a balanced nuclear equation to show the process.

W

Solution

1. Write out the equation, putting in letters for the unknowns:

$$^{228}_{90}\text{Th} \rightarrow \, ^{4}_{2}\text{He} + \, ^{A}_{Z}\text{X}$$

In these equations, the atomic numbers and mass numbers of each side must balance; this allows you to calculate the unknowns, A and Z and X.

2. Find the atomic number of the new element, X, using the atomic numbers shown by the equation.

$$_{90}\text{Th} \rightarrow {}_{2}\text{He} + {}_{Z}\text{X}$$

$$90 = 2 + Z$$

$$Z = 90 - 2 = 88$$

3. Use a Periodic Table to identify the new element:

element with atomic number 88 is radium, i.e. X = Ra.

4. Find the mass number of the particular isotope of radium, using the mass numbers in the equation.

$$^{228}\text{Th} \rightarrow {}^{4}\text{He} + {}^{A}\text{X}$$

$$228 = 4 + A \qquad \therefore A = 224$$

5. Write out the complete equation:

$$^{228}_{90}\text{Th} \rightarrow {}^{4}_{2}\text{He} + {}^{224}_{88}\text{Ra}$$

Note that ionic charges are not shown in nuclear equations. You may wonder, for example, why we do not write $^{4}_{2}\text{He}^{2+}$. In fact, the positive ions tend to pick up stray electrons and so the charges are usually omitted as a simplification. This also focuses attention on changes in the nucleus.

Now try some calculations yourself, by doing the next exercise:

Exercise 68 Write balanced nuclear equations to show the alpha decay of the following:

(a) $^{212}_{84}\text{Po}$ (b) $^{220}_{86}\text{Rn}$

You can adapt the method given in the last worked example to write an equation for β decay. The following equation illustrates how a nucleus consisting of protons and neutrons can emit electrons.

$$^{1}_{0}\text{n} \rightarrow {}^{1}_{1}\text{H} + {}^{0}_{-1}\text{e}$$

You make use of this idea in the next exercise.

Exercise 69 Write balanced nuclear equations to show the beta decay of the following:

(a) $^{212}_{84}\text{Po}$ (b) $^{24}_{11}\text{Na}$ (c) $^{108}_{47}\text{Ag}$

The next exercise summarises the effect of alpha and beta decay.

Exercise 70 (a) What happens to the atomic number and the mass number of a nucleus when it emits:

(i) an α-particle, (ii) a β-particle?

(b) Write the nuclear equations which represent:

(i) the loss of an α-particle by radium-226,

(ii) the loss of a β-particle by potassium-43,

(iii) the loss of an α-particle by the product of (ii).

We now briefly consider nuclear stability. In other words, why do some nuclides decay and not others?

Nuclear stability

To test your understanding of why certain nuclei are unstable, work through the following exercises, using Fig. 8, which shows the number of protons and neutrons in all stable nuclei, as a guide.

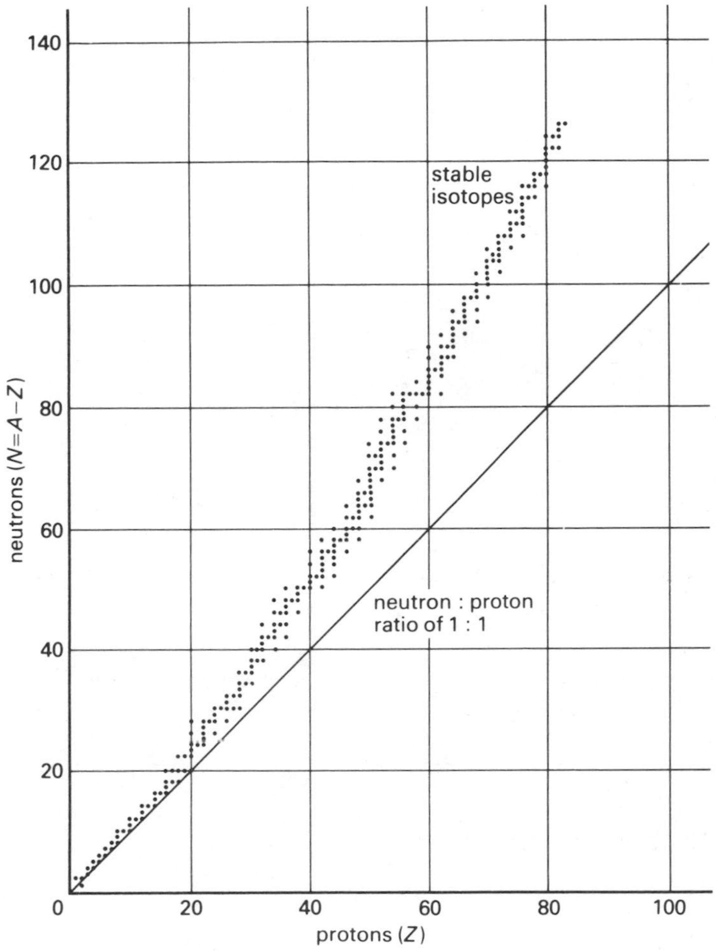

Fig.8.

Exercise 71

(a) From the graph, identify the heaviest known stable nucleus, using isotopic symbols.

(b) Calculate the neutron to proton ratios (N/Z) for these stable nuclei:

(i) the nucleus identified in (a)

(ii) $^{108}_{48}Cd$ (iii) $^{12}_{6}C$

(c) Describe how the ratio N/Z for stable nuclei changes with increasing atomic number.

(d) By calculating N/Z ratios, predict whether the nuclei listed below are likely to be stable or unstable. Check your prediction against the graph.

(i) $^{11}_{6}C$ (iii) $^{106}_{47}Ag$ (v) $^{206}_{82}Pb$

(ii) $^{60}_{30}Zn$ (iv) $^{143}_{56}Ba$

Exercise 72 $^{14}_{6}C$ is known to be a β emitter.

(a) Write a nuclear equation for the emission.

(b) Work out the *N*/*Z* ratio for reactant and product nuclei.

(c) In view of the *N*/*Z* ratios, what can you conclude about the stability of reactants and products? Summarise the way β-emission changes the stability of a nucleus.

Questions on the rate of decay of radioactive isotopes appear in Chapter 10.

END-OF-CHAPTER QUESTIONS

Exercise 73 Chlorine consists of isotopes of relative masses 34.97 and 36.96 with natural abundances of 75.77% and 24.23% respectively. Calculate the mean relative atomic mass of naturally-occurring chlorine.

Exercise 74 Calculate the relative atomic mass of natural lithium which consists of 7.4% of ^{6}Li (relative atomic mass 6.02) and 92.6% of ^{7}Li (relative atomic mass 7.02).

Exercise 75 A mixture of $^{2}_{1}H_2$ and $^{81}_{35}Br_2$ was analysed in a mass spectrometer. The following pattern of lines due to singly-charged ions was obtained.

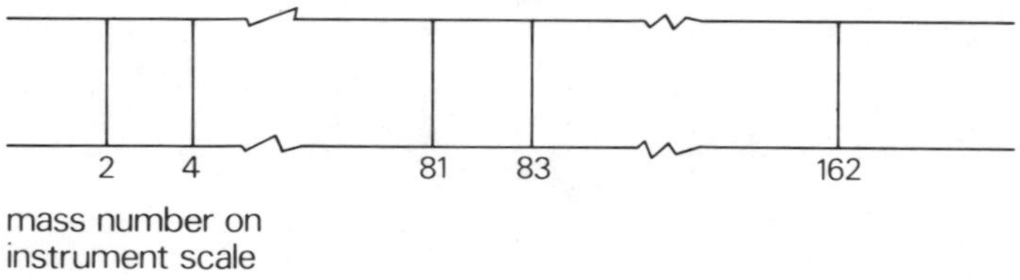

Fig.9.

(a) State which ions give rise to each of these lines.

(b) What apparent mass number would register on the instrument scale if the heaviest of these ions acquired a second charge?

Exercise 76 Using mass spectrometry, the element gallium has been found to consist of 60.4 per cent of an isotope of atomic mass 68.93 and 39.6 per cent of an isotope of atomic mass 70.92. Calculate, to three significant figures, the relative atomic mass of gallium.

Exercise 77 (a) Calculate the relative atomic mass of copper assuming it to contain 70% of ^{63}Cu and 30% of ^{65}Cu.

(b) Complete the following equations

$$^{220}_{86}Rn \rightarrow {}^{216}_{84}Po \ +$$

$$^{214}_{82}Pb \rightarrow {}^{214}_{83}Bi \ +$$

CHEMICAL ENERGETICS.

4

INTRODUCTION AND PRE-KNOWLEDGE

In any chemical reaction (and in many physical changes too) heat energy tends to be transferred, either from the system to the surroundings, or in the opposite direction. If we specify that all the reactants and products are in their standard states and at standard conditions, the quantity of energy transferred defines what is known as the standard enthalpy change of reaction, $\Delta H^\ominus$. A positive sign indicates transfer from the surroundings.

This information about a reaction can be summarised in two ways, by a thermochemical equation or by an energy (or enthalpy) level diagram. For instance, when two moles of hydrogen react with one mole of oxygen to produce two moles of liquid water, with both initial and final states at 298 K and 1 atm, 572 kJ of heat energy are released to the surroundings. The thermochemical equation for this reaction is:

$$2H_2(g) + O_2(g) \rightarrow 2H_2O(l); \quad \Delta H^\ominus = -572 \text{ kJ mol}^{-1}$$

Notice the use of the symbol $^\ominus$ to indicate standard conditions. If half the amounts are used, then the energy is halved also.

$$H_2(g) + \tfrac{1}{2}O_2(g) \rightarrow H_2O(l); \quad \Delta H^\ominus = -286 \text{ kJ mol}^{-1}$$

Notice that, in both examples, the unit for standard enthalpy change is kJ mol^{-1}. You can consider that the term 'per mole' refers to all the amounts specified in the equation. You may find it useful to think of the term 'per mole' as referring to 'a mole of equation' if, as in the two examples, the stoichiometric coefficients are not all unity.

It is important to include the state symbols (e.g. g, l, s, aq) because the energy change depends on the state of the substance. If the water produced were not allowed to condense, then we would write a thermochemical equation with a different value of the enthalpy change;

$$2H_2(g) + O_2(g) \rightarrow 2H_2O(g); \quad \Delta H^\ominus = -484 \text{ kJ mol}^{-1}$$

An enthalpy-level diagram corresponding to these equations is:

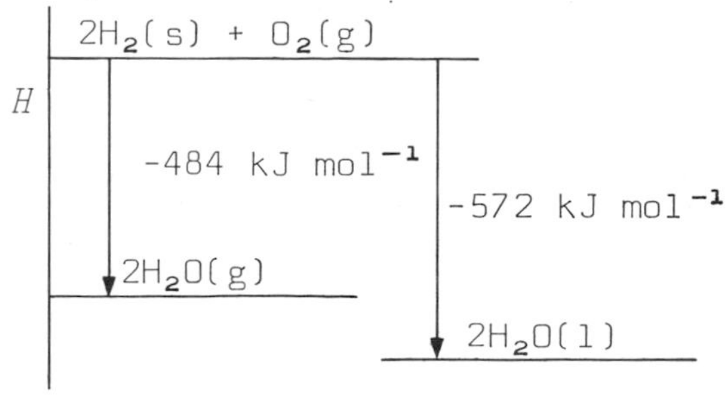

The standard enthalpy changes for particular types of reaction are often given special names, e.g. the standard enthalpy change of formation, $\Delta H_f^\ominus$, and the standard enthalpy change of combustion, $\Delta H_c^\ominus$. You should be able to write definitions for these based on this introductory section.

Try the next exercise to check that you have grasped the basic ideas.

Exercise 78 (a) Write a thermochemical equation showing that when 1.00 mol of carbon burns completely in oxygen, 394 kJ of heat are liberated.

(b) Calculate the enthalpy change on complete combustion of

(i) 10.0 mol of carbon, (iii) 18.0 g of carbon.

(ii) 0.25 mol of carbon,

(c) What mass of carbon would have to be burned to produce

(i) 197 kJ, (ii) 1000 kJ?

CALCULATING $\Delta H^{\ominus}$ FROM EXPERIMENTAL DATA

If we measure all the heat transferred between system and surroundings in order to return the system to 25 °C, this is the enthalpy change for the reaction. In practice, it is much easier to measure the enthalpy change of a reaction using a calorimeter in which the system is insulated from the surroundings. For an exothermic reaction, the energy which would otherwise be given to the surroundings results in an increase in the temperature of the system.

If the maximum temperature change of the system is recorded and if the heat capacity of the system is known, it is easy to calculate the quantity of heat which would have to be taken from the system in order to restore it to its initial temperature. This quantity of heat is the enthalpy change.

For an endothermic reaction, the temperature of the system would drop but, again, the enthalpy change can be calculated from the temperature change if the heat capacity of the system is known.

In calorimeters such as those shown in Fig. 10, the energy exchanged with the surroundings is usually small enough to be ignored.

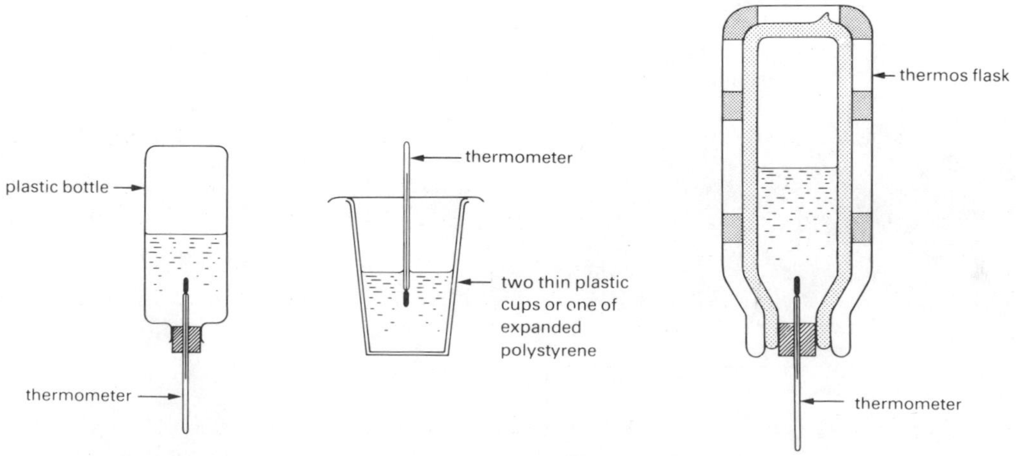

Fig.10.

We now give a Worked Example showing how the enthalpy change of reaction can be calculated from experimental data.

<u>Worked Example</u> An excess of zinc powder was added to 50.0 cm^3 of 0.100 M $AgNO_3$ in a polystyrene cup. Initially, the temperature was 21.10 °C and it rose to 25.40 °C. Calculate the enthalpy change for the reaction:

W

$$Zn(s) + 2Ag^+(aq) \rightarrow Zn^{2+}(aq) + 2Ag(s)$$

Assume that the density of the solution is 1.00 g cm^{-3} and its specific heat capacity is 4.18 kJ kg^{-1} K^{-1}. Ignore the heat capacity of the metals.

<u>Solution</u>

1. Since the polystyrene cup is an insulator and its heat capacity is almost zero, you can assume that no energy is exchanged between system and surroundings. All the chemical energy released in the reaction is transformed into heat energy which raises the temperature of the solution. The <u>total</u> energy change in the system is zero, so you can write:

$$\left[\begin{array}{l}\text{enthalpy change, } \Delta H\text{, due}\\ \text{to reaction (at constant } T)\end{array}\right] + \left[\begin{array}{l}\text{change in heat}\\ \text{energy of solution}\end{array}\right] = 0$$

Since $\left[\begin{array}{l}\text{change in heat}\\ \text{energy of solution}\end{array}\right] = \left[\begin{array}{l}\text{mass x specific heat capacity}\\ \text{x temperature change}\end{array}\right] = mc_p{}^* \ \Delta T$

you can now write: $\Delta H + mc_p\Delta T = 0$

$$\therefore \quad \Delta H = -mc_p\Delta T = \frac{50.0}{1000} \text{ kg} \times 4.18 \text{ kJ kg}^{-1} \text{ K}^{-1} \times 4.30 \text{ K} = -0.899 \text{ kJ}$$

*c_p is the symbol commonly used for specific heat capacity. The 'p' refers to constant pressure conditions (as does ΔH).

2. The value -0.899 kJ is the enthalpy change for the amounts used in the experiment. To obtain a value for the enthalpy change of reaction, compare the amounts used in the experiment with the amounts shown in the equation:

$$Zn(s) + 2Ag^+(aq) \rightarrow Zn^{2+}(aq) + Ag(s)$$

1 mol   2 mol

The amount of silver ion used = 0.0500 dm^3 x 0.100 mol dm^{-3} = 5.00 x 10^{-3} mol

$\therefore$ the enthalpy change using 2 mol of Ag^+

$$= -0.899 \text{ kJ} \times \frac{2.00 \text{ mol}}{5.00 \times 10^{-3} \text{ mol}} = -360 \text{ kJ}$$

3. Now write the complete thermochemical equation:

$$Zn(s) + 2Ag^+(aq) \rightarrow Zn^{2+}(aq) + Ag(s); \quad \Delta H = -360 \text{ kJ mol}^{-1}$$

Strictly speaking, you should not write $\Delta H^\ominus$ in this case because the conditions of the experiment were not standard, but the values of ΔH and $\Delta H^\ominus$ would be very close.

Note that enthalpy changes related to equations, which include all standard enthalpy changes, have the unit, kJ mol^{-1}.

Use a similar method in the next exercise.

Exercise 79 The following experiment was performed to determine the enthalpy change for the displacement reaction:

$Zn(s) + Cu^{2+}(aq) \rightarrow Cu(s) + Zn^{2+}(aq)$

25.0 cm^3 of 1.00 M copper(II) sulphate solution was placed in a polystyrene cup and the temperature of the solution was recorded to the nearest 0.2 °C every half-minute for 2½ minutes. At precisely 3 minutes, excess zinc powder was added, with continual stirring, and the temperature was recorded for an additional 6 minutes. The results of the experiment were as follows.

Time/min	0.0	0.5	1.0	1.5	2.0	2.5	3.0
Temperature/°C	27.0	27.0	27.2	27.2	27.2	27.2	-
Time/min	3.5	4.0	4.5	5.0	5.5	6.0	6.5
Temperature/°C	66.0	71.4	71.8	70.2	68.0	66.2	64.4
Time/min	7.0	7.5	8.0	8.5	9.0	9.5	
Temperature/°C	62.8	61.0	59.5	58.0	56.6	55.1	

(a) Plot the temperature (y-axis) against time (x-axis).

(b) Extrapolate the curve to 3.0 minutes to establish the maximum temperature rise as shown in Fig. 11.

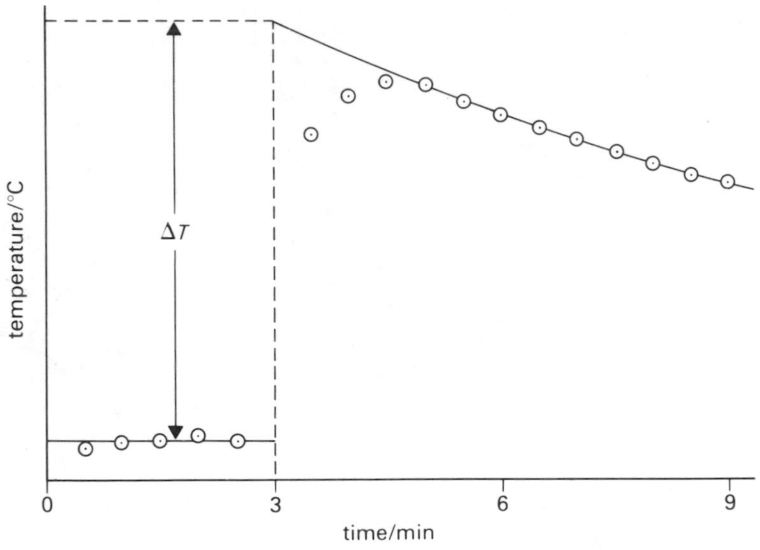

Fig.11.

(c) Calculate the enthalpy change for the quantities used, making the same assumptions as in the preceding exercise.

(d) Calculate the enthalpy change for one mole of Zn and $CuSO_4(aq)$, and write the thermochemical equation for the reaction.

Standard enthalpy change of solution

This is the enthalpy change which occurs when 1 mole of solute is dissolved in a specified amount of solvent under standard conditions. The abbreviation is $\Delta H^{\ominus}_{soln}$.

Strictly speaking, the resulting solution should be 'infinitely dilute', i.e. solvent is added until no further enthalpy change can be detected but, in practice, some convenient dilution is specified, such as 1 mol of solute to 100 mol of solvent. The following thermochemical equation represents 1 mol of copper(II) sulphate dissolving in 100 mol of water:

$$CuSO_4(s) + 100H_2O(l) \rightarrow CuSO_4(aq, 100H_2O); \quad \Delta H^\ominus = -67.4 \text{ kJ mol}^{-1}$$

Some data books list $\Delta H^\ominus_{soln}$ only at infinite dilution (-73.3 kJ mol^{-1} for $CuSO_4$), and others list only the heat of formation of the solution from water and the elements of the solute (-837.3 kJ mol^{-1} for $CuSO_4$). In the latter case, $\Delta H^\ominus_{soln}$ is obtained by subtracting $\Delta H^\ominus_f$ for the solute, i.e.
$\Delta H^\ominus_{soln}\,[CuSO_4(s)] = \Delta H^\ominus_f\,[CuSO_4(aq)] - \Delta H^\ominus_f\,[CuSO_4(s)]$.

We give a Worked Example to illustrate the type of calculations you might meet using $\Delta H^\ominus_{soln}$.

Worked Example

W

When 0.85 g of anhydrous lithium chloride, LiCl, was added to 36.0 g of water at 25.0 °C in a polystyrene cup, the final temperature of the solution was 29.7 °C. Calculate the enthalpy change of solution for one mole of lithium chloride.

Solution

1. In problems involving heat of solution, the amount of solvent is important so calculate the ratio: amount of LiCl/amount of H_2O. Then you can write the correct equation describing the process for one mole of the salt.

$$\text{amount of LiCl} = \frac{m}{M} = \frac{0.85 \text{ g}}{42.4 \text{ g mol}^{-1}} = 0.020 \text{ mol}$$

$$\text{amount of } H_2O = \frac{m}{M} = \frac{36.0 \text{ g}}{18.0 \text{ g mol}^{-1}} = 2.00 \text{ mol}$$

$$\therefore \quad \frac{\text{amount of LiCl}}{\text{amount of } H_2O} = \frac{0.020}{2.00} = \frac{1}{100}$$

2. Write the equation:

$$LiCl(s) + 100H_2O(l) \rightarrow LiCl(aq, 100H_2O)$$

3. Calculate the enthalpy change for the amounts used in the experiment:

$$\begin{bmatrix}\text{enthalpy change, } \Delta H,\\ \text{on dissolving LiCl}\end{bmatrix} + \begin{bmatrix}\text{change in heat energy}\\ \text{of the solution}\end{bmatrix} = 0$$

or $\quad \Delta H + mc_p\Delta T = 0$

$$\therefore \quad \Delta H = -mc_p\Delta T = -\frac{36.0}{1000} \text{ kg} \times 4.18 \text{ kJ kg}^{-1} \text{ K}^{-1} \times 4.7 \text{ K} = -0.71 \text{ kJ}$$

(Note that although the mass of the solution is greater than 36.0 g (36.85 g) its specific heat capacity is slightly <u>less</u> than 4.18 kJ kg^{-1} K^{-1}. The error arising from considering only the water in this type of calculation is therefore small.)

4. Scale up to the amounts shown in the equation, as in the last Worked Example:

$$\Delta H = -0.71 \text{ kJ} \times \frac{1 \text{ mol}}{0.020 \text{ mol}} = -35 \text{ kJ}$$

5. Write the complete thermochemical equation:

$$LiCl(s) + 100H_2O(l) \rightarrow LiCl(aq, 100H_2O); \quad \boxed{\Delta H_{soln} = -35 \text{ kJ mol}^{-1}}$$

Use a similar method, making the same assumptions, in the next exercise.

<u>Exercise 80</u>   When 2.67 g of ammonium chloride, NH_4Cl, was added to
90.0 g of water at 24.5 °C in a polystyrene cup, the
final temperature of the solution was 22.7 °C. Calculate
the enthalpy change of solution and write a thermochemical equation.

HESS'S LAW

Hess's law (sometimes called Hess's law of constant heat summation) is a
corollary to the law of conservation of energy. We use Hess's law to
calculate enthalpy changes which cannot be measured directly. It states
that the enthalpy change in converting reactants, A and B, to products,
X and Y, is the same, regardless of the route by which the chemical change
occurs, provided the initial and final conditions are the same.

The application of Hess's law is illustrated in the following Worked Example.

<u>Worked Example</u>   Calculate the standard enthalpy change for the
reaction

$$C(s) + \tfrac{1}{2}O_2(g) \rightarrow CO(g)$$

given the following information:

$$C(s) + O_2(g) \rightarrow CO_2(g); \quad \Delta H^{\ominus} = -394 \text{ kJ mol}^{-1}$$

$$CO(g) + \tfrac{1}{2}O_2(g) \rightarrow CO_2(g); \quad \Delta H^{\ominus} = -283 \text{ kJ mol}^{-1}$$

<u>Solution</u>

1.   First represent the direct combination of elements to give carbon dioxide:

$$\boxed{C(s) + O_2(g)} \longrightarrow \boxed{CO_2(g)}$$

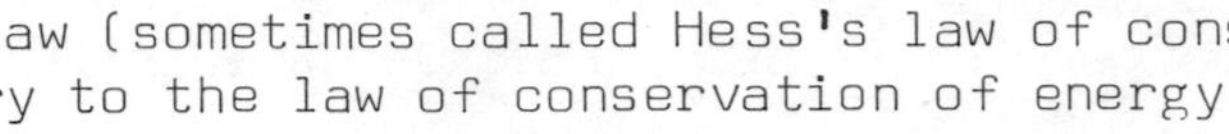

2.   Write the two-step process which gives the same product:

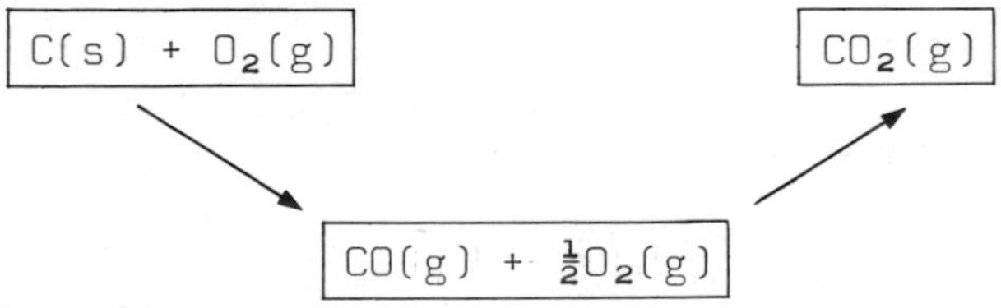

3.   Hess's law tells us that the enthalpy change via one route ('route 1')
must equal the enthalpy change via the two-step route ('route 2').

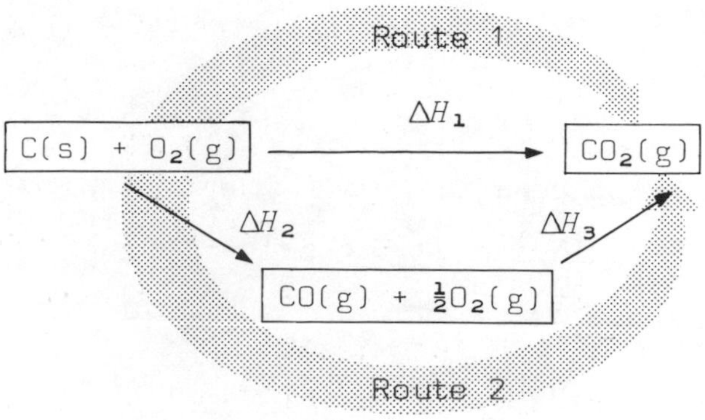

$$\Delta H_1 = \Delta H_2 + \Delta H_3$$

4. Calculate the unknown by substituting the known values:

$\Delta H_1 = \Delta H_2 + \Delta H_3$ or $\Delta H_2 = \Delta H_1 - \Delta H_3$

$\therefore \ \Delta H_f^\ominus[CO(g)] = -394 \text{ kJ mol}^{-1} - (-283 \text{ kJ mol}^{-1}) = \boxed{-111 \text{ kJ mol}^{-1}}$

5. The same result may be obtained by drawing an enthalpy level diagram to scale, as indicated here.

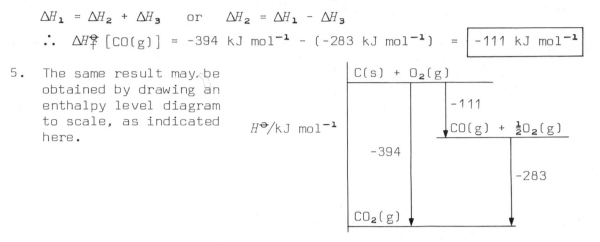

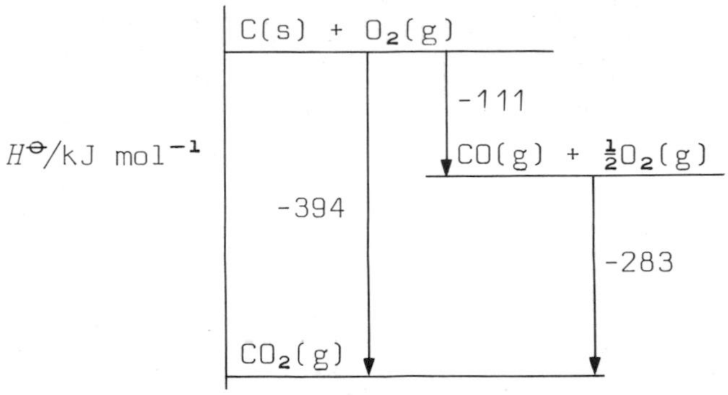

Both enthalpy cycles and enthalpy-level diagrams are very useful in solving problems. In the next exercise you use both methods: be careful to distinguish between the two. In an enthalpy-level diagram, the arrows should be drawn vertically, and preferably to scale, to represent both directions and extent of enthalpy change. In an enthalpy cycle, the arrows simply indicate a change from one state to another, and to avoid confusion we suggest you do not use vertical arrows in enthalpy cycles.

Exercise 81 Calculate the standard enthalpy change for the reaction

$$2NO_2(g) \rightarrow N_2O_4(g)$$

given the thermochemical equations:

$$N_2(g) + 2O_2(g) \rightarrow 2NO_2(g); \quad \Delta H^\ominus = +33.2 \text{ kJ mol}^{-1}$$

$$N_2(g) + 2O_2(g) \rightarrow N_2O_4(g); \quad \Delta H^\ominus = +9.2 \text{ kJ mol}^{-1}$$

by drawing an enthalpy-level diagram, and an enthalpy cycle.

In the next exercise, you use and justify a third way of applying Hess's law.

Exercise 82 Two of the possible methods of preparing a solution of ammonium chloride containing 1.00 mol of NH_4Cl in 200 mol of H_2O are summarised in the equations below:

Method 1

$$NH_3(g) + HCl(g) \rightarrow NH_4Cl(s); \quad \Delta H^\ominus = -175.3 \text{ kJ mol}^{-1}$$

$$NH_4Cl(s) + 200H_2O(l) \rightarrow NH_4Cl(aq, 200H_2O); \quad \Delta H^\ominus = +16.3 \text{ kJ mol}^{-1}$$

Method 2

$$NH_3(g) + 100H_2O(l) \rightarrow NH_3(aq, 100H_2O); \quad \Delta H^\ominus = -35.6 \text{ kJ mol}^{-1}$$

$$HCl(g) + 100H_2O(l) \rightarrow HCl(aq, 100H_2O); \quad \Delta H^\ominus = -73.2 \text{ kJ mol}^{-1}$$

$$NH_3(aq, 100H_2O) + HCl(aq, 100H_2O) \rightarrow NH_4Cl(aq, 200H_2O); \quad \Delta H^\ominus = -50.2 \text{ kJ mol}^{-1}$$

(a) Add the equations in Method 1, and simplify the result.

(b) Add the equations in Method 2, and simplify the result.

(c) How does this illustrate Hess's law?

(d) Draw an enthalpy-level diagram and an enthalpy cycle.

We usually use the enthalpy cycle method in the solutions we provide to exercises because it is the most general, but you may prefer to add equations as in Exercise 82. You should always be able to justify your method by means of an enthalpy level diagram.

In the next exercise, you apply Hess's law to another reaction for which the enthalpy change cannot be measured directly - the hydration of magnesium sulphate, $MgSO_4$.

Exercise 83 The following experiment was performed to determine the enthalpy change for the process

$$MgSO_4(s) + 7H_2O(l) \rightarrow MgSO_4{\cdot}7H_2O(s)$$

The experiment was divided into two parts. The first part measured the enthalpy change of solution for the process

$$MgSO_4(s) + 100H_2O(l) \rightarrow MgSO_4(aq, 100H_2O)$$

and the second part measured the enthalpy change of solution for

$$MgSO_4{\cdot}7H_2O(s) + 93H_2O(l) \rightarrow MgSO_4(aq, 100H_2O)$$

Part 1 Solution of $MgSO_4(s)$

45.00 g of water was placed in a polystyrene cup and the temperature of the water was found to be 24.1 °C. 3.01 g of $MgSO_4$ was dissolved in the water and after carefully stirring the maximum temperature was found to be 35.4 °C.

Part 2 Solution of $MgSO_4{\cdot}7H_2O(s)$

41.85 g of water was placed in a polystyrene cup and the temperature of the water was found to be 24.8 °C. 6.16 g of $MgSO_4{\cdot}7H_2O$ was dissolved in the water and the minimum temperature was found to be 23.4 °C.

(a) Calculate the enthalpy change of solution for one mole of $MgSO_4$. Assume c_p = 4.18 kJ kg^{-1} K^{-1}.

(b) Similarly, calculate the enthalpy change of solution for one mole of $MgSO_4{\cdot}7H_2O$.

(c) By means of an enthalpy cycle, calculate the enthalpy change for the reaction:

$$MgSO_4(s) + 7H_2O(l) \rightarrow MgSO_4{\cdot}7H_2O$$

(d) Plot the results on an enthalpy-level diagram.

Now we look at another application of Hess's law.

Calculating enthalpy of formation from enthalpy of combustion

Most compounds cannot be formed directly from the elements, so it is impossible to measure these enthalpy chages in a calorimeter. Here again, Hess's law helps us to calculate enthalpies of formation from enthalpies of combustion, which can be measured directly.

<u>Worked Example</u>   Calculate the standard enthalpy of formation of
ethane, C_2H_6, given:

$$C(s) + O_2(g) \rightarrow CO_2(g); \quad \Delta H^{\ominus} = -394 \text{ kJ mol}^{-1}$$

$$H_2(g) + \tfrac{1}{2}O_2(g) \rightarrow H_2O(l); \quad \Delta H^{\ominus} = -286 \text{ kJ mol}^{-1}$$

$$C_2H_6(g) + 2\tfrac{1}{2}O_2(g) \rightarrow 2CO_2(g) + 3H_2O(l); \quad \Delta H^{\ominus} = -1560 \text{ kJ mol}^{-1}$$

<u>Solution</u>

1. Starting with the reaction for which you want to calculate $\Delta H^{\ominus}_f$, begin
to construct an enthalpy cycle:

$$\boxed{2C(s) + 3H_2(g)} \xrightarrow{\Delta H_1} \boxed{C_2H_6(g)}$$

2. Put in the combustion reactions:

$$\boxed{2C(s) + 3H_2(g) + 3\tfrac{1}{2}O_2(g)} \xrightarrow{\Delta H_1} \boxed{C_2H_6(g) + 3\tfrac{1}{2}O_2(g)}$$

$$\Delta H_2 \searrow \qquad \swarrow \Delta H_3$$

$$\boxed{2CO_2(g) + 3H_2O(l)}$$

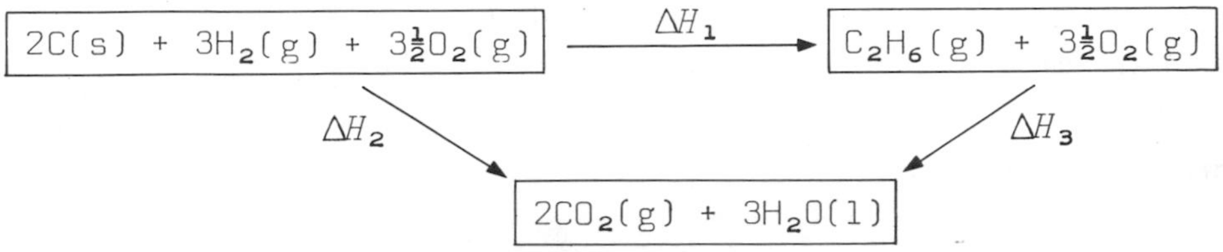

Note that the inclusion of $3\tfrac{1}{2}O_2(g)$, which is necessary for the combustions,
makes no different to ΔH_1.

3. Use Hess's law to equate the enthalpy changes.

$$\Delta H_2 = \Delta H_1 + \Delta H_3$$

or $\qquad \Delta H_1 = \Delta H_2 - \Delta H_3$

4. Substitute numerical values for ΔH_2 and ΔH_3

$$\Delta H_2 = 2 \times \Delta H^{\ominus}_C [C(s)] + 3 \times \Delta H^{\ominus}_C [H_2(g)]$$

$$= 2 \times (-394 \text{ kJ mol}^{-1}) + 3 \times (-286 \text{ kJ mol}^{-1})$$

$$= (788 - 858 \text{ kJ mol}^{-1}) = -1646 \text{ kJ mol}^{-1}$$

$$\Delta H_3 = \Delta H^{\ominus}_C [C_2H_6(g)] = -1560 \text{ kJ mol}^{-1}$$

$$\therefore \quad \Delta H_1 = \Delta H_2 - \Delta H_3$$

$$= -1646 \text{ kJ mol}^{-1} - (-1560 \text{ kJ mol}^{-1}) = \boxed{-86 \text{ kJ mol}^{-1}}$$

Exercises 84-85 test your understanding of this application of Hess's law.

<u>Exercise 84</u>   Calculate the standard enthalpy change of formation of
carbon disulphide, CS_2, given that

$$\Delta H^{\ominus}_f [CO_2(g)] = -393.5 \text{ kJ mol}^{-1},$$

$$\Delta H^{\ominus}_f [SO_2(g)] = -296.9 \text{ kJ mol}^{-1},$$

$$\Delta H^{\ominus}_C [CS_2(l)] = -1075.2 \text{ kJ mol}^{-1}.$$

<u>Exercise 85</u>   Calculate the standard enthalpy change of formation of
the following compounds:

(a) ethane, C_2H_6 (c) methylamine, CH_3NH_2
                              (Assume $N_2(g)$ is produced
(b) ethanol, C_2H_5OH on combustion.)

Having shown you how enthalpy changes of formation can be calculated, we now consider what use can be made of the values.

Calculating enthalpy of reaction from enthalpy of formation

In your data book you will find lists of enthalpy changes of formation for both organic and inorganic compounds. Many of the values have been calculated from experimental results in ways similar to those we have described.

These enthalpy changes of formation can be used to calculate the enthalpy changes in a reaction. The fact that we can predict the enthalpy change for any reaction is vitally important, for instance, to chemical engineers when planning chemical plant. They need to know how much heat will be generated or absorbed during the course of a particular reaction and make adequate provision for extremes of temperature in their designs.

Before you proceed, you should be familiar with the expression:

$$\Delta H^{\ominus}_{r} = \Sigma\Delta H^{\ominus}_{f}\,[\text{products}] - \Delta H^{\ominus}_{f}\,[\text{reactants}]$$

($\Delta H^{\ominus}_{r}$ is the standard enthalpy change for any reaction and the symbol Σ means 'the sum of'.)

This follows from the application of Hess's law to the enthalpy cycle shown below for a generalised reaction.

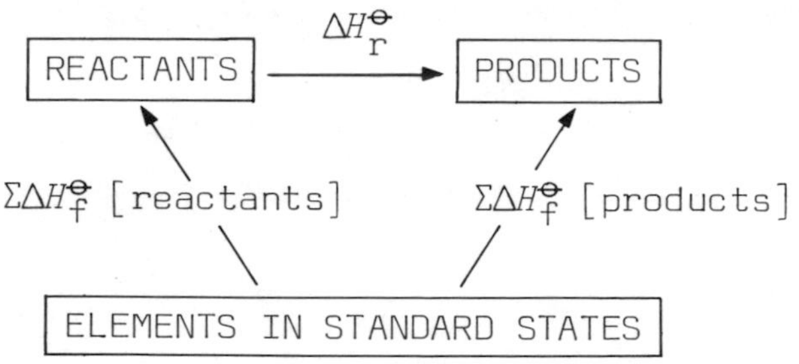

Once you have understood the derivation of the expression you can use it in calculations without drawing an enthalpy cycle, as we now show you in a Worked Example.

<u>Worked Example</u> Calculate the standard enthalpy change for the reaction

$$2H_2S(g) + SO_2(g) \rightarrow 3S(s) + 2H_2O(l)$$

using only standard enthalpy of formation data.

Compound	$\Delta H^{\ominus}_{f}$/kJ mol^{-1}
$H_2S(g)$	-20.6
$SO_2(g)$	-296.9
$H_2O(l)$	-285.9

Solution

1. Write an equation, leaving room to put the value of $\Delta H^\ominus_f$ under each formula.

$$2H_2S(g) + SO_2(g) \rightarrow 3S(s) + 2H_2O(l)$$

2. Calculate the standard enthalpy change of formation for the amount specified in the equation for each substance. In this equation, 2 mol of $H_2O(l)$ are produced.

$$2\ \Delta H^\ominus_f\ [H_2O(l)] = -571.8\ \text{kJ mol}^{-1}$$

Similarly, for $H_2S(g)$, $2\ \Delta H^\ominus_f\ [H_2S(g)] = -41.2\ \text{kJ mol}^{-1}$

and for $SO_2(g)$, $\Delta H^\ominus_f\ [SO_2(g)] = -296.9\ \text{kJ mol}^{-1}$

Because $\Delta H^\ominus_f$ for any element is zero, no value appears for sulphur.

3. Put these values under the compounds to which they refer.

	$2H_2S(g)$	+ $SO_2(g)$	$\rightarrow$ $3S(s)$	+ $2H_2O(l)$
$\Delta H^\ominus_f$/kJ mol^{-1}	-41.2	-296.9	0	-571.8

4. Add the values for the products.

$$\Sigma\Delta H^\ominus_f\ [\text{products}] = 0 - 571.8 = -571.8\ \text{kJ mol}^{-1}$$

Add the values for the reactants.

$$\Sigma\Delta H^\ominus_f\ [\text{reactants}] = (-41.2 - 296.9)\text{kJ mol}^{-1} = -338.1\ \text{kJ mol}^{-1}$$

5. Subtract:

$$\Delta H_r = \Sigma\Delta H^\ominus_f\ [\text{products}] - \Sigma\Delta H^\ominus_f\ [\text{reactants}]$$

$$= [-571.8 - (-338.1)]\ \text{kJ mol}^{-1} = \boxed{-233.7\ \text{kJ mol}^{-1}}$$

Now try some examples for yourself. Be very careful about positive and negative signs.

Exercise 86 Calculate the standard enthalpy changes for the following reactions:

(a) $CH_3OH(l) + 1\tfrac{1}{2}O_2(g) \rightarrow CO_2(g) + 2H_2O(l)$

(b) $2CO(g) + O_2(g) \rightarrow 2CO_2(g)$

(c) $ZnCO_3(s) \rightarrow ZnO(s) + CO_2(g)$

(d) $2Al(s) + Fe_2O_3(s) \rightarrow 2Fe(s) + Al_2O_3(s)$

Values of $\Delta H^\ominus_f$/kJ mol^{-1} are as follows:

CH_3OH	CO_2	CO	H_2O	ZnO	$ZnCO_3$	Al_2O_3	Fe_2O_3
-238.9	-393.5	-110.5	-285.9	-348.0	-812.5	-1675.7	-822.2

Exercise 87 The standard enthalpies of formation ($\Delta H^\ominus_f$) at 298 K for a number of compounds are, in kJ mol^{-1}:

$CH_4(g)$ -75; $H_2O(g)$ -242; $CO(g)$ -110

Calculate the enthalpy change for the reaction:

$$CH_4(g) + H_2O(g) \rightarrow CO(g) + 3H_2(g)$$

We now look at the enthalpy changes associated with the making and breaking of covalent bonds in molecules.

BOND ENTHALPY TERM (AVERAGE BOND ENERGY)

The energy required to break a particular bond depends not only on the nature of the two bonded atoms but also on the environment of those atoms. For instance, not all the C—H bonds in ethanol, C_2H_5OH, are equally strong because the oxygen atom has more influence on those near it than on those further away.

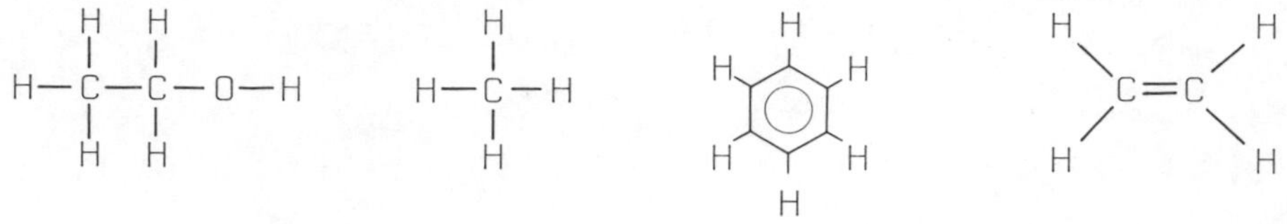

Similarly, C—H bonds in methane, CH_4, are not quite the same as those in benzene, C_6H_6, or ethane, C_2H_4. Fortunately, however, the differences are small enough for the concept of average bond energy (bond enthalpy term) to be very useful.

The enthalpy change associated with the breaking of a particular bond, i.e. the energy required to break it, is known as the bond dissociation enthalpy (expressed per mole of bonds). The bond in question is best specified by an equation, as shown in the table below for four successive dissociation reactions of methane.

Reaction	Bond dissociation enthalpy/kJ mol^{-1}
$CH_4(g) \rightarrow CH_3(g) + H(g)$	+435
$CH_3(g) \rightarrow CH_2(g) + H(g)$	+444
$CH_2(g) \rightarrow CH(g) + H(g)$	+440
$CH(g) \rightarrow C(g) + H(g)$	+343

The enthalpy change of complete dissociation of methane into gaseous atoms, by breaking all the C—H bonds, is obtained by adding the four equations:

$$CH_4(g) \rightarrow C(g) + 4H(g); \quad \Delta H^{\ominus} = +1662 \text{ kJ mol}^{-1}$$

The bond enthalpy term for C—H bonds in methane is the average of the four bond dissociation enthalpies and is denoted by $\overline{E}$(C—H),

i.e. $\overline{E}(\text{C—H}) = (1662/4)\text{kJ mol}^{-1} = 415 \text{ kJ mol}^{-1}$

We cannot always obtain bond enthalpy terms by the method just described because bond dissociation enthalpies are often not known. In the next section we look at another method of calculation.

Calculating bond enthalpy terms

Once again we can apply Hess's law to determine an enthalpy change which cannot be measured directly. We illustrate this method by a Worked Example in which we make use of tabulated values of enthalpy of formation of compounds and enthalpy of atomization of elements.

The <u>standard enthalpy (change) of atomization</u> of an element is the enthalpy change in the production of <u>one mole of gaseous atoms</u> from the element in its standard state. For example, it refers to processes such as

$$Na(s) \rightarrow Na(g) \quad \text{and} \quad \tfrac{1}{2}Cl_2(g) \rightarrow Cl(g)$$

<u>Worked Example</u> Calculate the C—H bond energy term in methane from standard enthalpies of atomization and formation.

W

<u>Solution</u>

1. Write the equation for the complete dissociation of methane.

$$CH_4(g) \rightarrow C(g) + 4H(g)$$

2. Construct an enthalpy cycle including the elements in their standard states.

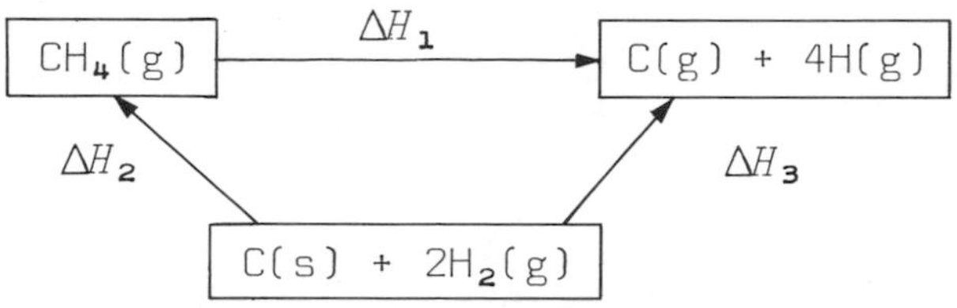

Here ΔH_2 is the standard enthalpy of formation of methane, $\Delta H^{\ominus}_{f}[CH_4(g)]$; and ΔH_3 is the standard enthalpy of atomization of carbon plus four times the standard enthalpy of atomization of hydrogen; i.e.

$$\Delta H_3 = \Delta H^{\ominus}_{at}[C(g)] + 4\,\Delta H^{\ominus}_{at}[H(g)]$$

3. Look up these values in your data book:

$$\Delta H^{\ominus}_{f}[CH_4(g)] = -74.8 \text{ kJ mol}^{-1}$$

$$\Delta H^{\ominus}_{at}[C(g)] = +715 \text{ kJ mol}^{-1}$$

$$4\,\Delta H^{\ominus}_{at}[H(g)] = 4(+218.0 \text{ kJ mol}^{-1}) = +872.0 \text{ kJ mol}^{-1}$$

4. Apply Hess's law

$$\Delta H_2 = \Delta H_1 - \Delta H_3$$

5. Substitute the appropriate values and solve for ΔH

$$\Delta H_1 = \Delta H_3 - \Delta H_2 = (715.0 + 872.0 - -74.8) \text{ kJ mol}^{-1} = 1661.8 \text{ kJ mol}^{-1}$$

6. This is the energy required to break four bonds. We want to know the average value for one bond.

$$\therefore \quad \overline{E}(C—H) = \frac{1661.8}{4} \text{ kJ mol}^{-1} = \boxed{415 \text{ kJ mol}^{-1}}$$

Now try two similar problems.

<u>Exercise 88</u> Determine the enthalpy change for the process:

$$H_2S(g) \rightarrow 2H(g) + S(g)$$

and so calculate the bond enthalpy term, $\overline{E}(H—S)$.

<u>Exercise 89</u> Calculate $\overline{E}(H—H)$, given the following data:

$$\tfrac{1}{2}N_2(g) + \tfrac{3}{2}H_2(g) \rightarrow NH_3(g); \quad \Delta H^{\ominus} = -46 \text{ kJ mol}^{-1}$$

$$\tfrac{1}{2}H_2(g) \rightarrow H(g); \quad \Delta H^{\ominus} = +218 \text{ kJ mol}^{-1}$$

$$\tfrac{1}{2}N_2(g) \rightarrow N(g); \quad \Delta H^{\ominus} = +473 \text{ kJ mol}^{-1}$$

Notice that in all these examples and exercises, reactants and products are in the gaseous state. If the substance we are decomposing exists as a liquid in its standard state, it must first be vaporized, and this requires energy.

$$CCl_4(l) \rightarrow CCl_4(g); \quad \Delta H^{\ominus}_{vap}[CCl_4(l)] = 30.5 \text{ kJ mol}^{-1}$$

Apply this idea in the next exercise.

Exercise 90 Calculate the enthalpy change for the process:

$$CCl_4(g) \rightarrow C(g) + 4Cl(g)$$

and calculate the bond enthalpy term, $\overline{E}$(C—Cl).

If we assume that molecules of very similar substances have very similar bonds, we can extend the calculations used above in order to determine two bond enthalpy terms by solving simultaneous equations.

Determining two bond enthalpy terms simultaneously

Work through the following Revealing Exercise, where we show you how to do this calculation. You need information obtained from enthalpies of formation and atomization as in the previous exercises:

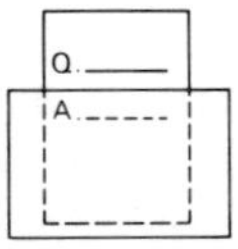

butane: $C_4H_{10}(g) \rightarrow 4C(g) + 10H(g); \quad \Delta H^{\ominus} = +5165 \text{ kJ mol}^{-1}$

pentane: $C_5H_{12}(g) \rightarrow 5C(g) + 12H(g); \quad \Delta H^{\ominus} = +6337 \text{ kJ mol}^{-1}$

Q1. How many bonds are broken in the atomization of butane and pentane?

A1. In butane, 3 C—C bonds and 10 C—H bonds are broken per molecule; in pentane, 4 C—C bonds and 12 C—H bonds are broken per molecule.

Q2. Express the enthalpy changes for the atomization of pentane and butane as the sums of the bond enthalpy terms.

A2. Butane: $3\overline{E}(\text{C—C}) + 10\overline{E}(\text{C—H}) = +5165 \text{ kJ mol}^{-1}$

Pentane: $4\overline{E}(\text{C—C}) + 12\overline{E}(\text{C—H}) = +6337 \text{ kJ mol}^{-1}$

Notice that for a compound, $\Delta H^{\ominus}_{at}$ refers to one mole of molecules, whereas for an element it refers to one mole of atoms.

Q3. Let $\overline{E}(\text{C—C}) = x \text{ kJ mol}^{-1}$ and $\overline{E}(\text{C—H}) = y \text{ kJ mol}^{-1}$ — Solve these equations simultaneously.

A3. $3x + 10y = 5165$ - - - - - - (1)

$4x + 12y = 6337$ - - - - - - (2)

Multiply equation (1) by 4 and equation (2) by 3 and subtract:

$$12x + 40y = 20660$$
$$12x + 36y = 19011$$
$$4y = 1649 \quad \therefore y = 412.25$$

Substitute this value for y into either equation (1) or (2).

$3x + 10(412.25) = 5165$

$3x = 5165 - 4122.5 = 1042.5 \quad \therefore x = 347.5$

$\therefore \bar{E}(C—C) = \boxed{347.5 \text{ kJ mol}^{-1}} \qquad \bar{E}(C—H) = \boxed{412.2 \text{ kJ mol}^{-1}}$

The next exercise can be done in the same way.

<u>Exercise 91</u> The enthalpies of complete dissociation (atomization) of gaseous hexane and heptane are 7512 kJ mol^{-1} and 8684 kJ mol^{-1} respectively. Calculate bond enthalpy terms for C—C and C—H bonds.

Now that we have shown you how bond enthalpy terms can be obtained we illustrate their use in estimating enthalpy changes in reactions for which there are no experimental data which can be applied to an enthalpy cycle.

Use of bond enthalpy terms to estimate enthalpy changes

In order to obtain a bond enthalpy term which can be tabulated and used generally, an average is taken for a particular type of bond in a large number of compounds. Values quoted in your data book should not be considered as accurately stating the bond strength in any particular molecular environment. Calculations using bond enthalpy terms can give only approximate answers.

Before presenting a Worked Example, we remind you that:

BOND BREAKING REQUIRES ENERGY (ENDOTHERMIC)
BOND MAKING RELEASES ENERGY (EXOTHERMIC)

<u>Worked Example</u> Use bond enthalpy terms to calculate the enthalpy change for the reaction:

$$CH_4(g) + Cl_2(g) \rightarrow CH_3Cl(g) + HCl(g)$$

<u>Solution</u>

1. Write the equation using structural formulae.

```
     H                              H
     |                              |
  H—C—H + Cl—Cl        →         H—C—Cl + H—Cl
     |                              |
     H                              H
```

2. List the bonds broken and bonds made under the equations as shown.

```
     H                              H
     |                              |
  H—C—H + Cl—Cl                  H—C—Cl + H—Cl
     |                              |
     H                              H
```

bonds broken: C—H and Cl—Cl

bonds made: C—Cl and H—Cl

3. Look up the bond enthalpy terms for the bonds broken and made. Add them up as shown, including negative signs for the exothermic bond-making.

$$\begin{array}{ccc}
& H & \\
& | & \\
H - C - H & + & Cl - Cl \\
& | & \\
& H &
\end{array}
\qquad \rightarrow \qquad
\begin{array}{ccc}
& H & \\
& | & \\
H - C - Cl & + & H - Cl \\
& | & \\
& H &
\end{array}$$

bonds broken

$\overline{E}(C-H)$ = 435 kJ mol^{-1}
$\overline{E}(Cl-Cl)$ = 242 kJ mol^{-1}
Total = 677 kJ mol^{-1}

bonds made

$-\overline{E}(C-Cl)$ = -339 kJ mol^{-1}
$-\overline{E}(H-Cl)$ = -431 kJ mol^{-1}
Total = -770 kJ mol^{-1}

4. Add the values for bond breaking and making (including the correct signs) to obtain the required enthalpy change.

ΔH = 677 kJ mol^{-1} - 770 kJ mol^{-1} = -93 kJ mol^{-1}

5. Write the complete thermochemical equation.

$CH_4(g) + Cl_2(g) \rightarrow CH_3Cl(g) + HCl(g);$ $\Delta H = \boxed{-93 \text{ kJ mol}^{-1}}$

The following enthalpy-level diagram summarises the enthalpy changes during the reaction we have just considered. Note that we have omitted the symbol $\ominus$ because the value is only an estimate and conditions are not standard.

$CH_3 + H + Cl + Cl$

Energy required to break a $Cl-Cl$ bond = +242 kJ mol^{-1}

$CH_3 + H + Cl_2$

H

Energy released in making a $C-Cl$ bond = -339 kJ mol^{-1}

Energy required to break a $C-H$ bond = 435 kJ mol^{-1}

$CH_3Cl + H + Cl$

$CH_4 + Cl_2$

Energy released in making an $H-Cl$ bond = -431 kJ mol^{-1}

ΔH_r = -93 kJ mol^{-1}

$CH_3Cl + HCl$

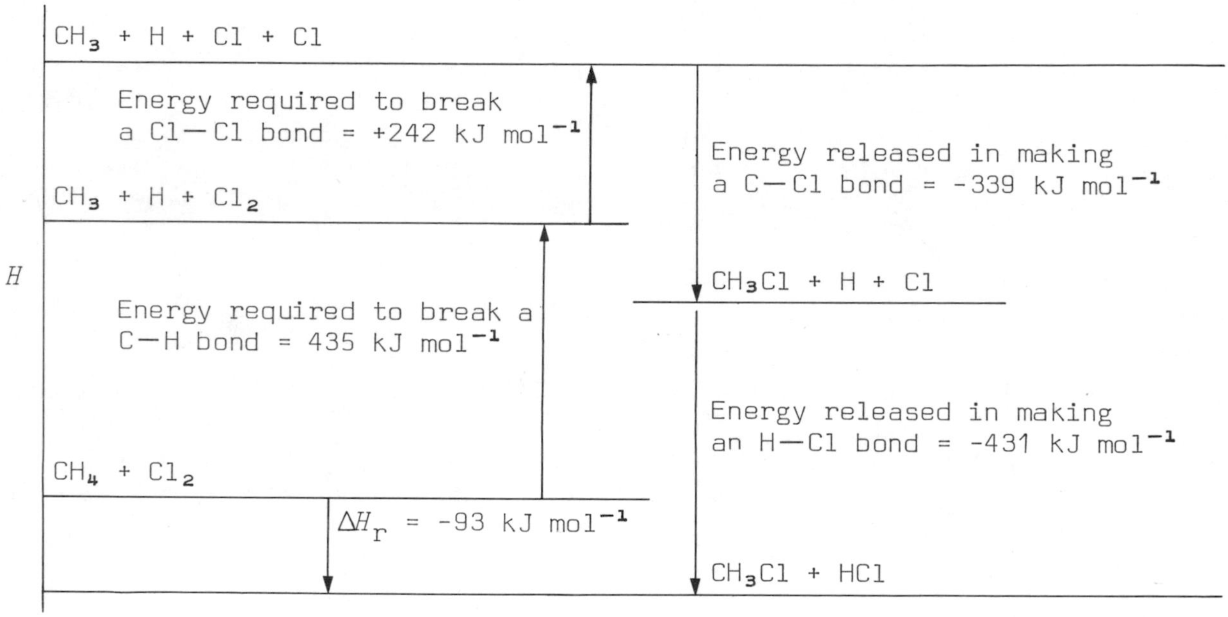

The following exercises are similar.

Exercise 92 Some bond enthalpy terms (in kJ mol^{-1}) are shown in the table below.

$\overline{E}(H-H)$	$\overline{E}(C-Cl)$	$\overline{E}(H-Cl)$	$\overline{E}(N\equiv N)$	$\overline{E}(N-H)$
+436	+242	+431	+945	+398

Calculate enthalpy changes for the following reactions:

(a) $H_2(g) + Cl_2(g) \rightarrow 2HCl(g)$

(b) $N_2(g) + 3H_2(g) \rightarrow 2NH_3(g)$

Exercise 93   Some bond energy terms are listed below:

Bond	H—H	C—H	C—Br	C—C	C=C	Br—Br
Bond energy /kJ mol^{-1}	435	415	284	356	598	193

(a)   What do you understand by <u>bond energy (enthalpy) term</u>?

(b)   Using the given data, calculate the enthalpies of formation, from gaseous atoms, of

    (i)   gaseous propene, (propylene),

$$\begin{array}{ccc} H & & H \\ \backslash & & | \\ C = C & - & C - H \\ / & | & | \\ H & H & H \end{array}$$

    (ii)   gaseous 1,2-dibromopropane

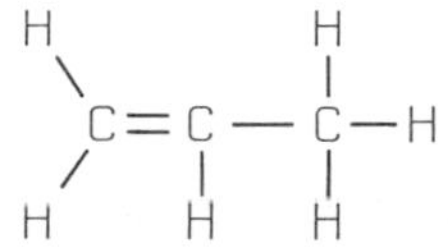

$$\begin{array}{ccc} H & Br & H \\ | & | & | \\ H - C - C - C - H \\ | & | & | \\ Br & H & H \end{array}$$

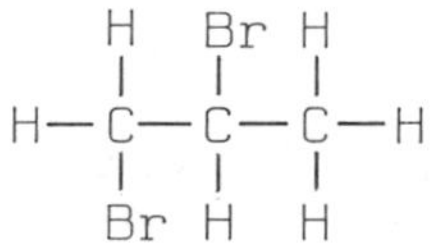

(c)   Calculate the enthalpy change, $\Delta H^{\ominus}$, for the reaction:

$$CH_2\!=\!CH\!-\!CH_3(g) + Br_2(g) \rightarrow CH_2BrCHBrCH_3(g)$$

Exercise 94   (a)   Use the values of bond enthalpy terms listed in your data book to calculate the standard enthalpy change for the reaction:

$$C_2H_6(g) + Cl_2(g) \rightarrow C_2H_5Cl(g) + HCl(g)$$

(b)   Calculate another value for this enthalpy change from enthalpies of formation.

(c)   Write a short account of the reasons why the two values you have calculated differ from each other.

So far, your work has been confined to covalent compounds. You have seen how to use bond energy terms to calculate the enthalpy change of formation of a molecular compound from its constituent gaseous atoms - an important step in the calculation of $\Delta H^{\ominus}$ for a reaction. Now we look at energy changes in the formation of ionic compounds.

LATTICE ENTHALPY

It is often useful to know the enthalpy of formation of an ionic compound from its constituent gaseous ions. This quantity is called the 'lattice enthalpy' (or lattice energy) of the compound, because ionic bonding always leads to a solid crystal lattice. It provides a measure of the strength of ionic bonds.

<u>The lattice enthalpy of an ionic compound</u> is the enthalpy change which occurs when one mole of it is formed, as a crystal lattice, from its constituent gaseous ions.

This definition of lattice enthalpy always gives a negative sign, e.g.

    $Na^+(g) + Cl^-(g) \rightarrow Na^+Cl^-(s)$; $\Delta H^{\ominus}_{lat} = -781$ kJ mol^{-1}

If we were considering the energy required to separate the salt into its separate ions then the enthalpy change would have a positive sign:

$$Na^+Cl^-(s) \rightarrow Na^+(g) + Cl^-(g); \quad \Delta H^{\ominus}_{lat} = +781 \text{ kJ mol}^{-1}$$

You may find that some books quote positive values of lattice enthalpy, but you will not go wrong if you always relate $\Delta H^{\ominus}$ to the appropriate equation.

Since it is impossible to determine lattice enthalpies directly by experiment we use an indirect method where we construct an enthalpy diagram called a Born-Haber cycle.

Calculating lattice enthalpy using a Born-Haber cycle

The Born-Haber cycle is yet another application of Hess's law but the alternative routes involve more steps than you have used so far.

To construct a Born-Haber cycle you need two enthalpy changes not previously mentioned in this chapter, ionization enthalpy (energy) and electron affinity.

The ionization enthalpy (energy) of an element is the enthalpy change which occurs when one mole of its gaseous atoms loses one mole of electrons to form one mole of gaseous positive ions, e.g.

$$Na(g) \rightarrow Na^+(g); \quad \Delta H^{\ominus}_i = +500 \text{ kJ mol}^{-1}$$

The electron affinity of an element is the enthalpy change which occurs when one mole of its gaseous atoms accepts one mole of electrons to form one mole of gaseous negative ions, e.g.

$$Br(g) + e^- \rightarrow Br^-(g); \quad \Delta H^{\ominus}_e = -342 \text{ kJ mol}^{-1}$$

We start with a Revealing Exercise concerned with a Born-Haber cycle.

Below is a generalised Born-Haber cycle for an ionic compound.

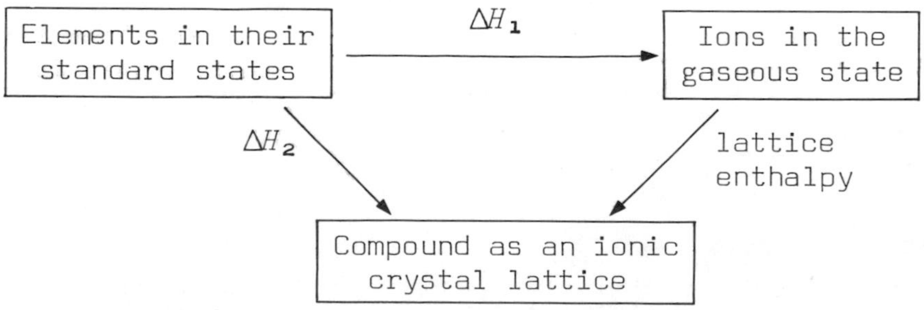

Q1. Redraw the cycle including the appropriate formulae for the formation of sodium chloride.

A1.

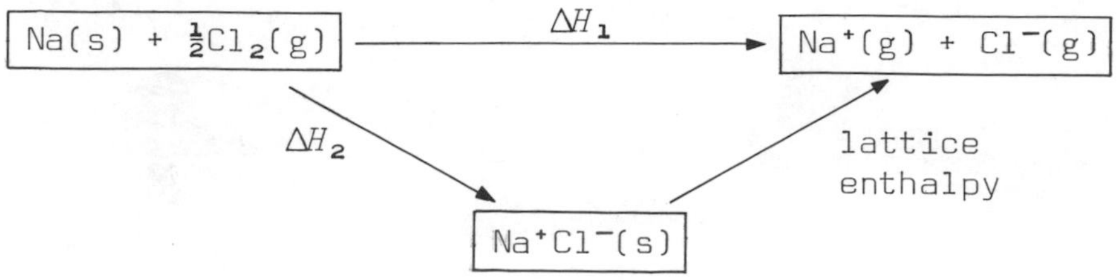

Q2. ΔH_1 is the sum of the enthalpy changes associated with four steps, two for sodium and two for chlorine. Write thermochemical equations for these four steps.

A2. $Na(s) \rightarrow Na(g)$; $\Delta H^{\ominus}_{at} = +108\ kJ\ mol^{-1}$

$Na(g) \rightarrow Na^+(g)$; $\Delta H^{\ominus}_{i} = +500\ kJ\ mol^{-1}$

$\frac{1}{2}Cl_2(g) \rightarrow Cl(g)$; $\Delta H^{\ominus}_{at} = +121\ kJ\ mol^{-1}$

$Cl(g) + e^- \rightarrow Cl^-(g)$; $\Delta H^{\ominus}_{e} = -364\ kJ\ mol^{-1}$

Q3. Calculate ΔH_1.

A3. $\Delta H_1 = \Delta H^{\ominus}_{at}(Na) + \Delta H^{\ominus}_{i}(Na) + \Delta H^{\ominus}_{at}(\frac{1}{2}Cl_2) + \Delta H^{\ominus}_{e}(Cl)$

So $\Delta H_1 = (+108 + 500 + 121 - 364)\ kJ\ mol^{-1} = +365\ kJ\ mol^{-1}$

Q4. What name is given to ΔH_2? Look up a value in your data book.

A4. ΔH_2 is the enthalpy change of formation of sodium chloride.

$\Delta H^{\ominus}_{f}[NaCl(s)] = -411\ kJ\ mol^{-1}$

Q5. Insert the values for ΔH_1 and ΔH_2 in the enthalpy cycle.

A5.

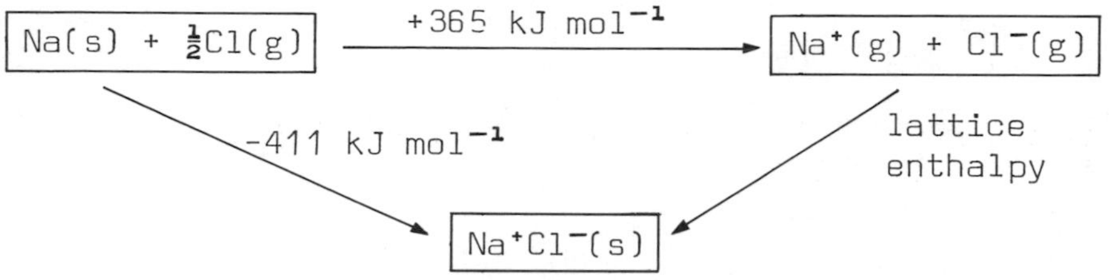

Q6. Use Hess's law to calculate the lattice enthalpy for sodium chloride.

A6. $365\ kJ\ mol^{-1}$ + lattice enthalpy = $-411\ kJ\ mol^{-1}$

$\therefore$ lattice enthalpy = $-411\ kJ\ mol^{-1} - 365\ kJ\ mol^{-1}$ = $\boxed{-776\ kJ\ mol^{-1}}$

Born-Haber cycles are often drawn as enthalpy level diagrams. Compare the following enthalpy-level diagram with the cycle you drew in the Revealing Exercise and trace the various steps.

This Born-Haber cycle is drawn approximately to scale, but it is quite acceptable to ignore the scale in order to save time and space.

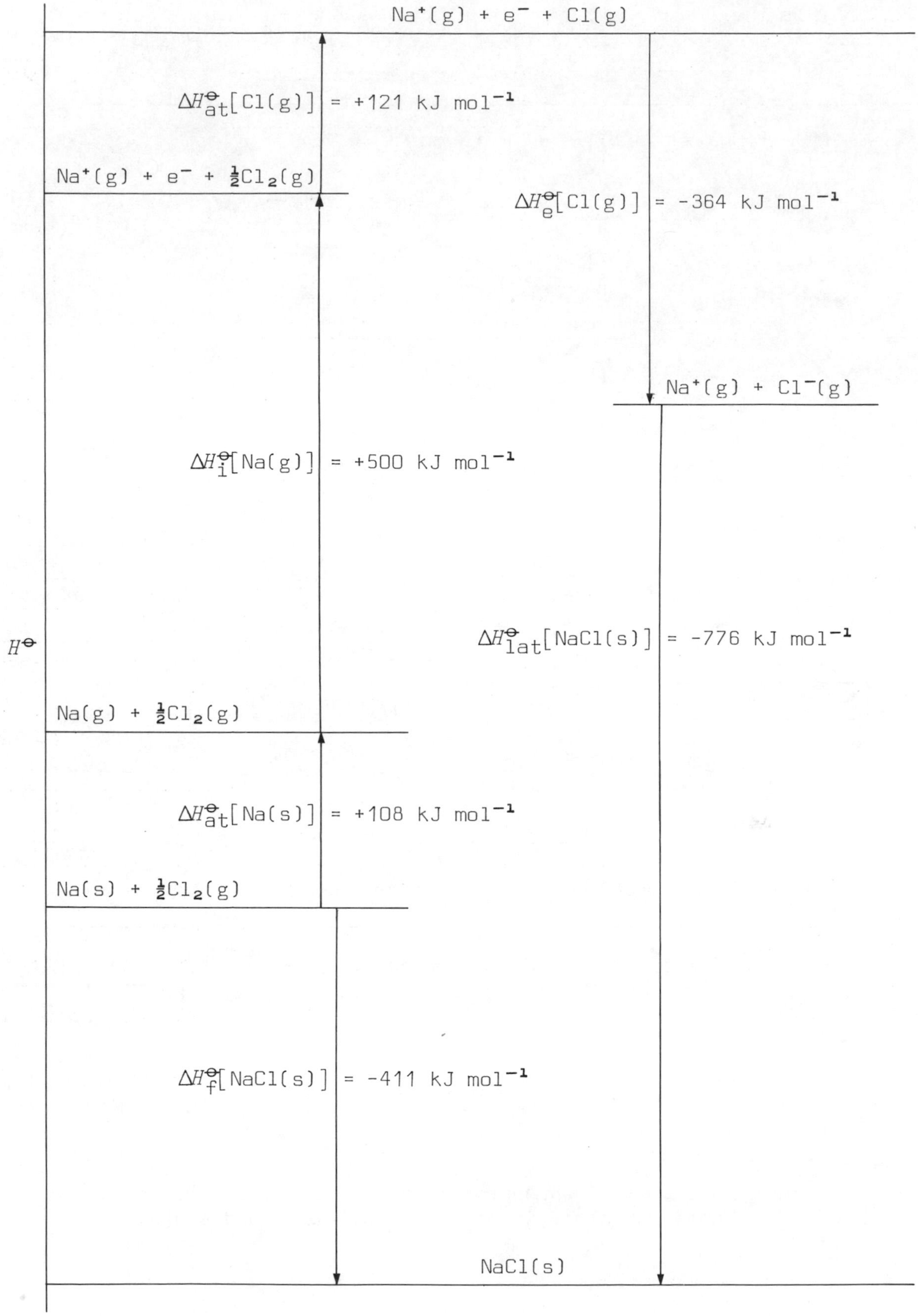

Born-Haber cycle for sodium chloride, drawn approximately to scale.

Try drawing some Born-Haber cycles for yourself in the following exercises, and use the cycles to calculate some lattice enthalpies.

Exercise 95 Draw a Born-Haber cycle for each of the following ionic compounds, and then calculate their lattice enthalpies. (Note that in sodium hydride, the hydrogen forms a negative ion.) The cycles need not be drawn to scale.

Compound	$\Delta H^\ominus_f$ /kJ mol^{-1}	Metal		Non-metal	
		$\Delta H^\ominus_{at}$ /kJ mol^{-1}	$\Delta H^\ominus_i$ /kJ mol^{-1}	$\Delta H^\ominus_{at}$ /kJ mol^{-1}	$\Delta H^\ominus_e$ /kJ mol^{-1}
KBr (K^+, Br^-)	-392	+ 89	+420	+112	-342
NaH (Na', H^-)	- 57	+108	+500	+218	- 72

Exercise 96 Complete the table below and draw Born-Haber cycles to obtain lattice enthalpies. Note that in this example you may have to combine successive values for $\Delta H^\ominus_i$ and $\Delta H^\ominus_e$. Also, whereas $\Delta H^\ominus_i$ is always positive, $\Delta H^\ominus_e$ may be positive or negative.

Compound	$\Delta H^\ominus_f$ /kJ mol^{-1}	Metal		Non-metal	
		$\Delta H^\ominus_{at}$ /kJ mol^{-1}	$\Delta H^\ominus_i$ /kJ mol^{-1}	$\Delta H^\ominus_{at}$ /kJ mol^{-1}	$\Delta H^\ominus_e$ /kJ mol^{-1}
$BaCl_2$ (Ba^{2+}, $2Cl^-$)	-860	+175	+ 500 +1000	+121	-364
SrO (Sr^{2+}, O^{2-})					

If lattice enthalpies are known (e.g. from theoretical calculations) Born-Haber cycles may be used to determine other enthalpy changes, particularly electron affinities, as in the next two exercises.

Exercise 97

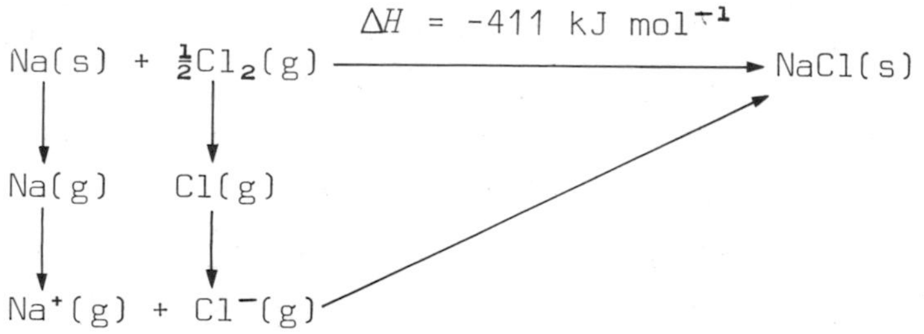

Use the above diagram and the following data to calculate ΔH for the reaction:

$Cl(g) + e^- \rightarrow Cl^-(g)$

$Na(s) \rightarrow Na(g)$; $\Delta H = 108$ kJ mol^{-1}

$Na(g) \rightarrow Na^+(g)$; $\Delta H = 500$ kJ mol^{-1}

$\frac{1}{2}Cl(g) \rightarrow Cl(g)$; $\Delta H = 121$ kJ mol^{-1}

$NaCl(s) \rightarrow Na^+(g) + Cl^-(g)$; $\Delta H = 776$ kJ mol^{-1}

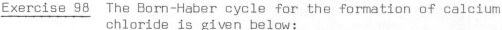

Exercise 98 The Born-Haber cycle for the formation of calcium chloride is given below:

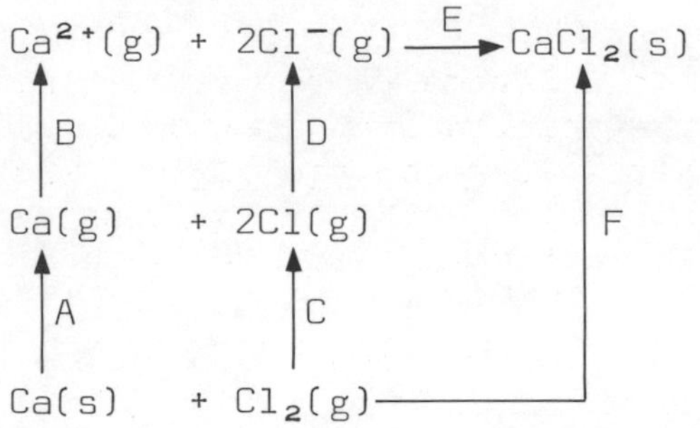

A = 177 kJ mol^{-1} C = 242 kJ mol^{-1} F = -795 kJ mol^{-1}

B = 1690 kJ mol^{-1} E = -2197 kJ mol^{-1}

(a) A is the enthalpy change of sublimation (atomization) of solid calcium. Similarly, define the following:

(i) B (ii) C (iii) E (iv) F

(b) Calculate the enthalpy change D.

Another use of Born-Haber cycles enables us to discuss the stoichiometry of compounds which might be formed by direct combination.

In the exercise which follows, you construct Born-Haber cycles for MgCl, $MgCl_2$ and $MgCl_3$ to determine enthalpies of formation, and then decide which would be the most likely formula for magnesium chloride.

Exercise 99 (a) Construct Born-Haber cycles for MgCl, $MgCl_2$ and $MgCl_3$, inserting all the values except $\Delta H^\ominus_f$. Since experimentally determined lattice enthalpies for MgCl and $MgCl_3$ are not available, use the theoretically calculated values:

$$\Delta H^\ominus_{lat}[MgCl] = -753 \text{ kJ mol}^{-1}$$

$$\Delta H^\ominus_{lat}[MgCl_3] = -5440 \text{ kJ mol}^{-1}$$

(b) Use the cycles to obtain values for

(i) $\Delta H^\ominus_f[MgCl]$ (ii) $\Delta H^\ominus_f[MgCl_2]$ (iii) $\Delta H^\ominus_f[MgCl_3]$

(c) Which of the three compounds MgCl, $MgCl_2$, $MgCl_3$ is/are energetically stable with respect to the elements?

(d) Calculate the enthalpy change for the hypothetical reaction

$$2MgCl(s) \rightarrow MgCl_2(s) + Mg(s)$$

using the $\Delta H^\ominus_f$ values you calculated in part (b).

(e) Discuss briefly the relative stabilities of MgCl and $MgCl_2$ in the light of your answer to (d). Does this explain why MgCl is not known?

We now give you another method for determining experimentally the enthalpy change of a reaction, i.e. using the technique of thermometric titration.

THERMOMETRIC TITRATIONS

The data below are taken from an experiment used to determine the concentrations of two acids, hydrochloric acid, HCl, and ethanoic acid, CH_3CO_2H, by thermometric titration. The temperature of the mixture was measured at regular intervals during the titration. The enthalpy change for each reaction can be calculated from the maximum temperature rise.

Titration of hydrochloric acid with 50.0 cm^3 of 1.00 M NaOH

Volume added/cm^3	0.0	5.0	10.0	15.0	20.0	25.0	30.0	35.0	40.0	45.0	50.0
Temperature/^oC	22.2	24.4	26.4	28.4	30.1	31.1	30.4	29.9	29.2	28.8	28.2

Titration of ethanoic acid with 50.0 cm^3 of 1.00 M NaOH

Volume added cm^3	0.0	5.0	10.0	15.0	20.0	25.0	30.0	35.0	40.0	45.0	50.0
Temperature ^oC	21.0	22.6	24.3	25.9	27.2	28.7	28.6	28.0	27.5	27.0	26.5

<u>Exercise 100</u> (a) Plot temperature (y-axis) against volume of acid added (x-axis) for each acid on the same graph.

(b) Extend the straighter portions of the curves near the top, as shown in Fig. 12. The point at which they meet corresponds to both the volume of acid required for neutralization and to the maximum temperature.

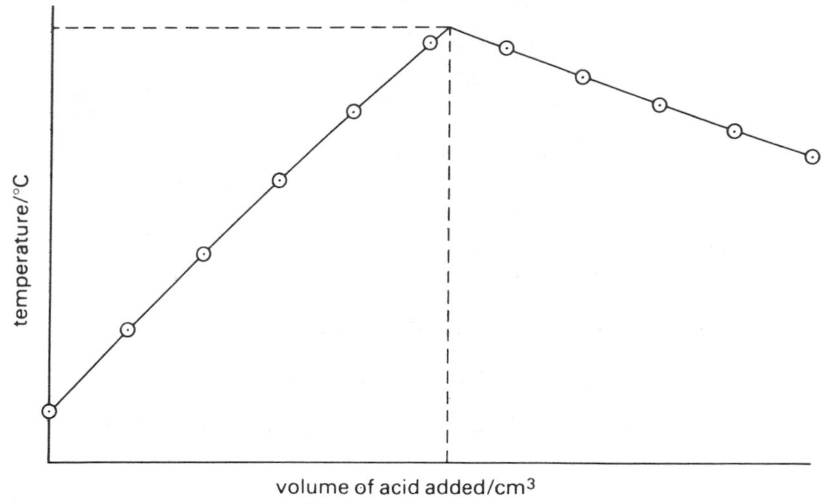

Fig.12.

(c) Calculate the concentration of each of the acids.

(d) From the maximum temperature rise, determine the quantity of energy released in each titration. Assume that the specific heat capacity of the solutions is the same as that for water, 4.18 kJ kg^{-1} K^{-1} and that the heat capacity of the cup is zero.

(e) Calculate the standard enthalpy change of neutralization for each reaction.

We have omitted calculations involving changes in free enthalpy and entropy, $\Delta G^\ominus$ and $\Delta S^\ominus$, since these topics are now excluded from most examination syllabuses at A-level. In any case, it is the underlying concepts that students find difficult, not the calculations themselves. Many of the calculations follow the same pattern as those using $\Delta H^\ominus$, while an exposition of the concepts is beyond the scope of this book.

For reversible reactions, $\Delta H^\ominus$ is related mathematically to the equilibrium constant. Some calculations involving this relationship are included in Chapter 6, and reference to the relationship between $\Delta H^\ominus$ and standard electrode potential, $E^\ominus$, is made in Chapter 8.

END-OF-CHAPTER QUESTIONS

Exercise 101 The standard enthalpies of combustion of carbon, hydrogen and methane are -394, -286 and -891 kJ mol^{-1} respectively.

Calculate the enthalpy of formation of methane.

Exercise 102 The enthalpy changes for two reactions are given by the equations:

$2Cr(s) + 1\frac{1}{2}O_2(g) \rightarrow Cr_2O_3(s)$; $\Delta H^\ominus$ = -1130 kJ mol^{-1}

$C(s) + \frac{1}{2}O_2(g) \rightarrow CO(g)$; $\Delta H^\ominus$ = -110 kJ mol^{-1}

What is the enthalpy change, in kJ, for the following reaction?

$3C(s) + Cr_2O_3(s) \rightarrow 2Cr(s) + 3CO(g)$

Exercise 103

	$CS_2(l)$	$NOCl(g)$	$CCl_4(l)$	$SO_2(g)$
$\Delta H_f^\ominus$/kJ mol^{-1}	88	53	-139	-296

From the data above, calculate the value of $\Delta H^\ominus$ in kJ mol^{-1} for the reaction:

$CS_2(l) + 4NOCl(g) \rightarrow CCl_4(l) + 2SO_2(g) + 2N_2(g)$

Exercise 104 Calculate the bond enthalpy term for the O—H bond in water from the following data:

$H_2(g) + \frac{1}{2}O_2(g) \rightarrow H_2O(g)$; $\Delta H^\ominus_{298}$ = -242 kJ mol^{-1}

$H_2(g) \rightarrow 2H(g)$; $\Delta H^\ominus_{298}$ = +436 kJ mol^{-1}

$O_2(g) \rightarrow 2O(g)$; $\Delta H^\ominus_{298}$ = +500 kJ mol^{-1}

Exercise 105 Calculate $\Delta H^\ominus$ at 25 °C for the reaction

$H(g) + Br(g) \rightarrow HBr(g)$

from the following data, which refer to 25 °C:

$H_2(g) \rightarrow 2H(g)$; $\Delta H^\ominus$ = +436 kJ mol^{-1}

$Br_2(g) \rightarrow 2Br(g)$; $\Delta H^\ominus$ = +193 kJ mol^{-1}

$H_2(g) + Br_2(g) \rightarrow 2HBr(g)$; $\Delta H^\ominus$ = -104 kJ mol^{-1}

Exercise 106 (a) Calculate the standard enthalpy of formation of ethanoic acid, $CH_3CO_2H(l)$, from the following data:

$\Delta H^{\ominus}_c$ [graphite] = -394 kJ mol^{-1}

$\Delta H^{\ominus}_c$ [$H_2(g)$] = -286 kJ mol^{-1}

$\Delta H^{\ominus}_c$ [$CH_3CO_2H(l)$] = -873.2 kJ mol^{-1}

Exercise 107 The standard enthalpy of formation at 298 K of carbon dioxide, CO_2, is -394 kJ mol^{-1} whilst that of carbon monoxide, CO, is -110 kJ mol^{-1}.

Calculate the enthalpy change for the reaction:

$$CO_2(g) \rightarrow CO(g) + \tfrac{1}{2}O_2(g)$$

Exercise 108 The following table lists the enthalpy changes of combustion of several monohydric alcohols.

alcohol	$\Delta H^{\ominus}_c$/kJ mol^{-1}
methanol	- 715
ethanol	-1367
propan-1-ol	-2017
butan-1-ol	-2675

(a) Draw a plot of $\Delta H^{\ominus}_c$ against the relative molecular mass of the alcohols.

(b) From your graph, estimate a value for the enthalpy change of combustion of pentan-1-ol.

(c) Given $\Delta H^{\ominus}_f[H_2O(l)]$ = -286 kJ mol^{-1} and $\Delta H^{\ominus}_f[CO_2(g)]$ = -394 kJ mol^{-1}, calculate the standard enthalpy change of formation of ethanol.

(d) The enthalpy change of combustion of ethane-1,2-diol is -1180 kJ mol^{-1}. What would you expect the corresponding value for propane-1,3-diol to be?

Exercise 109 (a) The standard enthalpy changes, in kJ mol^{-1}, of the following reactions refer to a temperature of 298 K.

$C(s) + O_2(g) \rightarrow CO_2(g)$; $\Delta H^{\ominus}$ = -394

$H_2(g) + \tfrac{1}{2}O_2(g) \rightarrow H_2O(l)$; $\Delta H^{\ominus}$ = -286

$C_2H_4(g) + 3O_2(g) \rightarrow 2CO_2(g) + 2H_2O(l)$; $\Delta H^{\ominus}$ = -1411

Calculate the standard enthalpy of formation of ethene, $CH_2{=}CH_2$.

(b) If, at 298 K, the enthalpy of sublimation of graphite is 713 kJ mol^{-1} and the bond energy terms H—H and C—H are 436 and 415 kJ mol^{-1} respectively, calculate the C=C bond energy term.

A

Exercise 110 The diagram shows in outline, and not to scale, the Born-Haber cycle for one mole of magnesium oxide. Some of the information has been left out.

The relevant data from which the cycle is constructed are:

Standard enthalpy of formation of magnesium oxide	= -610 kJ mol^{-1}
Sublimation energy of magnesium	= $+153$ kJ mol^{-1}
Sum of first two ionization energies of magnesium	= $+2178$ kJ mol^{-1}
Dissociation energy of oxygen (per mole of atoms formed)	= $+250$ kJ mol^{-1}
Electron affinity of oxygen (for two electrons gained)	= $+748$ kJ mol^{-1}

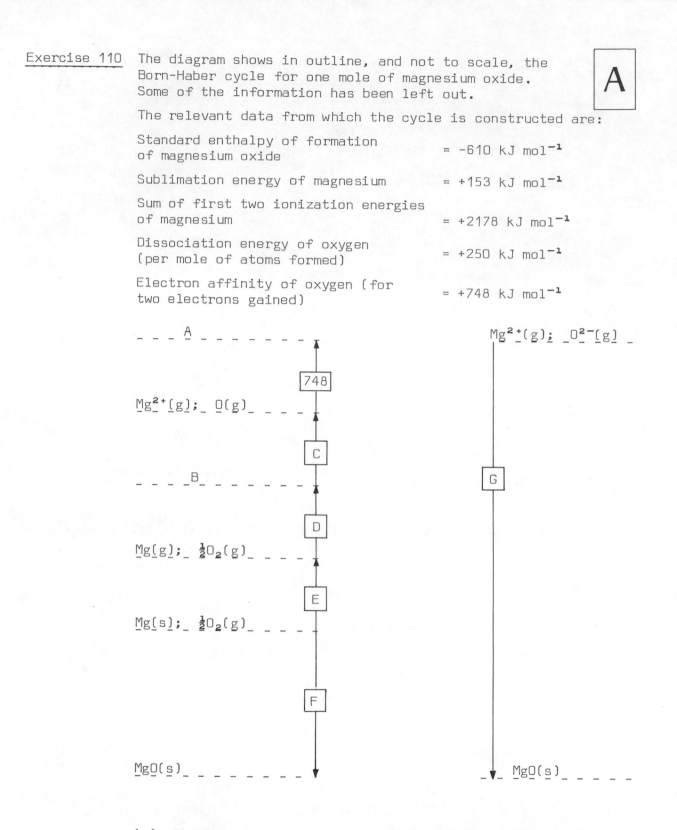

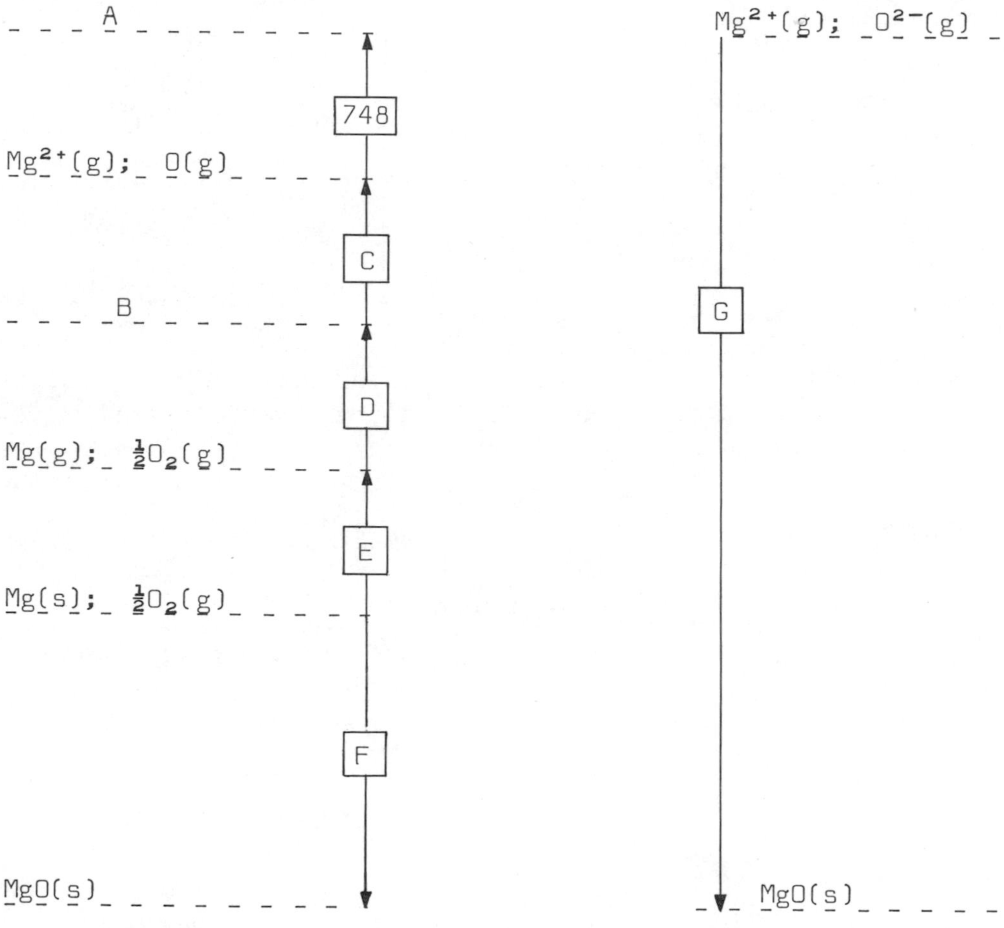

(a) Complete the diagram by writing the appropriate enthalpy changes for the labelled squares and the appropriate formulae for the dotted lines, A and B.

(b) State what is meant by the 'lattice energy' of magnesium oxide and compute its value.

Exercise 111 In an experiment to determine the enthalpy of neutralization of sodium hydroxide with sulphuric acid, 50 cm^3 of 0.40 M sodium hydroxide was titrated, thermometrically, with 0.50 M sulphuric acid. The results were plotted as follows:

A part

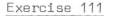

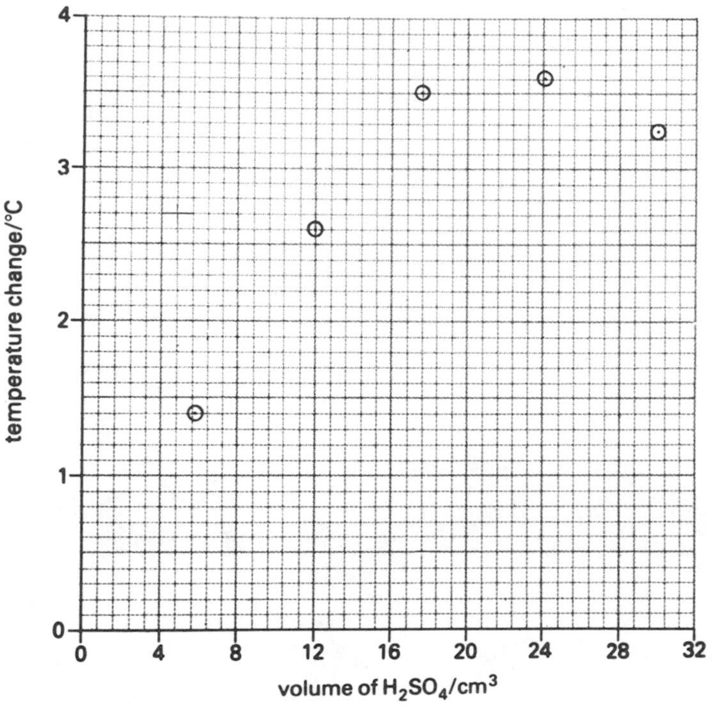

Fig.13.

Calculate a value for the enthalpy of neutralization of sodium hydroxide with sulphuric acid. (The specific heat capacity of water is 4.2 J K^{-1} g^{-1}.) What assumptions have you made in your calculations?

GASES

5

INTRODUCTION AND PRE-KNOWLEDGE

The gaseous state is characterised by the fact that a fixed amount varies readily in shape and size and generally has low density and high compressibility relative to solids and liquids.

Throughout this chapter, we assume that you are familiar with the concepts of 'standard temperature and pressure' (s.t.p.) and 'absolute temperature' and that you have used a variety of different units for pressure. While the IUPAC recommendation is that pressures should be expressed in Pascals (Pa = N m^{-2}) the officially obsolete units 'atmospheres' (atm) and 'millimetres of mercury' (mmHg) are likely to continue in use by chemists.

Partly because the pressure exerted by a gas is usually thought of in comparison with atmospheric pressure, and partly because barometers in schools are often graduated in millimetres of mercury, we have retained the use of the units atm and mmHg while also using Pa and kPa in some calculations.

Because the molecules of a gas are so far apart, the laws which describe their behaviour are remarkably simple.

THE GAS LAWS

We look in turn at several laws which describe quantitatively the behaviour of gases. These laws were discovered empirically (i.e. by studying the results of experiments) but were later shown to have the same theoretical basis.

Boyle's law and Charles' law

These laws were formulated by measuring how the volume of a sample of gas varies with temperature and pressure.

Boyle's law states that the pressure of a fixed mass of gas at constant temperature is inversely proportional to its volume, i.e.

$$p \propto \frac{1}{V} \quad \text{or} \quad pV = \text{constant} \quad \text{or} \quad p_1V_1 = p_2V_2$$

Charles' law states that the volume of a fixed mass of gas at constant pressure is directly proportional to the absolute temperature (K = ^{o}C + 273), i.e.

$$V \propto T \quad \text{or} \quad \frac{V}{T} = \text{constant} \quad \text{or} \quad \frac{V_1}{T_1} = \frac{V_2}{T_2}$$

Apply these laws in the three exercises following.

Exercise 112 (a) A meteorological balloon is to be filled with helium at atmospheric pressure. What will the volume of the balloon be if it is to hold all the gas from a 25.0 dm^3 gas cylinder at 150 atm? Assume constant temperature.

(b) The balloon will burst if its volume exceeds 4000 dm^3. If the filling temperature is 15 °C, what is the maximum temperature the balloon can stand at atmospheric pressure?

Exercise 113 100 cm^3 of hydrogen, at 1.00 atm, is compressed at constant temperature to 37.0 cm^3. What is the new pressure?

Exercise 114 73.0 cm^3 of nitrogen at 1456 mmHg is allowed to expand so that its pressure is 760 mmHg. What is its new volume at the same temperature?

We show you how to combine Boyle's law and Charles' law in the following Worked Example.

Worked Example A meteorological balloon has a volume of 6.15 m^3 when filled with helium at 14 °C and 762 mmHg. What will its volume be if the temperature rises to 19 °C and the pressure falls to 749 mmHg?

W

Solution

1. Calculate the effect on the volume of gas due to pressure change alone, using Boyle's law:

$$pV = \text{constant} \quad \text{or} \quad p_1V_1 = p_2V_2$$

$$\therefore V_2 = V_1 \times \frac{p_1}{p_2} = 6.15\ \text{m}^3 \times \frac{762\ \text{mmHg}}{749\ \text{mmHg}}$$

Notice that because a pressure decrease gives a volume increase, we multiply the volume by a factor greater than one, i.e. 762/749. Of course, the same factor arises if you use different units for pressure, e.g.

$$\frac{762\ \text{mmHg}}{749\ \text{mmHg}} = \frac{100.3\ \text{kPa}}{98.5\ \text{kPa}} = 1.017$$

2. Calculate the effect on this new volume of gas due to a temperature change alone, using Charles' law.

$$\frac{V}{T} = \text{constant} \quad \text{or} \quad \frac{V_2}{T_2} = \frac{V_3}{T_3}$$

$$\therefore V_3 = V_2 \times \frac{T_3}{T_2} \quad (V_2 \text{ is the volume calculated in step 1.})$$

$$\therefore V_3 = (6.15\ \text{cm}^3 \times \frac{762}{749}) \times \frac{(273 + 19)\text{K}}{(273 + 14)\text{K}}$$

Notice that because a temperature increase gives a volume increase we again multiply the volume by a factor greater than one, i.e. 292/287.

The final volume, V_3, is therefore given by

$$V_3 = 6.15\ \text{m}^3 \times \frac{762}{749} \times \frac{292}{287} = \boxed{6.37\ \text{m}^3}$$

Now try some exercises. You will soon find that, with a little practice, you can do both steps in one, but always check whether the factors should be greater or less than one.

New volume = original volume x pressure factor x temperature factor

Exercise 115 What is the volume at s.t.p. of a sample of gas which occupies 38.2 cm^3 at 18 °C and 765 mmHg?

Exercise 116 79.0 cm^3 of hydrogen were collected from a reaction between zinc and sulphuric acid, at 21 °C and 756 mmHg. What volume would the gas occupy at s.t.p?

The combination of Boyle's law and Charles' law which you have been using is often quoted separately as the combined gas law.

The combined gas law

In the previous exercises you applied Boyle's law and Charles' law to problems where p, V and T all vary. The final relationship in the worked example:

$$V_2 = V_1 \times \frac{p_1}{p_2} \times \frac{T_2}{T_1}$$

is a mathematical statement of the combined gas law, more commonly written as:

$$\frac{p_1 V_1}{T_1} = \frac{p_2 V_2}{T_2}$$

You will use the combined gas law most often for calculating volume changes, but it can, of course, be used to calculate pressure or temperature changes in a very similar way. Use it in the following exercises.

Exercise 117 The table below shows a number of gas volumes and the conditions under which they were collected and measured. Calculate the volumes the gases would occupy if the conditions were changed to those indicated.

	volume	conditions	new conditions
(a)	29.2 cm^3	25 °C, 762 mmHg	s.t.p.
(b)	5.13 dm^3	62 °C, 1.02 atm	s.t.p.
(c)	132 cm^3	s.t.p.	25 °C, 99.2 kPa
(d)	42.1 cm^3	31 °C, 98.7 kPa	20 °C, 27.5 kPa

(Standard pressure = 1.00 atm = 760 mmHg = 100 kPa)

Exercise 118 A sample of gas at 27 °C was heated so that both its pressure and its volume were doubled. What was the new temperature?

Exercise 119 A set of car tyres, each of volume 21.0 dm^3, were inflated till the pressure gauge read 28.5 lb in^{-2}. During a journey, the temperature of the tyres increased from 15 °C to 42 °C and their volume increased to 21.6 dm^3. What would the pressure gauge have read then? (The pressure of air in the tyre was the sum of the gauge pressure and atmospheric pressure, which may be taken as 14.7 lb in^{-2}.)

Having looked at some empirical laws, we now consider a very important theory which was developed to help explain them, particularly Gay-Lussac's law of combining volumes.

AVOGADRO'S THEORY AND ITS APPLICATIONS

Avogadro's theory (or hypothesis) states that equal volumes of gases, under the same conditions of temperature and pressure, contain equal numbers of molecules.

You have probably studied the way in which Avogadro's theory was used to explain Gay-Lussac's law of combining volumes and to help establish chemical formulae from measurements of reacting volumes of gases. However, we now often use the theory the other way round, i.e. to calculate reacting volumes of gases of known formulae.

Using Avogadro's theory to calculate reacting volumes

We show the procedure for this type of calculation in a Worked Example.

Worked Example What volume of oxygen would be required to burn completely 20 cm^3 of ethane, C_2H_6, and what volume of carbon dioxide would be produced? (Volumes measured at the same room temperature and pressure.)

Solution

1. Write the equation for the reaction:

$$C_2H_6(g) + 3\tfrac{1}{2}O_2(g) \rightarrow 2CO_2(g) + 3H_2O(l)$$

2. Write the amount of each gas under the equation:

$$C_2H_6(g) + 3\tfrac{1}{2}O_2(g) \rightarrow 2CO_2(g) + 3H_2O(l)$$

1 mol	$3\frac{1}{2}$ mol	2 mol

3. Write under the amounts the relative volume of each gas, by applying Avogadro's theory. Since equal volumes contain equal amounts, it follows that equal amounts occupy equal volumes.

$$C_2H_6(g) + 3\tfrac{1}{2}O_2(g) \rightarrow 2CO_2(g) + 3H_2O(l)$$

1 mol	$3\frac{1}{2}$ mol	2 mol
1 volume	$3\frac{1}{2}$ volumes	2 volumes
V	$3\frac{1}{2} \times V$	$2 \times V$

4. Use the volume given in the question to calculate the other volumes.

Volume of ethane = V = 20 cm^3

Volume of oxygen = $3\frac{1}{2} \times V = 3\frac{1}{2} \times 20\ cm^3$ = **70 cm^3**

Volume of carbon dioxide = $2 \times V = 2 \times 20\ cm^3$ = **40 cm^3**

Now try the following exercises in the same way.

Exercise 120 In each of the following reactions the volume of some of the gases involved is given. Calculate the volumes of the gases which react and are produced. All gas volumes are measured at the same temperature and pressure.

(a) $2N_2O(g) \rightarrow 2N_2(g) + O_2(g)$

15 cm^3 (under $2N_2(g)$)

(b) $4NH_3(g) + 3O_2(g) \rightarrow 2N_2(g) + 6H_2O(g)$

128 cm^3 (under $2N_2(g) + 6H_2O(g)$ together)

Exercise 121 What is the maximum volume of hydrogen chloride that can be obtained from 75 cm^3 of hydrogen at the same temperature and pressure?

Exercise 122 From the equation given, calculate the maximum volume of ammonia that can be oxidized by 15 cm^3 of oxygen at the same temperature and pressure. Also calculate the maximum volume of the products under the same conditions. What is the percentage change in volume during the reaction?

$$4NH_3(g) + 5O_2(g) \rightarrow 4NO(g) + 6H_2O(g)$$

Exercise 123 78.5 cm^3 of ethanol vapour, C_2H_5OH, at 110 °C was burnt in oxygen to carbon dioxide and water only. Calculate the volume of oxygen used and the volume of the products, all at the same temperature and pressure.

Another application of Avogadro's theory is in a method for finding the formula of a gaseous hydrocarbon by combustion.

Determining the formula of a gaseous hydrocarbon

We can easily measure the volumes of gases involved in the combustion of a hydrocarbon. From these volumes, and the equation for the reaction, we can obtain the formula of the hydrocarbon.

A Worked Example shows you how to work out a formula from the results of a combustion experiment. Then you can try some problems for yourself.

Worked Example — 10 cm^3 of a gaseous hydrocarbon at room temperature was mixed with 100 cm^3 of oxygen (an excess). After sparking the mixture and allowing it to cool to its original temperature, the total volume was found to be 95 cm^3. This contracted to 75 cm^3 when in contact with a concentrated solution of sodium hydroxide. What is the formula of the hydrocarbon?

W

Solution

1. Write an equation for the reaction in words:

   hydrocarbon(g) + oxygen(g) → carbon dioxide(g) + water(l)

2. Calculate the volume of carbon dioxide produced. This is equal to the contraction in volume as the sodium hydroxide absorbs carbon dioxide.

   ∴ volume of CO_2 formed = 95 cm^3 - 75 cm^3 = 20 cm^3

3. Calculate the volume of water vapour (if any) produced. In this example, the volumes are measured at room temperature. Thus, the water is liquid and has negligible volume. (In other problems, you may have to consider the volume of water vapour.)

4. Calculate the volume of oxygen used up in the reaction. The final volume, 75 cm^3, must be unused oxygen.

   ∴ volume of oxygen used = 100 cm^3 - 75 cm^3 = 25 cm^3

5. Write down the volume of each gas under the word-equation, and divide each by the smallest volume to obtain the relative volumes.

hydrocarbon(g) +	oxygen(g) →	carbon dioxide(g) +	water(l)
10 cm^3	25 cm^3	20 cm^3	-
1 volume	$2\frac{1}{2}$ volumes	2 volumes	

6. Apply Avogadro's theory to write down the relative amount of each gas. You may find it easier, though not strictly necessary, to convert these to whole numbers.

	hydrocarbon(g) +	oxygen(g) →	carbon dioxide(g) +	water(l)
	10 cm^3	25 cm^3	20 cm^3	-
	1 volume	$2\frac{1}{2}$ volumes	2 volumes	-
	1 mol	$2\frac{1}{2}$ mol	2 mol	? mol
or	2 mol	5 mol	4 mol	

7. Call the hydrocarbon C_xH_y and the amount of water z mol. Write a full equation in the form:

$$2C_xH_y(g) + 5O_2(g) \rightarrow 4CO_2(g) + zH_2O(l)$$

8. Calculate x, y and z by noting that the equation must balance, i.e. the same number of atoms of each sort must appear on each side of the equation. It is convenient to do this, one line at a time, in tabular form. Always consider atoms in the order C, O, H - it makes the calculation simpler.

	Left-hand side of equation	Right-hand side of equation	Calculation
C atoms	$2x$	4	$2x = 4 \therefore x = 2$
O atoms	$2 \times 5 = 10$	$(2 \times 4) + z = 8 + z$	$10 = 8 + z \therefore z = 2$
H atoms	$2y$	$2z$	$2y = 2z \therefore y = 2$

The formula of the hydrocarbon is therefore C_2H_2, and the equation is:

$$2C_2H_2(g) + 5O_2(g) \rightarrow 4CO_2(g) + 2H_2O(l)$$

Note that we could equally well have taken the ratio of volumes as $1:2\frac{1}{2}:2$, and arrived at the equivalent equation:

$$C_2H_2(g) + 2\tfrac{1}{2}O_2(g) \rightarrow 2CO_2(g) + H_2O(l)$$

Now you should try some similar problems yourself.

<u>Exercise 124</u> When 15 cm^3 of a gaseous hydrocarbon was exploded with 60 cm^3 of oxygen (an excess), the final volume was 45 cm^3. This decreased to 15 cm^3 on treatment with sodium hydroxide solution. What was the formula of the hydrocarbon? All measurements were made at the same room temperature and pressure.

The next exercise is more difficult because the information given is different. Also you cannot this time ignore the volume of water! However, the same general method should be used. If you cannot do it without help, look for clues in the answer page and then try again.

<u>Exercise 125</u> When 10 cm^3 of gaseous hydrocarbon, C_3H_x, was exploded with an excess of oxygen at 105 °C, the volume measured under the same physical conditions expanded by 5 cm^3. What is the value of x, and what volume of oxygen was used under these conditions?

<u>Exercise 126</u> 20 cm^3 of a gaseous hydrocarbon was mixed with 200 cm^3 of oxygen (an excess) and exploded. The final volume, at the same room temperature and pressure, was 150 cm^3, but this was reduced to 70 cm^3 after treatment with a concentrated alkali. Calculate the formula of the hydrocarbon.

<u>Exercise 127</u> 15 cm^3 of a gaseous hydrocarbon at 100 °C was mixed with an excess of oxygen and exploded by sparking. Cooling the products caused a reduction in volume equivalent to 30 cm^3 at the original temperature and pressure. A further reduction of 30 cm^3 occurred on exposure to concentrated alkali. What was the formula of the hydrocarbon?

<u>Exercise 128</u> 30 cm^3 of a gas, C_3H_x, was mixed with twice the minimum volume of oxygen needed for complete combustion and the mixture was exploded. After cooling to the same room temperature, and at the same pressure, the volume was 90 cm^3 less than before. Calculate the volume of oxygen used in the reaction, and the value of x.

In the next section, we show how Avogadro's theory leads to an important and useful statement about the volume of one mole of gas.

Molar volume of gases

The converse of Avogadro's theory (i.e. putting it the other way round) is that if we have samples of different gases at the same temperature and pressure and containing the same number of molecules, then they must occupy equal volumes.

If the fixed number of molecules referred to above is the number in one mole, then the corresponding fixed volume is known as the molar volume, V_m. Molar volume, like other volumes, varies with temperature and pressure but its value at s.t.p. is most often used.

An expression which is very useful in calculations involving molar volume is:

$$\boxed{\text{amount of gas} = \frac{\text{volume of gas}}{\text{molar volume}}} \quad \text{or} \quad \boxed{n = \frac{V}{V_m}}$$

where both volumes refer to the same conditions of temperature and pressure. This expression is very similar to the one you have already used often:

$$\text{amount of substance} = \frac{\text{mass}}{\text{molar mass}} \quad \text{or} \quad n = \frac{m}{M}$$

and is easily derived from it by dividing each mass by the appropriate density,

$$n = \frac{m/\rho}{M/\rho} = \frac{V}{V_m}$$

We use this expression in the following Worked Example which shows you how to calculate molar volume from experimental data for a gas of known formula.

Worked Example Calculate the molar volume at s.t.p. for carbon dioxide given that 2.50 g occupies 0.450 dm³ at 3.00 atm and 16 °C.

W

Solution

1. Calculate the volume at s.t.p. using the combined gas law:

$$\frac{p_1V_1}{T_1} = \frac{p_2V_2}{T_2} \quad \text{or} \quad V_2 = V_1 \times \frac{p_1}{p_2} \times \frac{T_2}{T_1}$$

i.e. $V_2 = 0.450\ \text{dm}^3 \times \frac{3.00\ \text{atm}}{1.00\ \text{atm}} \times \frac{273\ \text{K}}{(273 + 16)\text{K}} = 1.28\ \text{dm}^3$

2. Calculate the amount, n, of CO_2 using the expression:

$$n = \frac{m}{M}$$

i.e. $n = \frac{2.50\ \text{g}}{44.0\ \text{g mol}^{-1}} = 0.0568\ \text{mol}$

3. Calculate the molar volume, V_m, using the expression

$$n = \frac{V}{V_m}$$

i.e. $V_m = \frac{V}{n} = \frac{1.28\ \text{dm}^3}{0.0568\ \text{mol}} = \boxed{22.5\ \text{dm}^3\ \text{mol}^{-1}}$

Now you should be able to calculate some molar volumes for yourself.

Exercise 129 Given the following experimental results, calculate the molar volume of each gas at s.t.p.

(a) 0.122 g of hydrogen, $H_2(g)$, occupies 0.211 dm^3 at 7.00 atm and 20.0 °C.

(b) 1.10 g of butane, $C_4H_{10}(g)$, occupies 34.4 dm^3 at 600 °C and 30.0 mmHg. (1 atm = 760 mmHg.)

(c) 2.00 g of oxygen, $O_2(g)$, at 5.00 kPa and 500 K occupies 51.9 dm^3. (1.00 atm = 100 kPa.)

You will have noticed that the answers to the exercise are not equal. Even with the most accurate experimental data, slight differences in molar volumes occur. This is because the gases do not observe the gas laws precisely - only a hypothetical ideal gas does this.

However, we very often assume that gases are ideal and, therefore, assume that the molar volume of any gas is a constant. The value we use at s.t.p. is 22.4 $dm^3\ mol^{-1}$, a figure well worth remembering even though it is usually given in examinations when required.

$$V_m(\text{s.t.p.}) = 22.4\ \text{mol dm}^{-3}$$

The molar volume is often used to calculate amounts of gas from volume measurements, using almost the same method as in the last Worked Example.

Exercise 130 (a) Calculate the amount of gas contained in a 10.0 dm^3 globe at 27 °C and 350 mmHg.

(b) What is the volume of 8.00 g of oxygen at 23 °C and 0.200 atm?

We do not always take s.t.p. as our reference conditions. The next exercise refers to molar volume at 'room temperature and pressure'.

Exercise 131 If the pressure is 1.00 atm, at what temperature is the molar volume of a gas 24.0 $dm^3\ mol^{-1}$?

Exercise 132 Calculate the molar volume of

(a) ammonia at s.t.p. given that 0.0780 g occupies 86.1 cm^3 at 1.25 atm and 17 °C,

(b) methane, CH_4, at 25 °C and 1.00 atm, given that 0.0591 g occupies 90.2 cm^3 at s.t.p.

Exercise 133 (a) Calculate the amount of nitrogen in a 500 cm^3 flask at 15 °C and 1.01 atm.

(b) What mass of oxygen occupies 152 cm^3 at s.t.p.?

(c) What is the volume of 5.81 g of hydrogen at 19 °C and 92.8 kPa?

Molar volume can also be used in calculating the Avogadro constant.

A method for determining the Avogadro constant

There are many independent methods for determining the Avogadro constant, L, and we now outline one of the more accurate ones, which involves measurements of radioactivity. In the following exercise, you need only apply what you have already learned about the Avogadro constant (Chapter 1), radioactivity (Chapter 2) and molar volume.

Exercise 134 A sample of radium chloride was kept in a sealed tube and was found to produce helium gas at the rate of 4.13×10^{-4} cm^3 in one hour. Another sample of the same mass was found to emit α-particles at a rate of 3.07×10^{12} s^{-1}.

(a) How was the helium formed?

(b) Calculate a value for the Avogadro constant.

You should now be able to see the great value of Avogadro's theory in a variety of useful calculations. Next, we consider a very important relationship, known as the ideal gas equation, which embodies both Avogadro's theory and the combined gas law.

THE IDEAL GAS EQUATION

The ideal gas equation follows from the combined gas law and Avogadro's theory, as we now show. The combined gas law is written

$$\frac{p_1V_1}{T_1} = \frac{p_2V_2}{T_2} \quad \text{or} \quad \frac{pV}{T} = \text{constant}$$

For one mole of gas, the volume is written as V_m, while the constant is written as R, and is known as the gas constant.

i.e. $$\frac{pV_m}{T} = R$$

We have already used Avogadro's theory to show that the molar volume, V_m, is the same for all gases under the same conditions. It follows, therefore, that R is the same for all gases. Rearranging the expression, we have

$$pV_m = RT$$

This applies for one mole of gas, but for n mol it becomes:

$$\boxed{pV = nRT}$$

The gas constant, R

The gas constant, R, is a proportionality constant relating p, V, n and T. Its numerical value depends on the units chosen for the variables. T is always expressed in kelvin (K) and n in mol, but a variety of units are in use for p and V. Chemists usually express p in atmospheres (atm) and V in dm^3 - use these units in the next exercise.

Exercise 135 Use the fact that one mole of ideal gas occupies 22.4 dm^3 at s.t.p. to calculate a value for R, the gas constant.

In the next exercise, you calculate some other numerical values for R, using a variety of units. This involves using a conversion factor, as we show in a Worked Example.

Worked Example R is given as 0.0821 atm dm^3 K^{-1} mol^{-1}. What value should be used if the volume of gas were given in m^3?

W

Solution

1. Rearrange the expression for R to isolate the unit to be changed.

$$R = 0.0821 \text{ atm } (dm^3)\ K^{-1}\ mol^{-1}$$

2. Make the appropriate substitution, given that

$$1.00\ m^3 = 1.00 \times 10^3\ dm^3 \quad \text{or} \quad m^3 = 10^3\ dm^3$$

Dividing by 10^3, $1.00 \times 10^{-3}\ m^3 = 1.00\ dm^3$ or $dm^3 = 10^{-3}\ m^3$

$$\therefore \quad R = 0.0821 \text{ atm } (1.00 \times 10^{-3}\ m^3)\ K^{-1}\ mol^{-1}$$

$$= \boxed{8.21 \times 10^{-5} \text{ atm } m^3\ K^{-1}\ mol^{-1}}$$

Alternatively, the substitution can be made during the calculation using the ideal gas equation

$$pV = nRT$$

$$\therefore \quad R = \frac{pV}{nT} = \frac{1.00 \text{ atm} \times 22.4 \times 1.00 \times 10^{-3}\ m^3}{1.00 \text{ mol} \times 273\ K}$$

$$= \boxed{8.21 \times 10^{-5} \text{ atm } m^3\ K^{-1}\ mol^{-1}}$$

Note that when a quantity is expressed in a combination of units, those with positive indices are usually written first, in alphabetical order, followed by those with negative indices, also in alphabetical order. However, this is not a hard-and-fast rule.

Exercise 136 Obtain two more values of R, in different units.

(a) Substitute 1.00 atm = 760 mmHg in the expression

$$R = 0.0821 \text{ atm } dm^3\ K^{-1}\ mol^{-1}$$

(b) Repeat the calculation of R as in Exercise 135, using the following conversion factors.

$$1.00 \text{ atm} = 100 \text{ kPa} = 100 \text{ kN } m^{-2} \qquad 1.00 \text{ N m} = 1.00 \text{ J}$$

The most widely used values for R are 8.314 J K^{-1} mol^{-1} and 0.0821 atm dm^3 K^{-1} mol^{-1}. It is most important in calculations that the value for R is consistent with the units of the data you are using. You must choose a value for R in the exercises which follow, but first we give a Worked Example of a typical calculation involving the ideal gas equation.

Worked Example Assuming ideal behaviour, calculate the volume occupied by 2.00 g of carbon monoxide at 20 °C under a pressure of 6250 Pa.

W

Solution

1. Calculate the molar mass of carbon monoxide:

$$M = (12.0 + 16.0)\ \text{g mol}^{-1} = 28.0\ \text{g mol}^{-1}$$

2. Identify the values to be substituted in the ideal gas equation, $pV = nRT$.

$$p = 6250\ \text{Pa} = 6250\ \text{N m}^{-2}$$

$$n = 2.00\ \text{g} / 28.0\ \text{g mol}^{-1} = 0.0714\ \text{mol}$$

$$T = (273 + 20)\text{K} = 293\ \text{K}$$

$R = 8.314\ \text{J K}^{-1}\ \text{mol}^{-1}$ (This value for R is appropriate because the pressure is in N m^{-2}, and $1\ \text{J} = 1\ \text{N m}$.)

3. Substitute the values in the equation

$$pV = nRT \quad \text{in the form} \quad V = \frac{nRT}{p}$$

$$\therefore \quad V = \frac{0.0714\ \text{mol} \times 8.314\ \text{J K}^{-1}\ \text{mol}^{-1} \times 293\ \text{K}}{6250\ \text{N m}^{-2}}$$

$$= \frac{0.0714 \times 8.314\ \text{J} \times 293}{6250\ \text{N m}^{-2}}$$

but $1.00\ \text{J} = 1.00\ \text{N m}$

$$\therefore \quad V = \frac{0.0714 \times 8.314\ \text{N m} \times 293}{6250\ \text{N m}^{-2}} = \boxed{0.0278\ \text{m}^3 \quad \text{or} \quad 27.8\ \text{dm}^3}$$

Note that the units used in the calculation determine the unit of the answer. It would be possible to do the calculation using $R = 0.0821\ \text{atm dm}^3\ \text{K}^{-1}\ \text{mol}^{-1}$ but the answer would be expressed in a very peculiar unit of volume ($\text{atm dm}^5\ \text{N}^{-1}$)!

The next two exercises are straightforward applications of the ideal gas equation.

Exercise 137

(a) What is the volume of 0.500 mol of sulphur dioxide, SO_2, at s.t.p.?

(b) What is the volume of 1.50 g of hydrogen, H_2, at 15 °C and a pressure of 750 mmHg?

(c) At what temperature will 4.71 g of nitrogen occupy 12.0 dm^3 at 760 mmHg?

In the next exercise, you have to find reacting amounts from a chemical equation before applying the ideal gas equation.

Exercise 138 What mass of zinc is required to produce 2.00 dm^3 of hydrogen at 15 °C and 755 mmHg by reacting with acid?

In the next exercise, you use the ideal gas equation to get an idea of the pressure and volume changes which occur in an explosion. An explosion occurs when a chemical reaction produces large volumes of gas very rapidly. The gas cannot expand rapidly enough to avoid a very high pressure immediately around the explosive, and it is the spread of shock waves from this small region of high pressure which causes the noise and damage associated with explosions.

Exercise 139 A widely-used explosive is TNT* which has a formula $C_7H_5N_3O_6$. This is mixed with a solid oxidant so that oxygen required for combustion can be supplied rapidly.

(a) Write an equation for the combustion of TNT. Assume that C and H atoms are completely oxidized and that N atoms emerge as nitrogen gas, N_2.

(b) What amount of gas is produced from 1.00 mol of TNT, and what volume would it occupy at 1.00 atm and 400 °C?

(c) Assuming that 1.00 mol of TNT mixed with oxidant occupies 0.500 dm^3, what is the increase in volume expressed as a percentage?

(d) Assume that the reaction occurs so fast that the gaseous products occupy only 2.00 dm^3 at 600 °C. What would be the resulting pressure?

*TNT is an abbreviation of the old name trinitrotoluene. The modern systematic name is methyl-2,4,6-trinitrobenzene.

These calculations could also have been done using a known value for molar volume, but the use of the ideal gas equation is more general. Now try some further calculations.

Exercise 140 What is the volume of

(a) 0.250 mol of methane at 17 °C and 750 mmHg?

(b) 0.521 g of ammonia at 50 °C and 762 mmHg?

Exercise 141 (a) What volume of carbon dioxide, at 25 °C and 764 mmHg, could be obtained from 7.31 g of calcium carbonate?

(b) 1.52 g of pure zinc was dissolved in dilute acid and the resulting gas, after drying, occupied 557 cm^3 at 1.01 atm. What was the temperature?

(c) Excess magnesium was added to 50.0 cm^3 of 0.102 M hydrochloric acid. What volume of dry gas would be produced at 14 °C and 751 mmHg?

Exercise 142 5.00 cm^3 of water is introduced into an evacuated tube of volume 150 cm^3. The tube is sealed and heated to 300 °C. What is the resulting pressure?

One of the most important applications of the ideal gas equation is in the determination of molar mass. This is the subject of the next section.

Determining the molar mass of a gas or a volatile liquid

The ideal gas equation can be transformed into an equation involving the mass of the gas. You can use this form of the equation to determine the molar masses of a gas and of a volatile liquid, as in the following exercises.

Exercise 143 (a) Write down an expression for the amount, n, of a gas in terms of the mass, m, and the molar mass, M.

(b) Substitute your expression into the ideal gas equation to obtain an expression for the molar mass.

(c) Calculate the molar mass of a substance, given that 3.72 g of its vapour occupies 2.00 dm^3 at 740 mmHg and 100 °C.

Exercise 144 A flask of 109 cm^3 capacity weighed 78.521 g when evacuated and 78.719 g when filled with a gas at 21 °C and 759 mmHg. What was the molar mass of the gas?

Exercise 145 Calculate the molar masses of gases, A, B and C from the data given below:

	Mass of gas	Volume	Temperature	Pressure
A	0.672 g	257 cm^3	25 °C	1.00 atm
B	0.128 g	111 cm^3	17 °C	763 mmHg
C	1.60 g	526 cm^3	21 °C	103 kPa

(100 atm = 100 kPa)

In simple experiments to determine the molar mass of a gas, you are not likely to be able to weigh evacuated flasks. Instead, you deduct the mass of air from the mass of the 'empty' flask. Do this in the following exercise, which is based on the results of an experiment in an A-level class.

Exercise 146 Calculate the molar mass of carbon dioxide from the experimental results tabulated below.

Mass of flask filled with air	47.933 g
Mass of flask filled with CO_2	47.998 g
Mass of flask filled with water	152.8 g
Room temperature	26 °C
Atmospheric pressure	757 mmHg
Density of air under conditions of experiment	0.00118 g cm^{-3}

The method is equally applicable to volatile liquids. The experimental procedure is simpler because the mass of the vapour can be obtained by weighing the liquid before it is vaporized.

Exercise 147 0.25 cm^3 of a liquid of density 0.91 g cm^{-3} was injected into a syringe maintained at 160 °C. The liquid vaporized completely and occupied 81 cm^3 at 764 mmHg. What was the molar mass of the liquid?

Exercise 148 Calculate the molar masses of the volatile liquids D, E and F from the data given below:

	Mass of liquid	Volume of vapour	Temperature	Pressure
D	0.184 g	0.0792 dm^3	100 °C	0.984 atm
E	0.295 g	121 cm^3	140 °C	747 mmHg
F	0.163 g	65.0 cm^3	101 °C	105 kPa

(1.00 atm = 100 kPa)

Another empirical gas law which provides us with an alternative method for determining the molar mass of a gas is Graham's law of effusion.

GRAHAM'S LAW OF EFFUSION

Graham's law states that the rate of effusion (and the rate of diffusion) of a gas is inversely proportional to the square root of its density. Thus, if the rates of effusion of two different gases at the same conditions of temperature and pressure are compared, we have:

$$\frac{\text{rate}_1}{\text{rate}_2} = \sqrt{\frac{\text{density}_2}{\text{density}_1}}$$

Since the rate of effusion is inversely proportional to the time, t, taken for a fixed volume of gas to escape through a small hole, and the density, ρ, is directly proportional to the molar mass, M, other forms of the law are often used, e.g.

$$\frac{t_1}{t_2} = \sqrt{\frac{\rho_1}{\rho_2}} = \sqrt{\frac{M_1}{M_2}} \quad \text{or} \quad \left(\frac{t_1}{t_2}\right)^2 = \frac{\rho_1}{\rho_2} = \frac{M_1}{M_2}$$

Apply the appropriate form of the law in the following exercises.

Exercise 149 A volumetric flask was filled with hydrogen, H_2, and another with sulphur dioxide, SO_2, for an experiment. The flasks were stoppered, but not perfectly sealed, and put aside for use later.

(a) Use molar masses to calculate the ratio of the densities of hydrogen and sulphur dioxide under the same conditions.

(b) Assuming the flasks and the gaps between stoppers and flasks were identical, how much more rapidly would the hydrogen escape (to be replaced by air) than the sulphur dioxide?

Exercise 150 A gas syringe was filled first with hydrogen and then with carbon monoxide, and the gases were allowed to effuse through a tiny pinhole under the weight of the piston.

The time taken for 75 cm^3 of hydrogen to escape was 25 s, compared with 93 s for carbon monoxide. Given the molar mass of hydrogen as 2.02 g mol^{-1}, calculate the molar mass of carbon monoxide.

Exercise 151 The isotopes of uranium ^{235}U and ^{238}U can be separated by using the different rates of diffusion of the gaseous fluorides UF_6. What is the ratio of these rates of diffusion?

Exercise 152 Oxygen was allowed to escape from a syringe through a small hole under the weight of the piston. It took 47 s for 50 cm^3 of oxygen to escape. When the experiment was repeated for another gas, 50 cm^3 took 17 s to escape.

(a) What was the molar mass of the gas?

(b) How long would the same volume of ammonia take to escape?

Exercise 153 The times taken for the effusion of equal volumes of carbon dioxide and a mixture of carbon dioxide with carbon monoxide were 28 s and 24 s respectively. Calculate the apparent molar mass of the mixture and the proportion of each gas.

The gas laws you have studied so far relate mainly to pure gases. Now we consider a law which enables us to deal with mixtures of gases more fully.

DALTON'S LAW OF PARTIAL PRESSURES

You have learned that the pressure exerted by a gas is due to the combined effect of many collisions between individual molecules and the containing wall. Dalton invented the term 'partial pressure' to refer to the contribution to the total pressure made by the molecules of one particular sort in a mixture of gases.

Dalton's law of partial pressures states that, in a mixture of gases, the total pressure is the sum of the partial pressures of the component gases. The partial pressure of a component gas is equal to the pressure which it would exert if it alone occupied the total volume.

The first exercise in this section simply tests your understanding of the law.

Exercise 154 A globe contains oxygen at a pressure of 0.30 atm. Hydrogen is admitted until the total pressure is 0.80 atm, and then nitrogen is admitted until the total pressure is 0.90 atm. What is the partial pressure of each gas?

The last exercise was very simple. You are more likely to meet problems where you have to calculate partial pressures from the amounts of gas present. To do this, it is convenient to introduce a quantity known as 'mole fraction'.

In any mixture of substances (whether gaseous or not) X_A, the mole fraction of A, is given by the expressions:

$$\boxed{\text{mole fraction of A} = \frac{\text{amount of A}}{\text{total amount}}} \quad \text{or} \quad \boxed{X_A = \frac{n_A}{n}}$$

We now derive a simple and useful expression relating partial pressure to mole fraction. Dalton's law tells us that the partial pressure of a gas in a mixture is the same as if that gas alone occupied the entire volume. The ideal gas equation can then be applied to each gas in turn.

For example, in a mixture of three gases, A, B and C, with partial pressures p_A, p_B and p_C, we can write for the gas A:

$$p_A V = n_A RT \quad \text{or} \quad p_A = n_A \times \frac{RT}{V}$$

We can express the quantity $\frac{RT}{V}$ in terms of the total pressure, p, and the total amount of gas, n.

$$pV = nRT \quad \text{or} \quad \frac{RT}{V} = \frac{p}{n}$$

Substituting into the expression for p_A:

$$p_A = n_A \times \frac{RT}{V} = n_A \times \frac{p}{n} = p \times \frac{n_A}{n}$$

i.e. $$\boxed{p_A = p \times X_A}$$

in the same way, $p_B = p \times X_B$ and $p_C = p \times X_C$, i.e.

$$\boxed{\text{partial pressure} = \text{total pressure} \times \text{mole fraction}}$$

By calculating mole fractions from the equation for the reaction you should be able to do the next exercise.

Exercise 155 Some ammonia in a syringe is completely decomposed to nitrogen and hydrogen by passing it over heated iron wool. If the total pressure is then 760 mmHg, calculate the partial pressure of each gas.

If you can calculate partial pressures from amounts, then you should also be able to calculate amounts from partial pressures in the next exercise.

Exercise 156 If the amount of oxygen in the globe referred to in Exercise 154 were 0.25 mol, how much of the other gases would be present?

An application of Avogadro's theory enables us to calculate mole fractions and partial pressures from the volumes of gases which make up a mixture. Another way of stating Avogadro's theory is to say that the volume of a gas is proportional to its amount. Therefore, in a mixture of gases, A, B and C, we can express the mole fraction of A by:

$$X_A = \frac{\text{amount of A}}{\text{total amount}} = \frac{\text{volume of A}}{\text{total volume}}$$

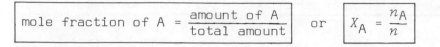

We have already shown that amount is proportional to partial pressure. You will find it helpful in the following exercises to remember these symmetrical relationships:

$$\frac{\text{partial pressure of A}}{\text{total pressure}} = \frac{\text{amount of A}}{\text{total amount}} = \frac{\text{volume of A}}{\text{total volume}}$$

Exercise 157 — 85.0 cm^3 of moist air, at 1.00 atm, was dried carefully and its volume, measured at the same temperature and pressure, fell to 82.0 cm^3. What was the partial pressure of water vapour in the moist air?

Exercise 158 — A diver should always breathe oxygen at a partial pressure of 2.0×10^4 N m^{-2}. What percentage of oxygen by volume should his breathing mixture contain when he dives to a depth of 40 m and the pressure on him rises to 5.0×10^5 N m^{-2}?

A

Exercise 159 — 75 cm^3 of hydrogen, 15 cm^3 of nitrogen, and 90 cm^3 of carbon dioxide, all measured at the same temperature and pressure, were mixed in a vessel. The total pressure was 2.4 atm. Calculate the partial pressure of each gas in the mixture.

Exercise 160 — A 500 cm^3 globe contains oxygen at 1.00 atm. 300 cm^3 of nitrogen, measured at the same temperature and pressure, is added under pressure, and then carbon dioxide is added till the total pressure is 3.10 atm. Calculate the partial pressure of each gas in the mixture, and the volume of carbon dioxide used.

Exercise 161 — A 600 cm^3 vessel contains 0.0232 g of nitrogen and 0.0417 g of carbon monoxide at 23 °C. Calculate the total pressure and the partial pressure of each gas.

Exercise 162 — A mixture of 82.2 g of helium and 27.4 g of oxygen was contained in a cylinder at 25.0 atm and 11 °C. What was the volume of the cylinder, and the partial pressure of each gas?

Exercise 163 — 15 cm^3 of a mixture of carbon monoxide and methane was mixed with excess oxygen and exploded. There was a contraction in volume of 21 cm^3 at the same room temperature and 1.0 atm pressure. Calculate the mole fraction of each gas in the mixture and their partial pressures.

The final section of this chapter concerns molecular speeds.

CALCULATING MOLECULAR SPEEDS

You have probably encountered two independent methods for calculating molecular speeds. One uses an equation derived from the kinetic theory of gases, which assumes 'ideal' behaviour. The other is based on the Zartman experiment, which was used to verify Maxwell's theoretical expression for the distribution of molecular speeds in a gas.

You compare these methods in the following exercises.

Exercise 164 By making simple assumptions about the nature and behaviour of gas molecules it is possible to derive a simple expression relating the pressure and volume of a gas containing N molecules each of mass m to the root-mean-square speed, c_{rms}. c_{rms} is the square root of the average value of c^2, where c is the speed of the molecule. The equation is

$$pV = \tfrac{1}{3}mN\overline{c^2}$$

(a) Combine the equation above with the ideal gas equation to derive an expression for c_{rms} in terms of the molar mass, M, the gas constant, R, and the absolute temperature, T.

(b) Calculate the root-mean-square speed of hydrogen molecules at 25 °C.

(c) Calculate the root-mean-square speed of bismuth atoms in vapour at 850 °C. (Use R = 8.31 J K^{-1} mol^{-1} and 1.00 J = 1.00 kg m^2 s^{-2}.)

Exercise 165 In the Zartman experiment, a 'beam' of bismuth atoms was directed at a rotating cylinder in a vacuum at 850 °C. A small slit in the cylinder allowed repeated pulses of atoms to cross inside the cylinder and reach a target area opposite the slit, as in Fig. 14.

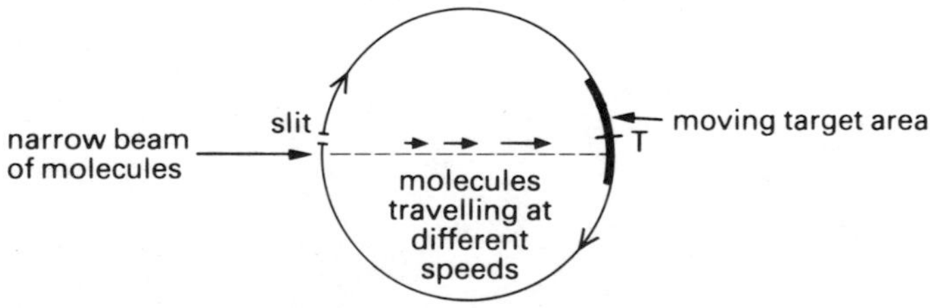

Fig.14. The Zartman experiment.

Atoms travelling at different speeds would strike different parts of the target area.

A layer of bismuth metal built up on the target area and the variation in its thickness was found to correspond with Maxwell's distribution curve.

Exercise 165 (continued)

It was found that the point T, where the largest number of atoms struck the target, was 1.50 cm from the point directly opposite the slit. The cylinder was 12.0 cm in diameter and was rotating at 125 revolutions per second.

(a) Calculate the speed (in m s^{-1}) at which T was travelling.

(b) How long would it have taken for T to travel 1.50 cm?

(c) In the interval of time calculated in (b) the atoms which struck point T had travelled a distance equal to the diameter of the cylinder. What, therefore, was their speed?

(d) Compare this speed with the root-mean-square speed calculated in the last exercise, and comment.

END-OF-CHAPTER QUESTIONS

Exercise 166

A sample of gas occupied 400 cm^3 at s.t.p. What would be its volume at 45 °C and 700 mmHg pressure?

Exercise 167

A

The following information refers to 5.60 g of a gas at a constant temperature of 0 °C.

Pressure /atm	2.25	1.90	1.60	1.12	0.75
Volume /dm^3	2.00	2.35	2.85	4.00	6.00

(a) Plot a graph of pressure against 1/volume.

(b) Is the gas behaving ideally? Give your reasons.

(c) Name and state the law which the information illustrates.

(d) Determine the gradient of the graph and write an expression relating the gradient, temperature and molar gas constant R.

(e) Hence, given that the molar volume is 22.4 dm^3 mol^{-1} at s.t.p., calculate the relative molecular mass of the gas.

(f) Calculate the density of the gas at 13 °C and 0.800 atm.

Exercise 168

A

When 30 cm^3 of a gaseous hydrocarbon were exploded with 350 cm^3 of oxygen, the volume of gas remaining, after cooling to room temperature, was 290 cm^3. This was reduced to 80 cm^3 by the addition of aqueous potassium hydroxide. Assume that all volumes are recorded at the same temperature and pressure.

(a) What volume of carbon dioxide is produced in the explosion?

(b) What volume of oxygen reacts with the hydrocarbon?

(c) What is the molecular formula of the hydrocarbon?

(d) State one chemical principle used in your calculation of the molecular formula of the hydrocarbon.

Exercise 169 1.75 g of chlorine occupies 0.545 dm^3 at s.t.p. At what temperature is its molar volume 25.0 dm^3 at 1.00 atm?

Exercise 170 Using the ideal gas equation, calculate the number of molecules of neon in a bulb of capacity 2.48 dm^3, evacuated to a pressure of 0.001 mmHg at 27 °C.

A part

1 mmHg = $13.60 \times 980.7 \times 10^{-2}$ N m^{-2} R = 8.31 J K^{-1} mol^{-1}

J = kg m^2 s^{-2} = N m L = 6.02×10^{23} mol^{-1}

Exercise 171 When 0.32 g of liquid bromine is volatilized, the volume of vapour formed, measured at s.t.p., is 45 cm^3. Calculate the relative molecular mass of bromine.

A part

S.t.p. is 273 K and 1 atm (i.e. 100 kN m^{-2})

R = 0.082 atm dm^3 K^{-1} mol^{-1} (i.e. 8.31 J K^{-1} mol^{-1}).

Exercise 172 In an experiment to determine the composition of a mixture of the gases propene, C_3H_6, and butene, C_4H_8, a student attempted to find the average relative molecular mass of the hydrocarbon mixture by finding the mass of a certain volume of it. The apparatus used was a standard volumetric flask, inverted and securely clamped in a a well-ventilated fume-cupboard.

A part

The results of the experiment were:

Mass of hydrocarbon mixture = 0.546 g

Volume of hydrocarbon mixture = 300 cm^3

Pressure = 1.00 atm

Temperature = 27 °C

(C = 12.0, H = 1.00, molar volume = 22.4 dm^3 mol^{-1})

(a) Calculate the volume of the hydrocarbon mixture at 0 °C and 1.00 atm pressure.

(b) Calculate the average mass of one mole of the particles in the hydrocarbon mixture.

(c) From your calculation, state which hydrocarbon is more abundant in the mixture.

Exercise 173 A flask contains 200 cm^3 of oxygen at 1.00 atm, and another flask contains 500 cm^3 of helium at 2.00 atm. The two flasks are connected to allow the gases to mix. What is the partial pressure of each gas in the mixture, and the total pressure?

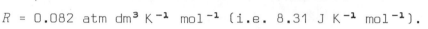

EQUILIBRIUM I – PRINCIPLES

6

INTRODUCTION AND PRE-KNOWLEDGE

Virtually all chemical reactions may be regarded as reversible and, in a closed system, a state of equilibrium will be reached in which both forward and reverse reactions are occurring simultaneously, but at equal rates.

Despite the dynamic nature of the equilibrium state, no change can be observed in macroscopic properties, particularly the concentrations of reactants and products. If the concentration of reactants is vanishly small compared with the products, we say that the reaction has 'gone to completion', whereas if the situation is the opposite we say that the reaction 'does not occur'.

However, for any particular reaction, the relative concentrations of reactants and products may be changed by altering the conditions of temperature and pressure and by the addition of other materials. The qualitative effects of such changes in conditions are described by Le Chatelier's principle:

> When a system at equilibrium is subjected to a change in conditions, the equilibrium position will shift in such a way as to relieve the immediate effect of the change, if that is possible.

We assume that you have observed some equilibrium shifts experimentally and discussed them qualitatively in terms of Le Chatelier's principle. In order to discuss them quantitatively, we now introduce the equilibrium law.

THE EQUILIBRIUM LAW

The equilibrium law applies to all equilibrium systems and we shall use it as a tool to make precise mathematical predictions and calculations. It was put forward as a result of careful experimental work, but has a firm basis in theoretical thermodynamics.

For any equilibrium system, which we can generalise by the equation

$$a\text{A} + b\text{B} \rightleftharpoons c\text{C} + d\text{D}$$

the concentrations of reactants and products are related by the expression

$$K_c = \frac{[\text{C}]^c_{\text{eqm}}\,[\text{D}]^d_{\text{eqm}}}{[\text{A}]^a_{\text{eqm}}\,[\text{B}]^b_{\text{eqm}}}$$

K_c is constant at constant temperature

(The suffix 'eqm' is frequently omitted but it must then be remembered that the concentration terms refer to the equilibrium state.)

We now show you, in a Worked Example, how to write an equilibrium law expression for a particular equilibrium system. We illustrate the conventions used in writing equilibrium law expressions and show you how to work out the unit of K_c.

<u>Worked Example</u> (a) Write an expression for K_c for the reaction

W

$$N_2O_4(g) \rightleftharpoons 2NO_2(g)$$

(b) What is the unit of K_c in this case? Assume the unit of concentration is mol dm^{-3}.

<u>Solution</u>

1. Compare the equation with the general equation given on the previous page, noting particularly the stoichiometric coefficients. This shows that:

$$K_c = \frac{[NO_2(g)]^2}{[N_2O_4(g)]}$$

2. To work out the unit, assume that

$$[NO_2(g)] = x \text{ mol dm}^{-3} \quad \text{and} \quad [N_2O_4(g)] = y \text{ mol dm}^{-3}$$

Substitution into the equilibrium law expression gives:

$$K_c = \frac{(x \text{ mol dm}^{-3})^2}{y \text{ mol dm}^{-3}} = \frac{x^2}{y} \times \frac{(\text{mol dm}^{-3})^2}{\text{mol dm}^{-3}} = \frac{x^2}{y} \text{ mol dm}^{-3}$$

The unit of K_c for this reaction is therefore mol dm^{-3}.

Now try writing some equilibrium law expressions by doing the next exercise.

<u>Exercise 174</u> (a) For each of the following reactions write an expression for K_c. Work out units for K_c and include them in your answers, assuming that concentrations are measured in mol dm^{-3}.

(i) $2HBr(g) \rightleftharpoons H_2(g) + Br_2(g)$

(ii) $2SO_2(g) + O_2(g) \rightleftharpoons 2SO_3(g)$

(iii) $Cu(NH_3)_4^{2+}(aq) \rightleftharpoons Cu^{2+}(aq) + 4NH_3(aq)$

(iv) $4PF_5(g) \rightleftharpoons P_4(g) + 10F_2(g)$

(v) $C_2H_5OH(l) + CH_3CO_2H(l) \rightleftharpoons CH_3CO_2C_2H_5(l) + H_2O(l)$

(b) Look at the examples in which K_c has no unit. What do all these reactions have in common?

As soon as you come to consider <u>numerical values</u> of K_c, you must remember two important points.

1. A particular value of K_c refers only to one particular temperature.
2. The form of the equilibrium law expression (and, therefore, the value of K_c) depends on the way the equation is written.

We illustrate these ideas in the following exercises.

<u>Exercise 175</u> The same equilibrium system may be represented by two different equations:

$$COCl_2(g) \rightleftharpoons CO(g) + Cl_2(g)$$

$$CO(g) + Cl_2(g) \rightleftharpoons COCl_2(g)$$

Write expressions for two equilibrium constants, K_c and K_c'. What is the mathematical relation between them?

Exercise 176 The equilibrium between dinitrogen tetroxide and nitrogen dioxide may be represented equally well by two different equations:

$$\tfrac{1}{2}N_2O_4(g) \rightleftharpoons NO_2(g)$$

$$N_2O_4(g) \rightleftharpoons 2NO_2(g)$$

(a) Write expressions for two equilibrium constants, K_c and K_c'.

(b) At 100 °C, $K_c' = 0.490$ mol dm^{-3} and at 200 °C, $K_c' = 18.6$ mol dm^{-3}

What are the values of K_c at these temperatures?

Calculating the value of an equilibrium constant

We illustrate the method of calculation by a Worked Example.

Worked Example Equilibrium was established at 308 K for the system

W

$$CO(g) + Br_2(g) \rightleftharpoons COBr_2(g)$$

Analysis of the mixture gave the following values of concentration:

$$[CO(g)] = 8.78 \times 10^{-3}\ \text{mol dm}^{-3}$$

$$[Br_2(g)] = 4.90 \times 10^{-3}\ \text{mol dm}^{-3}$$

$$[COBr_2(g)] = 3.40 \times 10^{-3}\ \text{mol dm}^{-3}$$

Calculate the value of the equilibrium constant.

Solution

1. Write the equilibrium law expression for this reaction in terms of concentration:

$$K_c = \frac{[COBr_2(g)]}{[CO(g)][Br_2(g)]}$$

2. Substitute the equilibrium concentrations:

$$K_c = \frac{3.40 \times 10^{-3}\ \text{mol dm}^{-3}}{8.78 \times 10^{-3}\ \text{mol dm}^{-3} \times 4.90 \times 10^{-3}\ \text{mol dm}^{-3}} = \boxed{79.0\ \text{dm}^3\ \text{mol}^{-1}}$$

Now try some similar problems.

Exercise 177 The equilibrium

$$N_2O_4 \rightleftharpoons 2NO_2$$

can be established in an inert solvent at 298 K. Analysis of an equilibrium mixture gave the concentration of N_2O_4 as 0.021 mol dm^{-3} and the concentration of NO_2 as 0.010 mol dm^{-3}. Calculate the value of the equilibrium constant at 298 K.

In some problems, you may have to calculate concentrations from amounts and total volume before you apply the equilibrium law. Try this in the next exercise.

Exercise 178 Some phosphorus pentachloride was heated at 250 °C in a sealed container until equilibrium was reached according to the equation

$$PCl_5(g) \rightleftharpoons PCl_3(g) + Cl_2(g)$$

Analysis of the mixture showed that it contained 0.0042 mol of PCl_5, 0.040 mol of PCl_3 and 0.040 mol of Cl_2. The total volume was 2.0 dm^3.

(a) Calculate the concentration of each component and hence determine the equilibrium constant, K_c.

(b) The value of K_c for this system is much greater than the one you calculated in Exercise 177. What can you say about the relative concentrations of reactants and products when

(i) K_c is very large, (ii) K_c is very small?

The next two exercises give you further practice.

Exercise 179 At 250 °C, equilibrium for the following system was established.

$$PCl_5(g) \rightleftharpoons PCl_3(g) + Cl_2(g)$$

Analysis revealed these equilibrium concentrations:

$$[PCl_3(g)] = 1.50 \times 10^{-2} \text{ mol dm}^{-3}$$

$$[Cl_2(g)] = 1.50 \times 10^{-2} \text{ mol dm}^{-3}$$

$$[PCl_5(g)] = 1.18 \times 10^{-3} \text{ mol dm}^{-3}$$

Calculate the value of K_c at 250 °C.

Exercise 180 Analysis of the equilibrium system

$$2SO_2(g) + O_2(g) \rightleftharpoons 2SO_3(g)$$

showed that

$[SO_2(g)] = 0.23 \text{ mol dm}^{-3}$, $[O_2(g)] = 1.37 \text{ mol dm}^{-3}$, and

$[SO_3(g)] = 0.92 \text{ mol dm}^{-3}$.

Calculate the value of K_c at the temperature of the system.

When volume can be omitted from an equilibrium law expression

In the next exercise you show that, for an equilibrium system where the equilibrium constant has no unit, amounts can be used directly in the equilibrium law expression without using the total volume.

Exercise 181 The table below shows the composition of two equilibrium mixtures at 485 °C.

	Amount of H_2/mol	Amount of I_2/mol	Amount of HI/mol
1	0.02265	0.02840	0.1715
2	0.01699	0.04057	0.1779

(a) Write the equation for the formation of hydrogen iodide from hydrogen and iodine.

(b) Write an expression for the equilibrium constant, K_c.

(c) Calculate a value of K_c for each mixture, assuming the volume of the equilibrium mixture is 1.00 dm^3. Include units in your working.

(d) For mixture 1, calculate a value for the equilibrium constant, assuming the volume is 2.00 dm^3.

(e) For mixture 2, calculate a value for the equilibrium constant, assuming the volume is V dm^3. Show that the volume cancels.

Exercise 181 should convince you that the volume of the reaction mixture made no difference to the value of the equilibrium constant. We can generalise this for gaseous reactions and reactions in solution: if the equation shows equal numbers of molecules on both sides, then the equilibrium law expression is independent of the volume, and K_c is unitless.

In these circumstances, to calculate an equilibrium constant you can use equilibrium amounts rather than concentrations; or you can include V as an unknown value which cancels as in Exercise 181(e).

In the next exercise you obtain an equilibrium constant from the results of an experiment in which measured quantities of ethyl ethanoate and water were mixed in the presence of a catalyst, hydrochloric acid, and allowed to reach equilibrium.

$$CH_3CO_2C_2H_5(l) + H_2O(l) \rightleftharpoons C_2H_5OH(l) + CH_3CO_2H(l)$$

(Note that the state symbol (l) is not strictly appropriate since there is only one liquid phase, not four separate ones. Nor is (aq) appropriate, since the liquid phase is not primarily water.)

At equilibrium, the total amount of acid present (CH_3CO_2H + HCl) was determined by titration with a standard alkali. The equilibrium concentrations of all four components of the mixture may then be calculated, leading to a value of K_c. Each step in the calculation is very simple, but you may find the flow chart in Fig. 15 helps you to see how the steps fit together.

The quantities given directly from experimental measurements are listed on the left-hand side and those required for the equilibrium constant are on the right. Intermediate calculations are indicated by the arrows. All the steps shown are necessary but you can choose, to some extent, the order in which you do them.

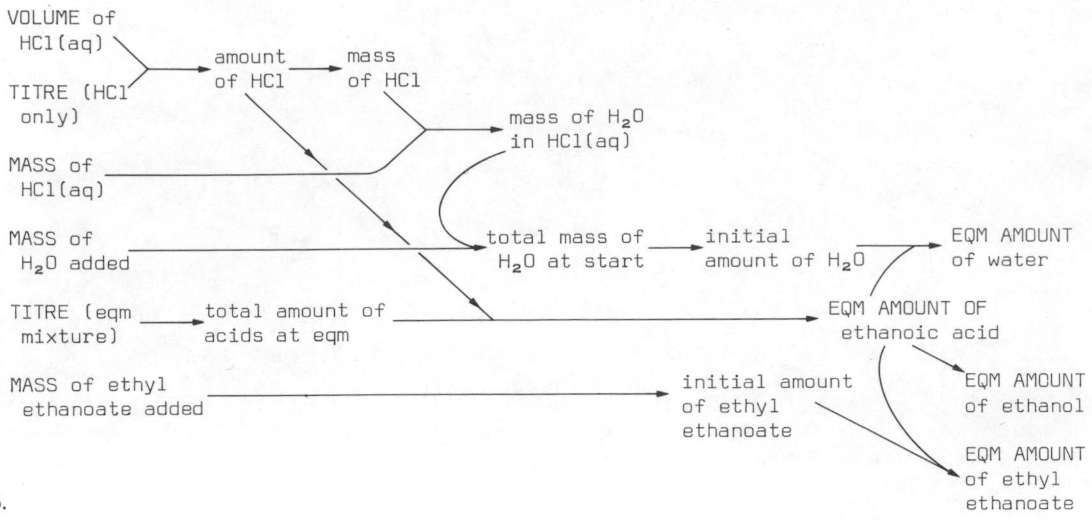

Fig.15.

Exercise 182 The following substances, in the quantities shown, were mixed in a specimen tube and allowed to reach equilibrium.

3.64 g of ethyl ethanoate, $CH_3CO_2C_2H_5$
0.99 g of water, H_2O
5.17 g of 2.00 M hydrochloric acid, HCl(aq)

Precisely the same mass of the hydrochloric acid was found to require 10.50 cm^3 of 0.974 M NaOH for neutralization. The equilibrium mixture required 39.20 cm^3 of the same alkali for neutralization. Calculate the equilibrium constant, K_c.

Various numerical problems involving equilibrium constants

We illustrate different types of problem by Worked Examples and Exercises.

Worked Example

W

For the equilibrium:

$$PCl_5(g) \rightleftharpoons PCl_3(g) + Cl_2(g)$$

K_c = 0.19 mol dm^{-3} at 250 °C

One equilibrium mixture at this temperature contains PCl_5 at a concentration of 0.20 mol dm^{-3} and PCl_3 at a concentration of 0.010 mol dm^{-3}.

Calculate the concentration of Cl_2 in this mixture.

Solution

1. Start by writing the equation (even if it is given in the question). Leave space where the equilibrium concentrations can be tabulated under the formulae of the compounds:

$$PCl_5(g) \rightleftharpoons PCl_3(g) + Cl_2(g)$$

Equilibrium
concn/mol dm^{-3}

2. Indicate the equilibrium concentrations under the equation. Let the concentration of chlorine, $[Cl_2(g)]$, be x mol dm^{-3}

	$PCl_5(g)$	$\rightleftharpoons$	$PCl_3(g)$	+	$Cl_2(g)$
Equilibrium concn/mol dm^{-3}	0.20		0.010		x

3. Write the equilibrium law expression in terms of concentration.

$$K_c = \frac{[PCl_3(g)][Cl_2(g)]}{[PCl_5(g)]}$$

4. Substitute all values into the expression (remember to include units).

$$0.19 \text{ mol dm}^{-3} = \frac{0.010 \text{ mol dm}^{-3} \times x \text{ mol dm}^{-3}}{0.20 \text{ mol dm}^{-3}}$$

5. Solve the equation for x:

$$x = \frac{0.19 \times 0.20}{0.010} = 3.8$$

$$\therefore \quad [Cl_2(g)] = \boxed{3.8 \text{ mol dm}^{-3}}$$

Exercise 183 Some N_2O_4 dissolved in chloroform was allowed to reach equilibrium at a known temperature.

$$N_2O_4 \rightleftharpoons 2NO_2$$

At this point the concentration of NO_2 was 1.85×10^{-3} mol dm^{-3}. What was the equilibrium concentration of N_2O_4?

$K_c = 1.06 \times 10^{-5}$ mol dm^{-3} at this temperature.

Exercise 184 At 1400 K, $K_c = 2.25 \times 10^{-4}$ mol dm^{-3} for the equilibrium:

$$2H_2S(g) \rightleftharpoons 2H_2(g) + S_2(g)$$

In an equilibrium mixture, $[H_2S(g)] = 4.84 \times 10^{-3}$ mol dm^{-3} and $[S_2(g)] = 2.33 \times 10^{-3}$ mol dm^{-3}.

Calculate the equilibrium concentration of H_2.

Exercise 185 In the following equilibrium:

$$H_2(g) + I_2(g) \rightleftharpoons 2HI(g)$$

$K_c = 54.1$, at a particular temperature. The equilibrium mixture was found to contain H_2 at a concentration of 0.48×10^{-3} mol dm^{-3}, and HI at a concentration of 3.53×10^{-3} mol dm^{-3}.

What is the equilibrium concentration of I_2?

Exercise 186 This question concerns the equilibrium system

$$CH_3CO_2H(l) + C_2H_5OH(l) \rightleftharpoons CH_3CO_2C_2H_5(l) + H_2O(l);$$

$K_c = 4.0$ at 25 °C

In a particular experiment, 0.33 mol of CH_3CO_2H, 0.66 mol of $CH_3CO_2C_2H_5$ and 0.66 mol of H_2O are found to be present. What amount of C_2H_5OH is present?

The next exercise is similar, but requires an extra step at the start; i.e. calculating concentrations from amounts and total volume.

Exercise 187 In another equilibrium mixture of the reaction

$$PCl_5(g) \rightleftharpoons PCl_3(g) + Cl_2(g)$$

at 250 °C in a 2.0 dm³ vessel, there is 0.15 mol of PCl_3 and 0.090 mol of Cl_2.

K_c = 0.19 mol dm^{-3} at 250 °C.

(a) Calculate the amount of PCl_5 present at equilibrium.

(b) Calculate the mass of PCl_5 present at equilibrium.

Don't forget that if the total number of molecules does not change during a homogeneous reaction, the volume doesn't make any difference to the calculation and you need not include it.

Now we show you how to apply the equilibrium law to a slightly different type of problem.

Worked Example 6.75 g of SO_2Cl_2 was put into a 2.00 dm³ vessel, the vessel was sealed and its temperature raised to 375 °C. At equilibrium, the vessel contained 0.0345 mol of Cl_2. Calculate the equilibrium constant for the reaction.

W

$$SO_2Cl_2(g) \rightleftharpoons SO_2(g) + Cl_2(g)$$

Solution

1. First, note that the number of molecules does change, so you will have to calculate concentrations from amounts.
2. Write the balanced equation, leaving room for initial concentrations above and equilibrium concentrations below:

Initial
concn/mol dm^{-3}

$$SO_2Cl_2(g) \rightleftharpoons SO_2(g) + Cl_2(g)$$

Equilibrium
concn/mol dm^{-3}

3. Calculate the initial concentrations and put the values above the appropriate formulae in the equation.

The amount, n, of SO_2Cl_2 $= \frac{m}{M} = \frac{6.75\ \text{g}}{135\ \text{g mol}^{-1}} = 0.0500$ mol

$$c = \frac{n}{V}$$

$$\therefore\ [SO_2Cl_2(g)] = \frac{0.0500\ \text{mol}}{2.00\ \text{dm}^3} = 0.0250\ \text{mol dm}^{-3}$$

The initial concentrations of SO_2 and Cl_2 are both zero. Write down the initial concentrations:

	$SO_2Cl_2(g)$	$\rightleftharpoons$	$SO_2(g)$	+	$Cl_2(g)$
Initial concn/mol dm^{-3}	0.025		0		0
Equilibrium concn/mol dm^{-3}	?		?		?

4. Now work out the equilibrium concentrations from the data and balanced equation:

(a) There is 0.0345 mol of Cl_2 present at equilibrium so there must also be 0.0345 mol of SO_2 present. The concentrations are:

$$[Cl_2(g)] = \frac{0.0345 \text{ mol}}{2.00 \text{ dm}^3} = 0.0173 \text{ mol dm}^{-3}$$

$$[SO_2(g)] = \frac{0.0345 \text{ mol}}{2.00 \text{ dm}^3} = 0.0173 \text{ mol dm}^{-3}$$

(b) The <u>amount</u> of SO_2Cl_2 left = initial amount - amount reacted

= (0.0500 - 0.0345) mol = 0.0155 mol

$$\therefore [SO_2Cl_2(g)] = \frac{0.0155 \text{ mol}}{2.00 \text{ dm}^3} = 0.00775 \text{ mol dm}^{-3}$$

Write these equilibrium concentrations under the equation:

Initial concn/mol dm^{-3}	0.025	0	0
	$SO_2Cl_2(g)$ $\rightleftharpoons$	$SO_2(g)$ +	$Cl_2(g)$
Equilibrium concn/mol dm^{-3}	0.00775	0.0173	0.0173

5. From here, proceed as in the first Worked Example. Write the equilibrium law expression and substitute the equilibrium concentrations:

$$K_c = \frac{[SO_2(g)][Cl_2(g)]}{[SO_2Cl_2(g)]}$$

$$= \frac{0.0173 \text{ mol dm}^{-3} \times 0.0173 \text{ mol dm}^{-3}}{0.00775 \text{ mol dm}^{-3}} = \boxed{0.0386 \text{ mol dm}^{-3}}$$

Now try the following three exercises, two of which are A-level questions.

<u>Exercise 188</u> Ethanoic acid, CH_3CO_2H, and pentene, C_5H_{10}, react to produce pentyl ethanoate in an inert solvent. A solution was prepared containing 0.020 mol of pentene and 0.010 mol of ethanoic acid in 600 cm^3 of solution. At equilibrium there was 9.0×10^{-3} mol of pentyl ethanoate. Calculate the value of K_c from these data.

$$CH_3CO_2H + C_5H_{10} \rightleftharpoons CH_3CO_2C_5H_{11}$$

<u>Exercise 189</u> A mixture of 1.90 mol of hydrogen and 1.90 mol of iodine was allowed to reach equilibrium at 710 K. The equilibrium mixture was found to contain 3.00 mol of hydrogen iodide. Calculate the equilibrium constant at 710 K for the reaction

A part

$$H_2(g) + I_2(g) \rightleftharpoons 2HI(g)$$

<u>Exercise 190</u> If a mixture of 6.0 g of ethanoic acid and 6.9 g of ethanol is allowed to reach equilibrium, 7.0 g of ethyl ethanoate is formed. Calculate K_c.

A part

$$CH_3CO_2H + C_2H_5OH \rightleftharpoons CH_3CO_2C_2H_5 + H_2O$$

Exercise 191 In the following equilibrium system:

$$H_2(g) + I_2(g) \rightleftharpoons 2HI(g)$$

20.57 mol of hydrogen and 5.22 mol of iodine were allowed to reach equilibrium at 450 °C. At this point, the mixture contained 10.22 mol of hydrogen iodide. Calculate the value of K_c at this temperature.

Exercise 192 The equilibrium

$$N_2O_4(l) \rightleftharpoons 2NO_2(l)$$

was established in a solvent at 10 °C starting with 0.1307 mol dm^{-3} of N_2O_4; the equilibrium mixture was found to contain 0.0014 mol dm^{-3} of NO_2. Calculate the value of K_c at this temperature.

The type of problem we present in the next section often requires the solution of a quadratic equation, although it is not always possible to tell just by reading the question. The chemistry is usually straightforward, but if you find the mathematics difficult, remember that in an examination you will secure a good proportion of the marks simply by applying the correct principle without actually solving the equation at the last step. In any case, some examining boards rarely set problems which require the solution of quadratic equations.

Problems involving quadratic equations

The general form of a quadratic equation is

$$ax^2 + bx + c = 0$$

where a, b and c are constants. The solution is given by the formula

$$x = \frac{-b \pm \sqrt{b^2 - 4ac}}{2a}$$

and it is easy to solve this using a calculator or tables of logarithms.

Worked Example Carbon monoxide and chlorine react to form phosgene, $COCl_2$. A mixture was prepared containing 0.20 mol of CO and 0.10 mol of Cl_2 in a 3.0 dm^3 vessel. At the temperature of the experiment, K_c = 0.410 dm^3 mol^{-1}. Calculate the concentration of $COCl_2$ at equilibrium.

W

Solution

1. In this case, it is convenient to work first in terms of amounts, and then divide them by the volume later when substituting concentrations into the equilibrium law. Write the initial amounts above the equation, leaving space below for the equilibrium amounts.

Initial amount/mol	0.20		0.10		0
	$CO(g)$	+	$Cl_2(g)$	$\rightleftharpoons$	$COCl_2(g)$
Equilibrium amount/mol					

2. Calculate the amounts at equilibrium in terms of one unknown amount.

(a) Let the amount of $COCl_2$ formed be x mol

(b) The balanced equation tells us that at equilibrium

$$\text{amount of CO} = \text{initial amount} - \text{amount reacted} = (0.20 - x)\ \text{mol}$$

(c) By similar reasoning, at equilibrium

$$\text{amount of } Cl_2 = (0.10 - x)\ \text{mol}$$

3. Put these equilibrium amounts in the appropriate spaces under the equation:

Initial amount/mol	0.20	0.10	0
	$CO(g)$ +	$Cl_2(g)$ ⇌	$COCl_2(g)$
Equilibrium amount/mol	$(0.20-x)$	$(0.10-x)$	x

4. Write the equilibrium law in terms of concentrations:

$$K_c = \frac{[COCl_2(g)]}{[CO(g)][Cl_2(g)]}$$

5. Put the equilibrium amounts in terms of concentrations by dividing by the volume; substitute these into the equilibrium law:

$$[COCl_2(g)] = \frac{x}{3.0}\ \text{mol dm}^{-3}$$

$$[CO(g)] = \frac{0.20 - x}{3.0}\ \text{mol dm}^{-3}$$

$$[Cl_2(g)] = \frac{0.10 - x}{3.0}\ \text{mol dm}^{-3}$$

$$\therefore\quad 0.410\ \text{dm}^3\ \text{mol}^{-1} = \frac{\frac{x}{3.0}\ \text{mol dm}^{-3}}{\frac{0.20 - x}{3.0}\ \text{mol dm}^{-3} \times \frac{0.10 - x}{3.0}\ \text{mol dm}^{-3}}$$

6. Simplify this equation:

$$0.410 = \frac{3x}{(0.20 - x) \times (0.10 - x)} = \frac{3x}{0.020 - 0.30x + x^2}$$

or $\quad 0.0082 - 0.123x + 0.410x^2 = 3x$

Put this quadratic equation in the standard form

$$0.410x^2 - 3.123x + 0.0082 = 0$$

7. Solve this quadratic equation. The solutions of this equation are:

$$x = \frac{-b \pm \sqrt{b^2 - 4ac}}{2a}$$

for which $a = 0.410$, $b = -3.123$, $c = 0.0082$.

$$x = \frac{3.123 \pm \sqrt{(3.123)^2 - (4 \times 0.410 \times 0.0082}}{(2 \times 0.410)}$$

$$= \frac{3.123 \pm \sqrt{9.7531 - 0.0134}}{0.820} = \frac{3.123 \pm \sqrt{9.7397}}{0.820} = \frac{3.123 \pm 3.121}{0.820}$$

$$\therefore\quad x = 7.6 \quad \text{or} \quad x = 2.4 \times 10^{-3}$$

Solving a quadratic equation produces two roots. Because the system began with 0.20 mol of CO and 0.10 mol of Cl_2, it is impossible to produce 7.6 mol of $COCl_2$. Therefore, the correct root is $x = 2.4 \times 10^{-3}$.

8. The amount of $COCl_2$ at equilibrium is 2.4×10^{-3} mol.

$$\text{Now, concentration} = \frac{\text{amount}}{\text{volume}}$$

$$\therefore \quad [COCl_2(g)] = \frac{2.4 \times 10^{-3}\ \text{mol}}{3.0\ \text{dm}^3} = \boxed{8.0 \times 10^{-4}\ \text{mol dm}^{-3}}$$

Now try the following exercises, some of which require the solution of a quadratic equation.

Exercise 193 Carbon monoxide will react with steam under the appropriate conditions according to the following reversible reaction:

$$CO(g) + H_2O(g) \rightleftharpoons CO_2(g) + H_2(g); \quad \Delta H^\ominus = -40\ \text{kJ mol}^{-1}$$

Calculate the number of moles of hydrogen in the equilibrium mixture when three moles of carbon monoxide and three moles of steam are placed in a reaction vessel of constant volume and maintained at a temperature at which the equilibrium constant has a numerical value of 4.00.

Exercise 194 For the equilibrium:

$$PCl_5(g) \rightleftharpoons PCl_3(g) + Cl_2(g)$$

$K_c = 0.19$ mol dm^{-3} at 250 °C. 2.085 g of PCl_5 was heated to 250 °C in a sealed vessel of 500 cm^3 capacity and maintained at this temperature until equilibrium was established. Calculate the concentrations of PCl_5, PCl_3 and Cl_2 at equilibrium.

Exercise 195 For the reaction:

$$H_2 + I_2 \rightleftharpoons 2HI$$

the equilibrium constant = 49.0 at 444 °C. If 2.00 mol of hydrogen and 2.00 mol of iodine are heated in a closed vessel at 444 °C until equilibrium is attained, calculate the composition in moles of the equilibrium mixture.

Exercise 196 For the equilibrium:

$$\underset{}{C_2H_5OH(l)} + \underset{\text{propanoic acid}}{C_2H_5CO_2H(l)} \rightleftharpoons \underset{\text{ethyl propanoate}}{C_2H_5CO_2C_2H_5(l)} + H_2O(l)$$

$K_c = 7.5$ at 50 °C. If 50.0 g of C_2H_5OH is mixed with 50.0 g of $C_2H_5CO_2H$, what mass of ethyl propanoate will be formed at equilibrium?

Exercise 197 In the following equilibrium:

$$CH_3CO_2H(l) + C_2H_5OH(l) \rightleftharpoons CH_3CO_2C_2H_5(l) + H_2O(l)$$

8.0 mol of ethanoic acid, and 6.0 mol of ethanol were placed in a 2.00 dm^3 vessel. What is the equilibrium amount of water? K_c at this temperature = 4.5.

Exercise 198 4.0 g of phosphorus pentachloride is allowed to reach equilibrium at 250 °C in a 750 cm^3 vessel. K_c = 0.19 mol dm^{-3} for the following reaction at 250 °C.

$$PCl_5(g) \rightleftharpoons PCl_3(g) + Cl_2(g)$$

Calculate the amount of phosphorus pentachloride at equilibrium.

So far in this Chapter, we have expressed the equilibrium law only in terms of the equilibrium concentrations of reactants and products. However, for gaseous systems, it is often convenient to use partial pressures rather than concentrations. The equilibrium law still applies, but gives a different equilibrium constant, K_p, which we now consider.

The equilibrium constant, K_p

You know, from Chapter 5 (Gases), that the partial pressure of a gas is proportional to its concentration. Therefore, the equilibrium law can be expressed in terms of partial pressures; the equilibrium constant is then given the symbol K_p.

To calculate K_p, we simply substitute equilibrium partial pressures into an equilibrium law expression. It is just like substituting equilibrium concentrations to calculate K_c. To see if you can do this, try the following exercises.

Exercise 199 In the equilibrium system

$$2SO_2(g) + O_2(g) \rightleftharpoons 2SO_3(g)$$

at 700 K, the partial pressures of the gases in an equilibrium mixture are:

$$p_{SO_2} = 0.090 \text{ atm}, \quad p_{SO_3} = 4.5 \text{ atm} \quad \text{and} \quad p_{O_2} = 0.083 \text{ atm}.$$

Calculate K_p for this system.

Exercise 200 In the equilibrium system:

$$N_2O_4(g) \rightleftharpoons 2NO_2(g)$$

at 55 °C, the partial pressures of the gases are:

$$p_{N_2O_4} = 0.33 \text{ atm} \quad \text{and} \quad p_{NO_2} = 0.67 \text{ atm}$$

Calculate K_p for this system.

Exercise 201 Consider the following reaction:

$$H_2(g) + I_2(g) \rightleftharpoons 2HI(g)$$

At a certain temperature, analysis of the equilibrium mixture of the gases yielded the following results:

$p_{H_2} = 0.25$ atm, $p_{I_2} = 0.16$ atm, $p_{HI} = 0.40$ atm

Calculate K_p for this reaction at the same temperature.

In the next exercise, you have to calculate the partial pressures before substituting in the equilibrium law expression. Remember that

partial pressure = mole fraction x total pressure

$$\text{mole fraction} = \frac{\text{amount of one component}}{\text{total amount of all components}}$$

Exercise 202 Analysis of the equilibrium system

$$N_2(g) + 3H_2(g) \rightleftharpoons 2NH_3(g)$$

showed 25.1 g of NH_3, 13.0 g of H_2 and 59.6 g of N_2.

(a) Calculate the mole fraction of each gas.

(b) The total pressure of the system was 10.0 atm. Calculate the partial pressure of each gas.

(c) Calculate K_p for the system.

Exercise 203 Phosphorus pentachloride dissociates on heating according to the equation:

$$PCl_5(g) \rightleftharpoons PCl_3(g) + Cl_2(g)$$

At a pressure of 10.0 atm and a temperature of 250 °C the amount of each gas present at equilibrium was: 0.33 mol of PCl_5, 0.67 mol of PCl_3 and 0.67 mol of Cl_2. Calculate the value of K_p.

Exercise 204 Ammonium aminomethanoate $NH_2CO_2NH_4$ decomposes according to the equation:

$$NH_2CO_2NH_4(s) \rightleftharpoons 2NH_3(g) + CO_2(g)$$

In a particular system at 293 K, there was 0.224 mol of CO_2 and 0.142 mol of NH_3 at a total pressure of 1.83 atm. Calculate K_p for this system.

Remember that the partial pressure of a solid is constant and can therefore be incorporated into the equilibrium constant. This is equivalent to taking it as unity.

Exercise 205 In the following reaction:

$$H_2(g) + I_2(g) \rightleftharpoons 2HI(g)$$

at a temperature of 485 °C and a pressure of 2.00 atm, the amount of each gas present at equilibrium was 0.56 mol of H_2, 0.060 mol of I_2 and 1.27 mol of HI. Calculate K_p.

You can use substitution into the expression for K_p to obtain other information. For instance, in the following exercises, you are given values for K_p and the amounts of gases present, and are required to calculate the total pressure.

Exercise 206 In the following reaction:

$$2NO_2(g) \rightleftharpoons 2NO(g) + O_2(g)$$

at a temperature of 700 K, the amount of each gas present at equilibrium is 0.96 mol of NO_2, 0.04 mol of NO, and 0.02 mol of O_2. If $K_p = 6.8 \times 10^{-6}$ atm, what must the total pressure have been to achieve this particular equilibrium mixture?

Exercise 207 K_p = 3.72 atm at 1000 K for the system:

$$C(s) + H_2O(g) \rightleftharpoons CO(g) + H_2(g)$$

What pressure must be applied to obtain a mixture containing 2.0 mol of CO, 1.0 mol of H_2 and 4.0 mol of H_2O?

K_p and K_c generally have different values and units (look, for example, at Exercises 178 and 203), but the two quantities are related because partial pressure is proportional to concentration.

Dalton's law of partial pressures enables us to say that, for a component, A, of a gaseous mixture

$$p_A V = n_A RT \quad \text{or} \quad p_A = \frac{n_A}{V}RT = [A(g)]RT$$

Substituting expressions like this into the expression for K_p for any reaction shows us that, in general

$$\boxed{K_p = K_c(RT)^{\Delta n}}$$

where Δn = (number of moles of gaseous products) - (number of moles of gaseous reactants). Clearly, if $\Delta n = 0$, $K_p = K_c$ and both are unitless.

Use this expression to do the following exercises. (Choose the value of R which has the most appropriate unit.)

Exercise 208 For the equilibrium system:

$$PCl_5(g) \rightleftharpoons PCl_3(g) + Cl_2(g)$$

K_p = 0.811 atm at 523 K. Calculate K_c at this temperature.

Exercise 209 In the equilibrium system:

$$N_2(g) + 3H_2(g) \rightleftharpoons 2NH_3(g)$$

$K_c = 2.0\ \text{mol}^{-2}\ \text{dm}^6$ at 620 K. What is the value of K_p?

Now we show you how to use an expression for K_p to investigate the effect of changing pressure.

The effect on an equilibrium system of changing pressure

We assume that you have already applied Le Chatelier's principle to determine the direction of any shift in equilibrium position caused by changing the total pressure. Now you can apply the equilibrium law to show why Le Chatelier's principle works in relation to pressure changes. You can also calculate the pressure change required to alter the equilibrium concentration to a given extent.

We show you how to achieve these objectives by means of a Revealing Exercise, followed by a Worked Example, both of which concern the equilibrium system

$$2SO_2(g) + O_2(g) \rightleftharpoons 2SO_3(g)$$

Q1. Write the equilibrium law expression for the system.

A1. $$K_p = \frac{p^2_{SO_3}}{p^2_{SO_2} \times p_{O_2}}$$

Q2. Write an expression for the partial pressure of each gas in terms of the mole fraction X and the total pressure p_T.

A2. $p_{SO_3} = X_{SO_3}p_T$ $\qquad p_{SO_2} = X_{SO_2}p_T$ $\qquad p_{O_2} = X_{O_2}p_T$

Q3. Substitute your answers to Q2 into the equilibrium law expression for K_p and simplify the equation.

A3. $$K_p = \frac{(X_{SO_3}p_T)^2}{(X_{SO_2}p_T)^2 \times (X_{O_2}p_T)} = \frac{1}{p_T} \times \frac{X^2_{SO_3}}{X^2_{SO_2} \times X_{O_2}}$$

Q4. If the pressure of the system is doubled (at constant temperature), which mole fractions must increase and which must decrease? (The new values must give the same value of K_p as before.)

A4. The mole fraction of SO_3 must increase and the mole fractions of SO_2 and O_2 must decrease.

Q5. In which direction must the equilibrium position shift in order to change the mole fractions? Is this in accordance with Le Chatelier's principle?

A5. The equilibrium shifts to the right, as predicted by Le Chatelier's principle.

The same method could be applied to any gaseous equilibrium system. We now show you, in a Worked Example, how to use the expression you derived in A3 of the Revealing Exercise.

Worked Example At 1100 K, $K_p = 0.13\ atm^{-1}$ for the system

$$2SO_2(g) + O_2(g) \rightleftharpoons 2SO_3(g)$$

W

If 2.00 mol of SO_2 and 2.00 mol of O_2 are mixed and allowed to react, what must the total pressure be to give a 90% yield of SO_3?

Solution

1. Write the balanced equation, leaving room above and below to put in initial amounts and equilibrium amounts. Put in the initial amounts:

Initial amount/mol	2.00	2.00	0
	$2SO_2(g)$ +	$O_2(g)$ ⇌	$2SO_3(g)$
Equilibrium amount/mol			

2. Calculate the amounts at equilibrium.

(a) The amount of SO_3 produced $= 2.00\ mol \times \frac{90}{100} = 1.80\ mol$

(b) The amount of SO_2 left at equilibrium

= initial amount - amount reacted

$= (2.00 - 1.80)\ mol = 0.20\ mol$

(c) Amount of oxygen reacting $= \frac{1}{2} \times$ amount of SO_2 reacting

$= \frac{1}{2} \times 1.80\ mol = 0.900\ mol$

The amount of oxygen remaining at equilibrium

= initial amount - amount reacting

$= (2.00 - 0.900)\ mol = 1.10\ mol$

3. Tabulate this equilibrium information.

Initial amount/mol	2.0	2.0	0
	$2SO_2(g)$ +	$O_2(g)$ ⇌	$2SO_3(g)$
Equilibrium amount/mol	0.20	1.10	1.80

4. Write an expression for the partial pressure of each gas.

The total amount of gas present $= (0.20 + 1.10 + 1.80)\ mol = 3.10\ mol$

$$p_{SO_3} = X_{SO_3}p_T = \frac{1.80}{3.10} \times p_T$$

Similarly for the other two partial pressures:

$$p_{SO_2} = \frac{0.20}{3.10} \times p_T \qquad p_{O_2} = \frac{1.10}{3.10} \times p_T$$

5. Write an expression for K_p, substitute the partial pressures and solve for p_T.

$$K_p = \frac{p^2_{SO_3}}{p^2_{SO_2} \times p_{O_2}}$$

$$\therefore\ 0.13\ atm^{-1} = \frac{\left(\frac{1.80}{3.10}\right)^2 p_T^2}{\left(\frac{0.20}{3.10}\right)^2 p_T^2 \times \left(\frac{1.10}{3.10}\right)} = \frac{0.337}{0.00416 \times 0.355\ p_T}$$

$$\therefore\ p_T = \frac{0.337}{0.13\ atm^{-1} \times 0.00416 \times 0.355} = \boxed{1.76 \times 10^3\ atm}$$

The pressure calculated in the Worked Example is far too high for economic production. In the next exercise, you calculate the pressure required for a more modest conversion.

Exercise 210 For the system described in the Worked Example, calculate the pressure necessary for 20% conversion of the SO_2.

Now try two similar exercises.

Exercise 211 Sulphur dioxide and oxygen in the ratio 2 mol:1 mol were mixed at a constant temperature of 1110 K and a constant pressure of 9 atmospheres in the presence of a catalyst. At equilibrium, one-third of the sulphur dioxide had been converted into sulphur trioxide:

$$2SO_2(g) + O_2(g) \rightleftharpoons 2SO_3(g)$$

Calculate the equilibrium constant, K_p, for this reaction under these conditions.

Exercise 212 A sample of dinitrogen tetroxide is 66% dissociated at a pressure of 98.3 kPa (747 mmHg) and a temperature of 60 °C. Standard pressure = 100 kPa (760 mmHg). Calculate the value of K_p for the equilibrium at 60 °C, stating the units.

A part

When considering mixtures of gases, a useful concept is 'average molar mass'. We deal with this in the next section.

AVERAGE MOLAR MASS OF A GASEOUS MIXTURE

You might be asked to calculate the average molar mass of a mixture of gases from the amounts present or, alternatively, to calculate the relative amounts present from the average molar mass of a mixture. We show you how to do this calculation in a Worked Example. You may like to try to solve the problem yourself before you look at our solution.

Worked Example Calculate the average molar mass of a mixture of 0.20 mol of SO_2, 1.10 mol of O_2 and 1.80 mol of SO_3.

W

Solution

Each component of the mixture contributes to the average molar mass, $\overline{M}$, in proportion to its mole fraction, X.

$$\overline{M} = X_{SO_2}M_{SO_2} + X_{O_2}M_{O_2} + X_{SO_3}M_{SO_3}$$

$$= \frac{0.20}{3.10} \times 64.1 \text{ g mol}^{-1} + \frac{1.10}{3.10} \times 32.0 \text{ g mol}^{-1} + \frac{1.80}{3.10} \times 80.1 \text{ g mol}^{-1}$$

$$= (4.1 + 11.4 + 46.5) \text{ g mol}^{-1} = \boxed{62.0 \text{ g mol}^{-1}}$$

Exercise 213 $2NO_2(g) \rightleftharpoons 2NO(g) + O_2(g)$

For this system, a particular equilibrium mixture has the composition 0.96 mol $NO_2(g)$, 0.040 mol $NO(g)$, 0.020 mol $O_2(g)$ at 700 K and 0.20 atm.

(a) Calculate the equilibrium constant, K_p, for this reaction under the stated conditions.

(b) Calculate the average molar mass of the mixture under the stated conditions.

Exercise 214 0.20 mol of carbon dioxide was heated with excess carbon in a closed vessel until the following equilibrium was attained:

$$CO_2(g) + C(s) \rightleftharpoons 2CO(g)$$

It was found that the average molar mass of the gaseous equilibrium mixture was 36 g mol^{-1}.

(a) Calculate the mole fraction of carbon monoxide in the equilibrium gaseous mixture.

(b) The pressure at equilibrium in the vessel was 12 atmospheres. Calculate K_p for the equilibrium at the temperature of the experiment.

(c) Calculate the mole fraction of carbon monoxide which would be present in the equilibrium mixture if the pressure were reduced to 2.0 atmospheres at the same temperature.

Exercise 215 For the equilibrium:

$$PCl_5(g) \rightleftharpoons PCl_3(g) + Cl_2(g)$$

a particular equilibrium mixture contains 0.20 mol PCl_5, 0.010 mol PCl_3 and 3.80 mol Cl_2, at a particular temperature and a total pressure of 3.0 atm.

(a) Calculate the equilibrium constant, K_p, for this reaction under the stated conditions.

(b) Calculate the average molar mass of the mixture under the stated conditions.

Exercise 216 For the equilibrium:

$$N_2O_4(g) \rightleftharpoons 2NO_2(g)$$

1.00 mol of N_2O_4 was introduced into a vessel and allowed to attain equilibrium at 308 K. It was found that the average molar mass of the mixture was 72.4 g mol^{-1}.

(a) Calculate the mole fraction of NO_2 in the equilibrium mixture.

(b) The pressure at equilibrium was 1.00 atm. Calculate K_p for the system at the temperature of the experiment.

(c) Calculate the mole fraction of NO_2 which would be present in the equilibrium mixture if the pressure were increased to 6.00 atm at the same temperature.

We now direct your attention from pressure changes to temperature changes.

THE VARIATION OF EQUILIBRIUM CONSTANT WITH TEMPERATURE

You already know, by applying Le Chatelier's principle, that the direction of a shift in equilibrium brought about by a temperature change depends on the sign of the standard enthalpy change, $\Delta H^{\ominus}$. You should not be surprised to find that the extent of the shift depends on the value of $\Delta H^{\ominus}$

Since a change in temperature does not, of itself, change the concentration of a substance, it follows that any shift in equilibrium position must be due to a change in the equilibrium constant. The effect on equilibrium of changing temperature is, therefore, different from the effects of changing concentration and pressure. Remember that:

> Equilibrium constants vary with temperature; they do not vary with pressure or concentrations.

> If $\Delta H^{\ominus}$ is positive, K increases with temperature (more products);
> if $\Delta H^{\ominus}$ is negative, K decreases with temperature (more reactants).

We now introduce an expression which relates changes in the equilibrium constant with temperature to the standard enthalpy change, $\Delta H^{\ominus}$, for the reaction. Since we have chosen not to discuss entropy and free energy, the derivation of this expression is beyond the scope of this book; we merely present it.

If K_1 and K_2 are the equilibrium constants for a particular reaction at temperatures T_1 and T_2, and the standard enthalpy change, $\Delta H^{\ominus}$, is taken as constant over a limited temperature range, then

$$\log\left(\frac{K_2}{K_1}\right) = \frac{\Delta H^{\ominus}}{2.30\ R}\left(\frac{1}{T_1} - \frac{1}{T_2}\right)$$

Thus, if four of the quantities $\Delta H^{\ominus}$, K_1, K_2, T_1, and T_2 are known, the fifth may be calculated. Note that if the reaction involves a change in the amount of gas present, K_p must be used in the expression. Otherwise, we may use K_c.

Use the expression derived above in the following exercises.

Exercise 217 At 1065 °C, K_p = 0.0118 atm for the reaction:

$$2H_2S(g) \rightleftharpoons 2H_2(g) + S_2(g); \quad \Delta H^{\ominus} = 177.3\ \text{kJ mol}^{-1}$$

Calculate the equilibrium constant for the reaction at 1200 °C. (Look up the appropriate value for the gas constant, R.)

Exercise 218 The value of K_p for a particular reaction is 2.44 atm at 1000 K and 3.74 atm at 1200 K. Calculate $\Delta H^{\ominus}$ for this reaction.

Exercise 219 The table below gives information about the values at different temperatures of the equilibrium constant for the reaction:

$$N_2(g) + O_2(g) \rightleftharpoons 2NO(g)$$

It also gives the partial pressures of NO in equilibrium with two different mixtures of nitrogen and oxygen at the given temperatures.

Temperature /K	$10^4 K_p$	Partial pressure of NO/atm x 10^2	
		p_{N_2} = 0.8 atm p_{O_2} = 0.2 atm	p_{N_2} = 0.8 atm p_{O_2} = 0.05 atm
1800	1.21	0.44	0.22
2000	4.08	0.81	0.40
2200	11.00	1.33	0.67
2400	25.10	2.00	1.00
2600	50.30	2.84	1.42

(a) Write an expression for the equilibrium constant K_p for the given reaction.

(b) Use the expression you have written to explain why the values in the fourth column are half those in the third.

(c) What condition of temperature and pressure should be used to obtain the best yield of NO? Justify your answer.

(d) Is the reaction exothermic or endothermic? Justify your answer.

Exercise 220 Calculate $\Delta H^{\ominus}$ from the data given in Exercise 219.

Most of the equilibrium systems in the preceding examples are homogeneous but you have also encountered the application of the equilibrium law to heterogeneous systems. In such systems, the concentration (or partial pressure) of a pure liquid or solid phase is constant and, therefore, included in the equilibrium constant.

We now reinforce this idea by applying the equilibrium law to an important group of heterogeneous sytems - sparingly soluble salts in contact with their saturated solutions.

SOLUBILITY PRODUCTS FOR SLIGHTLY SOLUBLE SALTS

Equilibria involving sparingly soluble salts deserve special attention because of their importance in chemical analysis. A simple application of the equilibrium law helps us to understand how precipitation can be controlled.

The solubility product, K_s, for a solution of a sparingly soluble salt in equilibrium with the solid is simply related to K_c, as shown in the following example:

$$Ag_3PO_4(s) \rightleftharpoons 3Ag^+(aq) + PO_4^{3-}(aq)$$

$$K_c = \frac{[Ag^+(aq)^3[PO_4^{3+}(aq)]}{[Ag_3PO_4(s)]}$$

$[Ag_3PO_4(s)]$ is constant and is combined with K_c to form a new constant, K_s.

$$K_c \times [Ag_3PO_4(s)] = K_s = [Ag^+(aq)]^3[PO_4^{3-}(aq)]$$

Numerical problems are often concerned with the relationship between solubility product and solubility, as we now show by a Worked Example.

Calculating solubility product from solubility

Worked Example The solubility of silver sulphide, Ag_2S, is 2.48×10^{-15} mol dm^{-3}. Calculate its solubility product.

W

Solution

1. Write the equation representing the dissolving of silver sulphide, leaving space above and below the equation for initial and equilibrium concentrations:

Initial concn/mol dm^{-3}		0	0
	$Ag_2S(s) \rightleftharpoons$	$2Ag^+(aq)$ +	$S^{2-}(aq)$
Equilibrium concn/mol dm^{-3}			

2. Calculate the concentrations of sulphide ion and silver ion.

The solubility indicates the amount of silver sulphide in a saturated solution. The balanced equation shows that every mole of Ag_2S which dissolves produces one mole of S^{2-} and two moles of Ag^+. Hence, at this temperature, in a saturated solution of Ag_2S,

$$[S^{2-}(aq)] = 2.48 \times 10^{-15} \text{ mol dm}^{-3}$$

and $[Ag^+(aq)] = 2 \times 2.48 \times 10^{-15}$ mol dm^{-3} = 4.96×10^{-15} mol dm^{-3}

Put these values below the equation:

Initial concn/mol dm^{-3}		0	0
	$Ag_2S(s) \rightleftharpoons$	$2Ag^+(aq)$ +	$S^{2-}(aq)$
Equilibrium concn/mol dm^{-3}		4.96×10^{-15}	2.48×10^{-15}

3. Write the expression for K_s, substitute the equilibrium concentrations and do the arithmetic.

$$K_s = [Ag^+(aq)]^2[S^{2-}(aq)]$$
$$= (4.96 \times 10^{-15} \text{ mol dm}^{-3})^2 \times (2.48 \times 10^{-15} \text{ mol dm}^{-3})$$
$$= \boxed{6.10 \times 10^{-44} \text{ mol}^3 \text{ dm}^{-9}}$$

In this type of calculation, some students wonder why a concentration term is both doubled and squared. The explanation is as follows: firstly, the stoichiometry tells us that $[Ag^+(aq)]$ is twice the solubility; secondly, the form of the equilibrium law tells us that the silver ion concentration is squared. The same principle applies in the following exercises.

Exercise 221 Calculate the solubility products of the following substances from their solubilities.

Substance		Solubility/mol dm^{-3}
(a)	$CdCO_3$	1.58×10^{-7}
(b)	CaF_2	2.15×10^{-4}
(c)	$Cr(OH)_3$	1.39×10^{-8}

Exercise 222 In an experiment to determine the solubility product of calcium hydroxide, $Ca(OH)_2$, 25.0 cm^3 of a saturated solution required 11.45 cm^3 of 0.100 M HCl for neutralization.

Calculate the concentration of hydroxide ions and, hence, the solubility and solubility product of calcium hydroxide at the temperature of the experiment.

Now that you know how to calculate solubility product from solubility, we show you how to do the reverse process.

Calculating solubility from solubility product

Worked Example Calculate the solubility, in mol dm^{-3}, of calcium sulphate.

K_s for calcium sulphate = 8.64×10^{-8} mol^2 dm^{-6}.

W

Solution

1. Write the equation for the solution of the salt, leaving room above and below for concentrations. Initially, there is no $Ca^{2+}(aq)$ or $SO_4^{2-}(aq)$.

Initial concn/mol dm^{-3}		0	0
	$CaSO_4(s) \rightleftharpoons$	$Ca^{2+}(aq)$ +	$SO_4^{2-}(aq)$
Equilibrium concn/mol dm^{-3}			

2. From the balanced equation we know that the equilibrium concentration of $Ca^{2+}(aq)$ is equal to that of $SO_4^{2-}(aq)$. Let this be x mol dm^{-3}.

Initial concn/mol dm^{-3}		0	0
	$CaSO_4(s) \rightleftharpoons$	$Ca^{2+}(aq)$ +	$SO_4^{2-}(aq)$
Equilibrium concn/mol dm^{-3}		x	x

3. Write the expression for K_s

$$K_s = [Ca^{2+}(aq)][SO_4^{2-}(aq)]$$

4. Substitute K_s and the equilibrium concentrations.

$$8.64 \times 10^{-8}\ \text{mol}^2\ \text{dm}^{-6} = x\ \text{mol dm}^{-3} \times x\ \text{mol dm}^{-3}$$

$$x^2 = 8.64 \times 10^{-8}$$

$$\therefore x = 2.94 \times 10^{-4} \text{ and solubility } = \boxed{2.94 \times 10^{-4}\ \text{mol dm}^{-3}}$$

Try the next exercise, using a similar method.

Exercise 223 Calculate the solubilities of the following slightly soluble substances from their solubility products.

Substance	Solubility product
(a) CuS	6.3×10^{-36} mol^2 dm^{-6}
(b) $Fe(OH)_2$	6.0×10^{-15} mol^3 dm^{-9}
(c) Ag_3PO_4	1.25×10^{-20} mol^4 dm^{-12}

The idea of solubility product strictly applies only to slightly soluble electrolytes, i.e. solutions in which the concentration of ions is very low. At higher concentrations, the ions interact with one another so that their effective concentrations differ from their actual concentrations. In this respect, very dilute solutions may be regarded as 'ideal solutions' which obey simple laws, just as gases at low pressure observe the ideal gas law.

Calcium hydroxide is a border-line case; strictly, it is too soluble for its solution to be regarded as ideal, but the error is small enough for us to use calcium hydroxide as an example in a simple experiment. Obviously, titration would not be a good method for determining solubility products for less soluble salts; you will use another method in Chapter 8 (Redox Reactions).

So far in this section we have considered only pure solutions. Now we deal with mixtures of solutions with a common ion. The effect is often demonstrated in concentrated solutions: calculations can only be done for dilute solutions.

The common ion effect

Provided that the total ion concentration is no more than about 0.1 mol dm^{-3}, you can use solubility products in calculations involving mixtures of solutions. We illustrate this in a Worked Example.

Worked Example Calculate the solubility of silver chloride, AgCl, in

(a) pure water, (b) 0.10 M NaCl.

$K_s(\text{AgCl}) = 2.0 \times 10^{-10}$ mol^2 dm^{-6}

W

Solution

(a) This is not a new type of calculation but you need the result for comparison. Refer to the last Worked Example, if necessary.

$$\text{Solubility} = \sqrt{K_s} = \sqrt{2.0 \times 10^{-10}\ \text{mol}^2\ \text{dm}^{-6}} = \boxed{1.4 \times 10^{-5}\ \text{mol dm}^{-3}}$$

(b) 1. Write the equation and indicate the concentrations present initially and at equilibrium.

	AgCl(s)	⇌	$Ag^+(aq)$	+	$Cl^-(aq)$
Initial concn/mol dm^{-3}			0		0.10
Equilibrium concn/mol dm^{-3}			x		$(x + 0.10)$

If x mol dm^{-3} of AgCl dissolves, then x mol dm^{-3} of $Ag^+(aq)$ and x mol dm^{-3} of $Cl^-(aq)$ are produced. The total concentration of chloride ions equals that produced by the AgCl dissolving plus the 0.10 mol dm^{-3} initially present.

2. Write the equilibrium law expression.

$$K_s = [Ag^+(aq)][Cl^-(aq)]$$

3. Substitute the equilibrium concentrations and the solubility product.

$$2.0 \times 10^{-10} \text{ mol}^2 \text{ dm}^{-6} = x \times (x + 0.10) \text{ mol}^2 \text{ dm}^{-6}$$

4. Now you can make a very useful approximation. In otherwise pure water, the concentration of $Ag^+(aq)$ from dissolved AgCl is 1.4×10^{-5} mol dm^{-3}. Applying Le Chatelier's principle, addition of Cl^- will drive the equilibrium to the left, thereby reducing the amount dissolved still further. So you can assume that the concentration of chloride ion from the AgCl is negligible compared to the 0.10 mol dm^{-3} solution of Cl^- present initially.

Expressed mathematically, $(x + 0.10)$ mol $dm^{-3} \simeq 0.10$ mol dm^{-3}

(You can test this approximation after the calculation.)

5. The equilibrium law expression now becomes

$$2.0 \times 10^{-10} = x \times 0.10$$

$$\therefore x = \frac{2.0 \times 10^{-10}}{0.10} = 2.0 \times 10^{-9}$$

6. The solubility in 0.10 M NaCl = $[Ag^+(aq)]$

$$= x \text{ mol dm}^{-3} = \boxed{2.0 \times 10^{-9} \text{ mol dm}^{-3}}$$

Note the dramatic effect of the common ion; this solubility is less than one-thousandth of the solubility in water!

You can quite simply check that the approximation in step 4 is valid:

$$x + 0.10 = (2.0 \times 10^{-9}) + 0.10 = 0.100000002 \simeq 0.10$$

A better, but more tedious, check is to calculate x from the quadratic equation in step 3:

$$2.0 \times 10^{-10} = x \times (x + 0.10)$$

$$\therefore x^2 + 0.10x - 2.0 \times 10^{-10} = 0$$

$$\therefore x = \frac{-0.10 \pm \sqrt{0.010 + 8.0 \times 10^{-10}}}{2} = 2.0 \times 10^{-9}$$

You will find it very useful to make such approximations because they often make calculations much simpler, as you can see in the following exercises.

Exercise 224 At 20 °C, the solubility product of strontium sulphate is 4.0×10^{-7} mol^2 dm^{-6}, and that of magnesium fluoride is 7.2×10^{-9} mol^3 dm^{-9}. Estimate, to two significant figures, the solubility at 20 °C in mol dm^{-3} of

(a) strontium sulphate in a 0.1 M solution of sodium sulphate;

(b) magnesium fluoride in a 0.2 M solution of sodium fluoride.

Exercise 225 A saturated solution of strontium carbonate was filtered. When 50 cm^3 of the filtrate was added to 50 cm^3 of 1.0 M sodium carbonate solution, some strontium carbonate was precipitated. Calculate the concentration of strontium ions remaining in the solution. All the work was done at 25 °C.

You can also apply the solubility product principle to a consideration of the conditions under which precipitation may occur.

Will precipitation occur?

To predict whether precipitation will occur when solutions are mixed, you need to recognise that solubility product is a value which the product of ion concentrations in solution can never exceed at equilibrium.

We show you how to apply this idea in a Worked Example followed by two exercises.

Worked Example If 50.0 cm^3 of 0.050 M $AgNO_3$ is mixed with 50.0 cm^3 of 0.010 M $KBrO_3$, will a precipitate of $AgBrO_3$ form? $K_s(AgBrO_3) = 6 \times 10^{-5}$ mol^2 dm^{-6}.

Solution

1. Work out what the concentrations of Ag^+ and BrO_3^- would be after mixing but before any reaction:

   (a) Amount of Ag^+ = 0.050 dm^3 × 0.050 mol dm^{-3}

   $$[Ag^+(aq)] = \frac{0.050\ \text{dm}^3 \times 0.050\ \text{mol dm}^{-3}}{0.100\ \text{dm}^3} = 0.025\ \text{mol dm}^{-3}$$

   (The volume is doubled, so the concentration is halved.)

   (b) Similarly, $[BrO_3^-(aq)] = \frac{1}{2} \times 0.010\ \text{mol dm}^{-3} = 0.0050\ \text{mol dm}^{-3}$

2. Calculate the product of the ion concentrations (called the ion product).

   Ion product = $[Ag^+(aq)][BrO_3^-(aq)]$

   $= 0.025\ \text{mol dm}^{-3} \times 0.0050\ \text{mol dm}^{-3} = 1.25 \times 10^{-4}\ \text{mol}^2\ \text{dm}^{-6}$

3. Compare the value of the ion product with the solubility product.

   If the ion product is greater than K_s, precipitation will occur.
   If the ion product is less than K_s, precipitation will not occur.

∴ a silver bromate precipitate will appear because the ion product is greater than the solubility product.

Exercise 226 Will a precipitate of BaF_2 form if 150 cm^3 of 0.1 M $Ba(NO_3)_2$ is mixed with 50.0 cm^3 of 0.050 M KF?

K_s = 1.7 x 10^{-6} mol^3 dm^{-9}.

Exercise 227 You are given the following numerical values only for the solubility products of various salts at 25 °C.

K_s[silver chloride] = 2 x 10^{-10}

K_s[lead(II) bromide] = 3.9 x 10^{-5}

K_s[silver bromate(V)] = 6.0 x 10^{-5}

K_s[magnesium hydroxide] = 2.0 x 10^{-11}

(a) State the units for each of the above solubility products.

(b) Which of the following pairs of solutions (all of concentration 1.0 x 10^{-3} mol dm^{-3}) will form a precipitate when equal volumes are mixed at 25 °C? Give reasons for your answers.

(i) Silver nitrate and sodium chloride.

(ii) Lead(II) nitrate and sodium bromide.

(iii) Silver nitrate and potassium bromate(V).

(iv) Magnesium sulphate and sodium hydroxide.

The final application of the equilibrium law in this chapter refers to the distribution of a solute between two immiscible solvents.

DISTRIBUTION EQUILIBRIA

When a solute is distributed between two immiscible solvents, there is a dynamic equilibrium at the interface, where the solute is transferred from one solvent to the other and back at equal rates. For example, the distribution of ammonia between water and 1,1,1-trichloroethane (tce) may be represented:

$$NH_3(tce) \rightleftharpoons NH_3(aq)$$

Application of the equilibrium law gives an expression for the equilibrium constant which, in this situation, is known as the distribution coefficient, K_d, or the partition coefficient.

$$K_d = \frac{[NH_3(aq)]}{[NH_3(tce)]}$$

Use similar expressions in the following exercises.

Exercise 228 Calculate the value of the distribution coefficient for the distribution equilibrium of butanedioic acid between water and ethoxyethane (ether) from the following data. All experiments were performed at the same temperature.

Concn of acid in water /mol dm^{-3}	0.0759	0.108	0.158	0.300
Concn of acid in ether /mol dm^{-3}	0.0114	0.0162	0.0237	0.0451

The distribution of a solute between two solvents is sometimes used in extraction and purification procedures. For instance, you could remove most of the ammonia from a solution in 1,1,1-trichloroethane by shaking it with water in several portions, as you discover in the next exercise.

Exercise 229 This exercise concerns the removal of ammonia from a solution in 1,1,1-trichloroethane (0.10 mol dm^{-3}) by shaking with water. $K_d = 290$.

How much ammonia remains in the organic layer after shaking 100 cm^3 of solution with

(a) 100 cm^3 of water,

(b) four successive 25 cm^3 portions of water?

In the distribution equilibria you have considered so far, the solute is in the same molecular form in both solvents. For example, ammonia in 1,1,1-trichloroethane exists entirely as molecules of NH_3, and in water almost entirely as molecules of NH_3. There is a slight reaction in water:

$$NH_3(aq) + H_2O(l) \rightleftharpoons NH_4^+(aq) + OH^-(aq)$$

but K_c for this is about 10^{-7}, so virtually all the ammonia remains as molecules.

By contrast, ethanoic acid exists in aqueous solution almost entirely as CH_3CO_2H molecules and in 1,1,1-trichloroethane almost entirely as $(CH_3CO_2H)_2$ molecules, i.e. as dimers. You need not concern yourself with the derivation of the expression for K_d, but the simple result is that a squared term appears:

$$K_d = \frac{[(CH_3CO_2H)(tce)]}{[CH_3CO_2H(aq)]^2}$$

The next exercise concerns a similar equilibrium system.

Exercise 230 Trichloromethane was added to a series of aqueous solutions of ethanoic acid (acetic acid) and the mixtures shaken at laboratory temperature. By titration, the following concentrations of ethanoic acid were found in the two layers:

Trichloromethane/g dm^{-3}	17.5	43.5	84.6
Water/g dm^{-3}	292	479	642

By neglecting any dissociation of ethanoic acid in water, deduce its molecular formula in trichloromethane.

You may use graph paper if you wish.

END-OF-CHAPTER QUESTIONS

Exercise 231 At a certain temperature, the equilibrium constant for the following reaction is equal to 4.

$$CH_3CO_2H(l) + C_2H_5OH(l) \rightleftharpoons CH_3CO_2C_2H_5(l) + H_2O(l)$$

If one mole of ethanoic acid (CH_3CO_2H) is added to one mole of ethanol and the mixture is allowed to reach equilibrium, how many moles of ethanoic acid will be present in the equilibrium mixture?

Exercise 232 At a certain temperature, for the reaction:

$$CO(g) + Cl_2(g) \rightleftharpoons COCl_2(g)$$

the equilibrium partial pressures of carbon monoxide, chlorine and carbonyl chloride are 2, 4 and 48 atm, respectively. What is the value of K_p?

Exercise 233 If the solubility of bismuth trisulphide, Bi_2S_3, is 3.6×10^{-7} mol dm^{-3}, what is the value of its solubility product?

Exercise 234 The solubility product of silver chromate is 1.0×10^{-12} mol^3 dm^{-9}. In a solution in which $[CrO_4^{2-}]$ is 1.0×10^{-4} mol dm^{-3}, what is the maximum $[Ag^+]$, in mol dm^{-3}?

Exercise 235 For the equilibrium

$C_5H_{10}(l)$ +	$CH_3CO_2H(l)$	$\rightleftharpoons$	$CH_3CO_2C_5H_{11}(l)$
(pentene)	(ethanoic acid)		(pentyl ethanoate)

K_c = 540 dm^3 mol^{-1} at a certain temperature.

An equilibrium mixture at this temperature contains 5.66×10^{-3} mol dm^{-3} of pentene and 2.55×10^{-3} mol dm^{-3} of ethanoic acid.

Calculate the concentration of pentyl ethanoate in the mixture.

Exercise 236 For the equilibrium:

$$N_2(g) + 3H_2(g) \rightleftharpoons 2NH_3(g)$$

at a particular temperature and total pressure of 2 atm, an equilibrium mixture consists of 1.0 mol of NH_3, 3.6 mol of H_2 and 13.5 mol of N_2.

(a) Calculate the equilibrium constant, K_p, for this reaction under the stated conditions.

(b) Calculate the average molar mass of the mixture under the stated conditions.

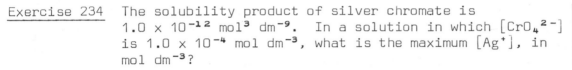

Exercise 237 In the following reaction:

$$CO_2(g) + H_2(g) \rightleftharpoons CO(g) + H_2O(g)$$

at a temperature of 100 °C and a pressure of 2.0 atm, the amounts of each gas present at equilibrium were 6.2×10^{-3} mol of CO, 6.2×10^{-3} mol of H_2O, 0.994 mol of CO_2 and 0.994 mol of H_2.

Calculate the value of K_p.

Exercise 238 At 1000 K, K_p = 1.9 atm for the system:

$$C(s) + CO_2(g) \rightleftharpoons 2CO(g)$$

What must the total pressure be to have an equilibrium mixture containing 0.013 mol of CO_2 and 0.024 mol of CO?

Exercise 239 In the following equilibrium:

$$2HI(g) \rightleftharpoons H_2(g) + I_2(g)$$

1.0 mol of I_2 and 2.0 mol of H_2 are allowed to react in a 1.0 dm^3 vessel at 440 °C. What are the equilibrium concentrations of HI, H_2 and I_2 at this temperature, given that K_c = 0.020 at 440 °C?

Exercise 240 A part

(a) Write an expression for the solubility product of lead(II) chloride.

(b) The solubility product of lead(II) chloride is 1.6×10^{-5} mol^3 dm^{-9} at a given temperature.

(i) What is the solubility in mol dm^{-3} of lead(II) chloride in water at the same temperature?

(ii) How many moles of chloride ion must be added to a 1.0 M solution of lead(II) nitrate at the same temperature in order just to cause a precipitate of lead(II) chloride? Assume that no change in volume occurs on adding the chloride ion.

Exercise 241 A

In the vapour phase, the following equilibrium exists between dinitrogen tetroxide and nitrogen dioxide:

$$N_2O_4(g) \rightleftharpoons 2NO_2(g)$$

At atmospheric pressure and a given temperature T, 20.0 per cent of the molecules in the vapour are NO_2.

(a) Using the atmosphere as the unit of pressure, calculate the equilibrium constant, K_p, at temperature T.

(b) Calculate the percentage of NO_2 molecules in the vapour at temperature T when the (total) pressure of the vapour is 0.50 atmosphere.

Exercise 242 When lead sulphate is added to an aqueous solution of sodium iodide, the following equilibrium is obtained.

A

$$PbSO_4(s) + 2I^-(aq) \rightleftharpoons PbI_2(s) + SO_4^{2-}(aq)$$

The equilibrium constant for this reaction may be determined by adding an excess of lead sulphate to a known volume of a standard solution of sodium iodide and allowing the mixture to equilibrate in a water-bath thermostatically controlled at the desired temperature. Cold water is then added to the reaction mixture to 'freeze' the equilibrium and the mixture is then titrated with standard silver nitrate solution. In a typical experiment using 50.0 cm^3 of 0.100 M sodium iodide, a titre of 31.0 cm^3 of 0.100 M silver nitrate was obtained.

(a) Give an expression for the equilibrium constant, K, of the reaction.

(b) Why is it not necessary to know the mass of lead sulphate used in the experiment?

(c) From the data given above, calculate

(i) the concentration of iodide ions present initially,

(ii) the concentration of iodide ions present at equilibrium,

(iii) the concentration of iodide ions which have reacted,

(iv) the concentration of sulphate ions formed,

(v) a value for K.

Exercise 243 The following results were obtained in an experiment to determine the partition coefficient of ammonia between water and trichloromethane ($CHCl_3$):

A part

10.0 cm^3 of the aqueous layer required 43.2 cm^3 of 0.250 mol dm^{-3} hydrochloric acid for neutralization;

25.0 cm^3 of the trichloromethane layer required 21.6 cm^3 of 0.050 mol dm^{-3} hydrochloric acid for neutralization.

Calculate the partition coefficient at the laboratory temperature.

EQUILIBRIUM II – ACIDS AND BASES

7

INTRODUCTION AND PRE-KNOWLEDGE

In this chapter we extend your study of equilibrium by applying the principles developed in Chapter 5 to equilibria involving weak acids and weak bases, including buffer solutions and acid-base indicators.

We assume that you are already familiar with Le Chatelier's principle and can apply the equilibrium law in calculations involving equilibrium constants. You have already met several types of equilibrium constant (K_c, K_p, K_s, and K_d): in this chapter we introduce some more, the first of which concerns the ionization of water.

THE IONIZATION OF WATER

Water is a special case of an amphiprotic substance; it can react with itself.

$$H_2O(l) + H_2O(l) \rightleftharpoons H_3O^+(aq) + OH^-(aq)$$

The evidence that this process takes place in even the purest water is that it has a slight electrical conductivity. A few ions must be present to carry a current and these must have come from the ionization of the water molecules themselves. A simpler way of writing the equation is:

$$H_2O(l) \rightleftharpoons H^+(aq) + OH^-(aq)$$

Application of the equilibrium law gives:

$$K_c = \frac{[H^+(aq)][OH^-(aq)]}{[H_2O(l)]}$$

So few water molecules are ionized that their concentration remains effectively constant (proportional to the density of the water). We can, therefore, include $[H_2O(l)]$ with the constant K_c.

$$K_c \times [H_2O(l)] = [H^+(aq)]\,[OH^-(aq)]$$

Let $K_c \times [H_2O(l)]$ be another constant, K_w, the <u>ionic product of water</u>

i.e. $K_w = [H^+(aq)][OH^-(aq)]$

In your reading, you may also see the ionic product written in the form:

$$K_w = [H_3O^+(aq)][OH^-(aq)]$$

This means the same and can be used in exactly the same way; its only difference is that it has been derived using the 'full' equation for the ionization of water:

$$H_2O(l) + H_2O(l) \rightleftharpoons H_3O^+(aq) + OH^-(aq)$$

The value of the ionic product has been measured experimentally and found to be 1.0×10^{-14} mol^2 dm^{-6} at 25 °C. Since it is directly related to the equilibrium constant for the self-ionization of water, this value is, of course, temperature dependent. Summarising:

$$\boxed{K_w = [H^+(aq)][OH^-(aq)] = 1.0 \times 10^{-14}\ \text{mol}^2\ \text{dm}^{-6}\ \text{at 25 °C}}$$

Exercise 244 Use the expression above to calculate the fraction of molecules in pure water which are ionized at 25 °C. (Take the concentration of un-ionized molecules as 55.6 mol dm^{-3}.)

The aim of the following Revealing Exercise is to examine what happens when the equilibrium in pure water is upset by the addition of hydrogen ions from an acid or hydroxide ions from an alkali. Also, since K_W is constant, we derive a definition of acidic, neutral and basic solutions in terms of the concentration of hydrogen ions.

Consider the equilibrium:

$$H_2O(l) \rightleftharpoons H^+(aq) + OH^-(aq); \quad K_W = 1.0 \times 10^{-14}\ mol^2\ dm^{-6}$$

Q1. Calculate $[H^+(aq)]$ and $[OH^-(aq)]$ in pure water.

A1. $K_W = [H^+(aq)][OH^-(aq)] = 1.0 \times 10^{-14}\ mol^2\ dm^{-6}$

Since $[H^+(aq)] = [OH^-(aq)]$, then

$[H^+(aq)]^2 = 1.0 \times 10^{-14}\ mol^2\ dm^{-6}$

$\therefore\ [H^+(aq)] = \sqrt{1.0 \times 10^{-14}\ mol^2\ dm^{-6}} = 1.0 \times 10^{-7}\ mol\ dm^{-3}$

and $[OH^-(aq)] = 1.0 \times 10^{-7}\ mol\ dm^{-3}$

Q2. Acid is added to the solution to increase the concentration of hydrogen ions. How does the equilibrium shift?

A2. It shifts to the left.

Q3. Is the concentration of hydroxide ions less than or greater than 1.0×10^{-7} mol dm^{-3} now?

A3. It is less than 1.0×10^{-7} mol dm^{-3}.

Q4. Does the concentration of hydroxide ions ever decrease to zero? Explain.

A4. No. The ionization of water is an equilibrium system. The product of the concentrations of hydrogen ions and hydroxide ions must always be 1.0×10^{-14} mol^2 dm^{-6}. The only way for one value to drop to zero would be for the other to be infinity, which is impossible.

Q5. What is the effect of adding alkali to the equilibrium system existing in pure water?

A5. It shifts the equilibrium to the left.

Q6. Is the concentration of hydrogen ions greater than or less than 1.0×10^{-7} mol dm^{-3} now?

A6. It is less than 1.0×10^{-7} mol dm^{-3}.

Q7. Does the hydrogen ion concentration ever decrease to zero? Explain.

A7. No, for similar reasons to those given in A4.

To summarise, in a <u>neutral</u> solution,

$[H^+(aq)] = [OH^-(aq)]$

$[H^+(aq)] = [OH^-(aq)] = 1.0 \times 10^{-7}$ mol dm^{-3} (at 25 °C)

In an <u>acidic</u> solution,

$[H^+(aq)] > [OH^-(aq)]$

and $[H^+(aq)] > 1.0 \times 10^{-7}$ mol dm^{-3}

In a <u>basic</u>, or <u>alkaline</u>, solution,

$[H^+(aq)] < [OH^-(aq)]$

and $[H^+(aq)] < 1.0 \times 10^{-7}$ mol dm^{-3}

Using the expression for the ionic product of water

Since the product of the concentrations of hydrogen ions and hydroxide ions is always constant for any dilute solution at a given temperature, we can calculate the concentration of one of these ions given the concentration of the other. We show you how to do this in a Worked Example.

<u>Worked Example</u> Calculate the hydroxide ion concentration in an acid solution which has a concentration of hydrogen ions of 0.10 mol dm^{-3} given that $K_W = 1.0 \times 10^{-14}$ mol^2 dm^{-6} at 25 °C.

W

<u>Solution</u>

1. Rearranging the expression:

$$K_W = [H^+(aq)][OH^-(aq)]$$

gives $$[OH^-(aq)] = \frac{K_W}{[H^+(aq)]}$$

2. Substitute the given values of K_W and $[H^+(aq)]$,

$$[OH^-(aq)] = \frac{1.0 \times 10^{-14}\ \text{mol}^2\ \text{dm}^{-6}}{0.10\ \text{mol dm}^{-3}} = \boxed{1.0 \times 10^{-13}\ \text{mol dm}^{-3}}$$

Notice that even in strongly acidic solutions, there are hydroxide ions present.

<u>Exercise 245</u> Given that $K_W = 1.0 \times 10^{-14}$ mol^2 dm^{-6}, calculate the hydroxide ion concentrations of these acid solutions at 25 °C. Assume that both acids are fully dissociated into ions.

(a) 0.010 M HCl (b) 0.10 M H_2SO_4

By rearranging the expression for the ionic product of water to solve for $[H^+(aq)]$, you can calculate the hydrogen ion concentration given the hydroxide ion concentration.

Exercise 246 Given that K_W = 1.0 x 10 mol^2 dm^{-6} at 25 °C, calculate the hydrogen ion concentrations of the following solutions. Assume that both alkalis are fully dissociated in solution.

(a) 0.010 M KOH (b) 0.050 M $Ba(OH)_2$

In the next section, we show you how to express the concentrations of hydrogen ion and hydroxide ion more conveniently as pH and pOH.

pH - THE HYDROGEN ION EXPONENT

In your pre-A-level study you have probably used pH numbers to indicate whether a solution is acidic (pH < 7), neutral (pH = 7) or alkaline (pH > 7). Here we give a precise definition of pH, and show how it is related to the hydrogen ion concentration.

Chemists are interested in hydrogen ion concentrations covering a very large range, say from 10 mol dm^{-3} right down to 1.0 x 10^{-15} mol dm^{-3}, or even smaller. It is very convenient, particularly for small concentrations, to use a logarithmic scale and to work with positive numbers, and this is the basis of the pH scale as a measure of acidity.

$[H^+(aq)] = 1.0 \times 10^{-pH}$ or, taking logarithms, $pH = -\log_{10}[H^+(aq)]$

Strictly speaking, since we can only take the logarithm of a pure number, we should write $\log_{10}([H^+(aq)]/\text{mol dm}^{-3})$ but this practice is not usually observed.

The following table should clarify the meaning of pH and illustrate the convenience of its use. Notice that an increase of 1 in pH corresponds to a decrease to one-tenth of the hydrogen ion concentration.

$[H^+(aq)]$ mol dm^{-3}		pH
1.0	= 1.0 x 10^0	0
0.10	= 1.0 x 10^{-1}	1
0.0010	= 1.0 x 10^{-3}	3
0.0000010	= 1.0 x 10^{-6}	6
0.0000000010	= 1.0 x 10^{-9}	9

The term pOH is similarly defined, i.e. $pOH = -\log_{10}[OH^-(aq)]$, but is less often used.

Now try the following exercise to check that you understand the definition.

Exercise 247 Calculate the pH of 10 M HCl solution, assuming it to be fully dissociated.

You have already made use of the ionic product of water to determine the concentration of hydrogen ions in alkaline solution. Now we take you through an exercise making use of the ionic product of water to calculate the pH of an alkaline solution.

Worked Example What is the pH of a 1.0 M NaOH solution? ($K_W = 1.0 \times 10^{-14}$ mol^2 dm^{-6}.)

W

Solution

Assume that NaOH is fully dissociated in solution so that $[OH^-(aq)]$ = 1.0 mol dm^{-3}.

1. Starting with the ionic product of water, solve for $[H^+(aq)]$

$$[H^+(aq)][OH^-(aq)] = 1.0 \times 10^{-14} \text{ mol}^2 \text{ dm}^{-6}$$

$$[H^+(aq)] = \frac{1.0 \times 10^{-14} \text{ mol}^2 \text{ dm}^{-6}}{[OH^-(aq)]}$$

2. Calculate the hydrogen ion concentration:

$$[H^+(aq)] = \frac{1.0 \times 10^{-14} \text{ mol}^2 \text{ dm}^{-6}}{1.0 \text{ mol dm}^{-3}} = 1.0 \times 10^{-14} \text{ mol dm}^{-3}$$

3. Substitute this value into the definition of pH:

$$pH = -\log_{10}[H^+(aq)]$$

$$= -\log(1.0 \times 10^{-14}) = -(0 - 14) = \boxed{14}$$

An alternative solution makes use of pOH:

1. $pOH = -\log_{10}[OH^-(aq)] = -\log 1.0 = 0$
2. $[H^+(aq)][OH^-(aq)] = 1.0 \times 10^{-14}$ mol^2 dm^{-6}

   Taking logarithms

   $\log[H^+(aq)] + \log[OH^-(aq)] = -14$

   i.e. $pH + pOH = +14$

   $\therefore \quad pH = 14 - 0 = \boxed{14}$

The relationships between $[H^+(aq)]$, $[OH^-(aq)]$, pH and pOH are summarised in Fig. 16.

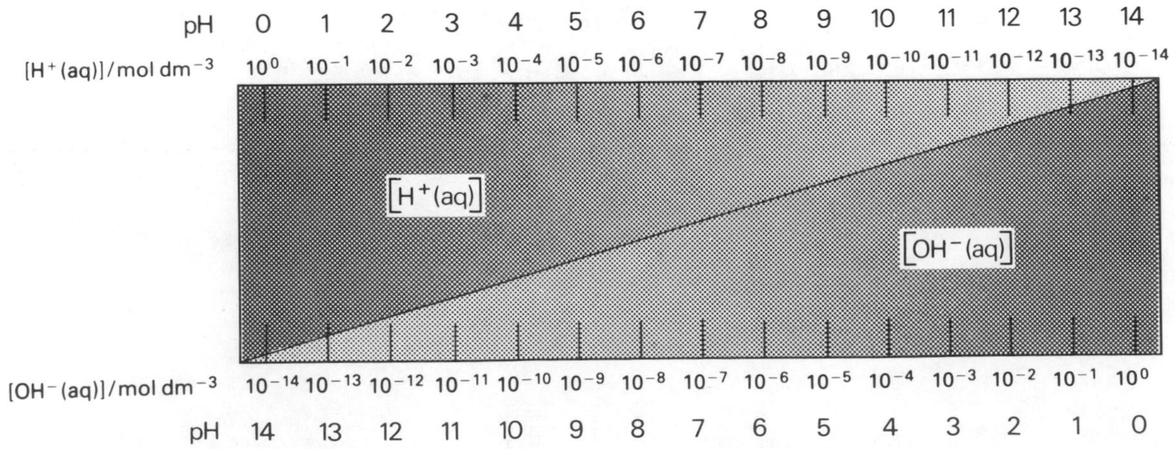

Fig.16.

In the next exercise, use either of the methods shown in the Worked Example.

Exercise 248 Calculate the pH, at 25 °C, of solutions in which the hydroxide ion concentrations are as follows:

(a) 1.0×10^{-8} mol dm^{-3} (c) 1.0×10^{-4} mol dm^{-3}

(b) 0.10 mol dm^{-3} (d) 10 mol dm^{-3}

Up to now, we have only considered pH values which are whole numbers. We now show you how to do calculations in which this is not the case.

Calculating pH from hydrogen ion concentration

Worked Example Calculate the pH of a solution whose hydrogen ion concentration is 4.0×10^{-3} mol dm^{-3}.

Solution

Substitute the hydrogen ion concentration into the expression:

$$pH = -\log[H^+(aq)]$$

$$pH = -\log(4 \times 10^{-3}) = -(-3.00 + 0.60) = -(-2.40) = \boxed{2.40}$$

Now try the following exercise. Of course, if you have a scientific calculator, you can let the machine do the work.

Exercise 249 Calculate the pH of the solutions in which the hydrogen ion concentrations are:

(a) 7.0×10^{-7} mol dm^{-3} (d) 2.62×10^{-8} mol dm^{-3}

(b) 3.2×10^{-9} mol dm^{-3} (e) 0.54 mol dm^{-3}

(c) 8.34×10^{-3} mol dm^{-3} (f) 2.23×10^{-3} mol dm^{-3}

You have already seen that the ionic product of water is temperature dependent. In working through the following exercise, you learn how temperature affects the pH of pure water or a neutral solution.

Exercise 250 At 10 °C, $K_w = 0.30 \times 10^{-14}$ mol^2 dm^{-6} and at 50 °C, $K_w = 5.47 \times 10^{-14}$ mol^2 dm^{-6}. Calculate the pH of pure water at these temperatures.

Now we show you how to calculate the hydrogen ion concentration, given the pH of the solution.

Calculating hydrogen ion concentration from pH

Worked Example What is the hydrogen ion concentration of a solution whose pH is 3.70?

W

Solution

1. Substitute in the expression:

$$pH = -\log[H^+(aq)]$$

$$3.70 = -\log[H^+(aq)]$$

2. Rearrange the expression for $\log[H^+(aq)]$

$$\log[H^+(aq)] = -3.70$$

3. Take antilogs:

$$[H^+(aq)] = \text{antilog}\ (-3.70)$$

$$= \text{antilog}\ (-4 + 0.30) = \boxed{2.0 \times 10^{-4}\ \text{mol dm}^{-3}}$$

If you are using a calculator, refer to the instruction booklet.

Use the same method in the next exercise.

Exercise 251 Calculate the hydrogen ion concentration in solutions which have the following pH values.

(a) 9.20	(c) 11.11	(e) 5.64
(b) 2.632	(d) 1.11	(f) 7.71

Some indication of the strength of an acid is given by the pH of its aqueous solution. However, pH varies with concentration so that, to compare the strengths of different acids, we need another measure. This is the subject of the following sections.

THE RELATIVE STRENGTHS OF ACIDS

Consider again the equation for the dissociation in water of a general acid which we represent as HA.

$$HA(aq) \rightleftharpoons H^+(aq) + A^-(aq)$$

The further the position of equilibrium lies to the right, the more hydrogen ions are present and the stronger the acid is considered to be.

Acids such as HCl, H_2SO_4 and HNO_3 are classified as strong acids because they dissociate virtually completely. Thus, we could write the equation for the dissociation of HCl in water as

$$HCl(aq) \rightarrow H^+(aq) + Cl^-(aq)$$

Notice that we have removed the reverse arrow indicating that the reaction 'goes to completion'. Acids which are not fully dissociated in water are classified as weak acids.

Note that the terms 'weak' and 'dilute' have different meanings in chemistry; the same applies to the terms 'strong' and 'concentrated'. Make sure that you understand the difference.

The dissociation constant of an acid

The pH of a solution is not a very reliable measure of the strength of an acid because pH varies with concentration. There is another value we could use which is constant at all dilutions - the equilibrium constant for the dissociation of an acid in water.

Consider the dissociation of ethanoic acid in water.

$$CH_3CO_2H(aq) + H_2O(l) \rightleftharpoons CH_3CO_2^-(aq) + H_3O^+(aq)$$

The expression for the equilibrium constant for this reaction is

$$K_c = \frac{[CH_3CO_2^-(aq)][H_3O^+(aq)]}{[H_2O(l)][CH_3CO_2H(aq)]}$$

In practice, we always omit the $[H_2O(l)]$ term in all expressions which refer to the dissociation of an acid in water. The water is present in such great excess that its concentration remains virtually constant and can be combined in the expression with the constant, K_c.

Putting the $[H_2O(l)]$ term on the other side of the expression above gives:

$$K_c[H_2O(l)] = \frac{[CH_3CO_2^-(aq)][H_3O^+(aq)]}{[CH_3CO_2H(aq)]}$$

Since constant x constant = another constant, we let $K_c \times [H_2O(l)] = K_a$

We call K_a the acid dissociation constant, which we can now express as:

$$K_a = \frac{[CH_3CO_2^-(aq)][H_3O^+(aq)]}{[CH_3CO_2H(aq)]}$$

This can also be written in the form:

$$K_a = \frac{[CH_3CO_2^-(aq)][H^+(aq)]}{[CH_3CO_2H(aq)]}$$

This last expression could be derived directly from the equation:

$$CH_3CO_2H(aq) \rightleftharpoons CH_3CO_2^-(aq) + H^+(aq)$$

in which the water is omitted. This form of writing the expression for the dissociation constant, K_a, of an acid is the one most commonly used in text-books and is the form we have chosen to adopt in this chapter. So, for a general acid HA dissociating in water, we will normally write:

$$HA(aq) \rightleftharpoons H^+(aq) + A^-(aq)$$

Hence, we will usually write the equilibrium law expression for the dissociation of an acid in the form:

$$K_a = \frac{[H^+(aq)][A^-(aq)]}{[HA(aq)]}$$

K_a values, unlike pH values, are unaffected by concentration changes and are influenced only by changes in temperature. So the value of K_a provides an accurate measure of the extent to which an acid dissociates, i.e. the strength of that acid.

We now show you how to calculate K_a for a weak acid solution from pH, and the reverse process. For each type of calculation, read the Worked Example and then try the exercises which follow it.

Calculating K_a from hydrogen ion concentrations (or from pH)

We can use the expression for the dissociation of an acid, HA:

$$K_a = \frac{[H^+(aq)][A^-(aq)]}{[HA(aq)]}$$

to calculate a value of K_a for an acid solution if we know the initial concentration of the acid and its hydrogen ion concentration (or pH). You can see how this is done by working through the following Worked Example.

Worked Example A 0.100 M solution of an acid, HA, has a hydrogen ion concentration of 7.94×10^{-6} mol dm^{-3} at 25 °C. Calculate the dissociation constant of the acid at this temperature.

W

Solution

1. Write the equation and above it insert the initial concentrations.

Initial concn/mol dm^{-3}	0.100	0	0
	$HA(aq) \rightleftharpoons$	$H^+(aq)$ +	$A^-(aq)$

2. Write the equilibrium concentrations under the equation. There must be the same concentration of A^- ions produced as hydrogen ions, i.e. 7.94×10^{-6} mol dm^{-3}.

   The equilibrium concentration of HA = initial concentration minus concentration reacted = $(0.100 - 7.94 \times 10^{-6})$ mol dm^{-3}

Initial concn/mol dm^{-3}	0.100	0	0
	$HA(aq)$	$\rightleftharpoons H^+(aq)$	$+ A^-(aq)$
Equilibrium concn/mol dm^{-3}	$(0.100 - 7.94 \times 10^{-6})$	7.94×10^{-6}	7.94×10^{-6}

Note that we have ignored the hydrogen ions which arise from the ionization of water. We can do this because, even in pure water, $[H^+(aq)]$ is as small as 1.0×10^{-7} mol dm^{-3}, and the addition of an acid suppresses the ionization still further.

3. Write the expression for K_a for this acid and substitute the equilibrium concentrations:

$$K_a = \frac{[H^+(aq)][A^-(aq)]}{[HA(aq)]} = \frac{(7.94 \times 10^{-6} \text{ mol dm}^{-3})^2}{(0.100 - 7.94 \times 10^{-6}) \text{ mol dm}^{-3}}$$

   But $7.94 \times 10^{-6} << 0.100 \quad \therefore \quad (0.100 - 7.94 \times 10^{-6}) \simeq 0.100$

   i.e. the ionization makes virtually no difference to the concentration of HA.

$$\therefore \quad K_a = \frac{(7.94 \times 10^{-6} \text{ mol dm}^{-3})^2}{(0.100 \text{ mol dm}^{-3})} = \boxed{6.3 \times 10^{-10} \text{ mol dm}^{-3}}$$

The following exercises are similar: you must decide for yourself whether you can make the approximation which we made in the Worked Example.

Exercise 252 The pH of 0.010 M carbonic acid solution, H_2CO_3, is 4.17 at 25 °C. Calculate the value of the dissociation constant, K_a, at this temperature.

Exercise 253 The pH of 5.0×10^{-3} M benzoic acid solution, $C_6H_5CO_2H$, is 3.28 at 25 °C. Calculate the value of the dissociation constant at this temperature.

Exercise 254 The pH of 0.010 M ethanoic acid solution is 3.40 at a particular temperature. What is the dissociation constant, K_a, of the acid at this temperature?

Exercise 255 The pH of 0.10 M methanoic acid (HCO_2H) is 2.40. What is the dissociation constant, K_a, of this acid?

$$HCO_2H(aq) \rightleftharpoons HCO_2^-(aq) + H^+(aq)$$

Calculating pH from the dissociation constant

Now we take you through a calculation where you are given a value for the dissociation constant of an acid and you have to determine its pH at a particular concentration, i.e. the reverse of our previous exercise.

Worked Example What is the pH of a 0.100 M solution of ethanoic acid at 25 °C? ($K_a = 1.7 \times 10^{-5}$ mol dm^{-3})

W

Solution

1. Write the equation for the dissociation of ethanoic acid, leaving space above and below for initial and equilibrium concentrations.

Initial concn/mol dm^{-3}			
	$CH_3CO_2H(aq)$ $\rightleftharpoons$	$CH_3CO_2(aq)$ +	$H^+(aq)$
Equilibrium concn/mol dm^{-3}			

2. Indicate the initial concentrations above the equation.

Initial concn/mol dm^{-3}	0.100	0	0
	$CH_3CO_2H(aq)$ $\rightleftharpoons$	$CH_3CO_2^-(aq)$ +	$H^+(aq)$

3. Let $[H^+(aq)] = x$ mol dm^{-3}. There must also be the same concentration of ethanoate ions produced.

   And if x mol dm^{-3} of ions are produced, this must leave $(0.100-x)$ mol dm^{-3} of ethanoic acid undissociated.

4. Now write the equilibrium concentrations under the equation

Initial concn/mol dm^{-3}	0.100	0	0
	$CH_3CO_2H(aq)$ $\rightleftharpoons$	$CH_3CO_2^-(aq)$ +	$H^+(aq)$
Equilibrium concn/mol dm^{-3}	$0.100 - x$	x	x

5. Write the expression for K_a for this reaction and substitute the equilibrium concentrations in the expression:

$$K_a = \frac{[CH_3CO_2^-(aq)][H^+(aq)]}{[CH_3CO_2H(aq)]}$$

$$1.7 \times 10^{-5}\ \text{mol dm}^{-3} = \frac{(x\ \text{mol dm}^{-3})^2}{(0.100 - x)\text{mol dm}^{-3}}$$

6. Decide whether x is small enough to put $(0.100 - x) \simeq 0.100$

(If this is the case, your calculation is simplified: if not, you must solve a quadratic equation, as in Chapter 6.)

Clearly, x is small because K_a is small, but is it small enough to be ignored relative to 0.100? A simple 'rule-of-thumb' is that if the initial acid concentration, c_0, divided by K_a, is greater than 1000, then $c_0 - x \simeq c_0$.

$$\text{i.e. if } \frac{c_0}{K_a} > 1000 \quad \text{then} \quad c_0 - x \simeq c_0$$

7. Here, $\dfrac{c_0}{K_a} = \dfrac{0.10}{1.7 \times 10^{-5}} = 5880 \quad \therefore c_0 = x \simeq c_0$ so you can write:

$$1.7 \times 10^{-5} = \frac{x^2}{0.100}$$

$$x^2 = 1.7 \times 10^{-5} \times 0.100$$

$$x = \sqrt{(1.7 \times 10^{-6})} = 1.3 \times 10^{-3}$$

$$\therefore [H^+(aq)] = 1.3 \times 10^{-3}\ \text{mol dm}^{-3}$$

8. Substitute this value of $[H^+(aq)]$ into the expression which defines pH.

$$\text{pH} = -\log[H^+(aq)] = -\log(1.3 \times 10^{-3}) = -(-3 + 0.114) = \boxed{2.9}$$

(The validity of the simplifying assumption made in step 6 can be shown by solving the quadratic equation at the end of step 5 - this gives the same answer, but by a tedious route.

A simpler check is to compare the calculated value of x with the initial concentration:

$$c_0 - x = (0.100 - 0.0013)\text{mol dm}^{-3} = 0.099\ \text{mol dm}^{-3}$$

i.e. $c_0 - x \simeq c_0$ and the assumption was justified.)

In the following exercises, you must judge whether or not to make the simplifying assumption made in the Worked Example.

<u>Exercise 256</u> (a) Given that the ionic product of water, $[H^+][OH^-]$, is $1.00 \times 10^{-14}\ \text{mol}^2\ \text{dm}^{-6}$ at 298 K, calculate to three significant figures the pH at this temperature of a 0.0500 molar aqueous solution of sodium hydroxide.

A

(b) The dissociation constant of ethanoic acid at 5 °C is $1.69 \times 10^{-5}\ \text{mol dm}^{-3}$. Calculate to two significant figures the pH of a 0.100 molar solution of this acid at 5 °C.

<u>Exercise 257</u> Calculate the pH of a 0.100 M iodic(V) acid solution, HIO_3, at 25 °C. (K_a = 0.17 mol dm^{-3})

$$HIO_3(aq) \rightleftharpoons H^+(aq) + IO_3^-(aq)$$

Much of what you have learned about weak acids can be applied, in a slightly modified form, to weak bases.

THE RELATIVE STRENGTHS OF BASES

A base is defined as a proton acceptor. The strengths of different bases may be compared by looking at the equilibrium constants for reactions in which the bases accept protons from water. For example, ammonia is a weak base which reacts with water as follows.

$$NH_3(aq) + H_2O(l) \rightleftharpoons NH_4^+(aq) + OH^-(aq)$$

$$K_c = \frac{[NH_4^+(aq)][OH^-(aq)]}{[NH_3(aq)][H_2O(l)]}$$

A new constant is formed by combining the two constants K_c and $[H_2O(l)]$. This is called the base dissociation constant, K_b.

$$\text{i.e.} \quad K_b = K_c \times [H_2O(l)] = \frac{[NH_4(aq)][OH^-(aq)]}{[NH_3(aq)]}$$

Note that the conjugate acid, NH_4^+, has its own acid dissociation constant which is simply related to K_b as follows.

$$NH_4^+(aq) \rightleftharpoons NH_3(g) + H^+(aq)$$

$$K_a = \frac{[NH_3(aq)][H^+(aq)]}{[NH_4^+(aq)]}$$

But $K_w = [H^+(aq)][OH^-(aq)]$ or $[H^+(aq)] = K_w/[OH^-(aq)]$

Substituting for $[H^+(aq)]$ in the expression for K_a gives

$$K_a = \frac{[NH_3(aq)]\,K_w}{[NH_4^+(aq)][OH^-(aq)]} = \frac{K_w}{K_b} \quad \text{or} \quad K_b = \frac{K_w}{K_a}$$

Use these relationships in the next exercise to calculate $[OH^-(aq)]$ and, hence, the pH.

Exercise 258 Calculate the pH of the following solutions.

(a) 0.25 M NH_3 ($K_b = 1.8 \times 10^{-5}$ mol dm^{-3})

(b) 0.15 M $C_4H_9NH_2$

(K_a for $C_4H_9NH_3^+ = 1.7 \times 10^{-11}$ mol dm^{-3})

If the dissociation constants for the acid and base are known, it is a fairly simple matter to calculate the pH at any stage of an acid-base titration.

pH CHANGES DURING TITRATIONS

You can calculate the pH at any stage during the titration by calculating the amount of excess acid before the equivalence point, or the amount of excess alkali after the equivalence point. From this amount you can calculate the pH as previously described. We begin with the simplest case, in which both acid and base are strong.

Worked Example Calculate the pH of the solution resulting from adding 7.00 cm^3 of 0.100 M NaOH to 25.0 cm^3 of 0.100 M HCl.

Solution

Amount of acid initially = cV = 0.100 mol dm^{-3} x 0.0250 dm^3 = 2.50 x 10^{-3} mol

Amount of alkali added = cV = 0.100 mol dm^{-3} x 0.00700 dm^3 = 7.00 x 10^{-4} mol

According to the equation:

$$NaOH(aq) + HCl(aq) \rightarrow NaCl(aq) + H_2O(l)$$

the amount of acid reacting = the amount of alkali added.

The amount of acid left = initial amount - amount reacted

= (2.50 x 10^{-3} - 7.00 x 10^{-4})mol = 1.80 x 10^{-3} mol

Total volume of solution = 25.0 cm^3 + 7.0 cm^3 = 32.0 cm^3

$$\therefore \quad [H^+(aq)] = [HCl(aq)] = n/V = \frac{0.00180 \text{ mol}}{0.0320 \text{ dm}^3} = 0.0563 \text{ mol dm}^{-3}$$

$$\therefore \quad pH = -\log[H^+(aq)] = -(-2 + 0.751) = +2 - 0.75 = \boxed{1.25}$$

Worked Example Calculate the pH of the solution resulting from adding 25.2 cm^3 of 0.100 M NaOH to 25.0 cm^3 of 0.0500 M H_2SO_4.

Solution

Amount of acid initially = cV = 0.0500 mol dm^{-3} x 0.0250 dm^3 = 1.25 x 10^{-3} mol

Amount of alkali added = cV = 0.100 mol dm^{-3} x 0.0252 dm^3 = 2.52 x 10^{-3} mol

According to the balanced equation:

$$H_2SO_4(aq) + 2NaOH(aq) \rightarrow Na_2SO_4(aq) + 2H_2O(l)$$

the amount of alkali reacting = twice the initial amount of acid.

The amount of excess alkali = amount added - amount reacted

= 2.52 x 10^{-3} mol -(2 x 1.25 x 10^{-3}) mol = 2.00 x 10^{-5} mol

Total volume of solution = 25.0 cm^3 + 25.2 cm^3 = 50.2 cm^3

$$\therefore \quad [OH^-(aq)] = [NaOH(aq)] = n/V = \frac{2.00 \times 10^{-5} \text{ mol}}{0.0502 \text{ dm}^3} = 3.98 \times 10^{-4} \text{ mol dm}^{-3}$$

But K_W = 1.00 x 10^{-14} mol^2 dm^{-6} = $[H^+(aq)][OH^-(aq)]$

$$\therefore \quad [H^+(aq)] = \frac{K_W}{[OH^-(aq)]} = \frac{1.00 \times 10^{-14} \text{ mol}^2 \text{ dm}^{-6}}{3.98 \times 10^{-4} \text{ mol dm}^{-3}} = 2.51 \times 10^{-11} \text{ mol dm}^{-3}$$

$$\therefore \quad pH = -\log[H^+(aq)] = -(-11 + 0.400) = +11 - 0.400 = \boxed{10.6}$$

Note that a difference of only 0.2 cm^3 of alkali between this point and the equivalence point at 25.0 cm^3 gives a large difference in pH, 7.0 → 10.6.

Use a method similar to that in the Worked Examples to do the next exercise.

Exercise 259 25.0 cm^3 of 0.100 M HCl was titrated with 0.100 M NaOH. Calculate the pH of the solution for each of the following volumes of NaOH:

0.0, 5.0, 10.0, 20.0, 24.0, 25.0, 26.0, and 30.0 cm^3

You are less likely to meet such calculations involving weak acids and/or weak bases but the principle is the same. First, calculate the concentration of weak acid or base corresponding to the excess, and then calculate the pH as in previous exercises.

Exercise 260 25.0 cm^3 of 0.20 M CH_3CO_2H (ethanoic acid) was titrated with 0.20 M NaOH. Calculate the pH of the solution for each of the following volumes of added NaOH. (For CH_3CO_2H, K_a = 1.7 x 10^{-5} mol dm^{-3}.)

5.0, 10.0, 20.0, 24.0, 24.5, 25.5, 26.0 and 30.0 cm^3.

Exercise 261 25.0 cm^3 of 0.15 M HCl was titrated with 0.15 M NH_3. Calculate the pH of the solution for each of the following volumes of added NH_3. (For NH_3, K_b = 1.8 x 10^{-5} mol dm^{-3}.)

5.0, 10.0, 20.0, 24.0, 24.5, 25.5, 26.0 and 30.0 cm^3

The pH at the equivalence point, i.e. the pH of salt solutions

At the equivalence point the reaction between acid and base is complete, so that only a solution of the salt remains. If the salt is derived from a strong acid and a strong base, e.g.

$$HCl(aq) + NaOH(aq) \rightarrow NaCl(aq) + H_2O(l)$$

then the ions present, in this case Na^+ and Cl^-, show no tendency to react with water to change the pH from the value of 7 for pure water, i.e. for NaCl(aq), pH = 7.

However, a salt derived from a weak acid, e.g. ethanoic acid, changes the pH because the ions react with water to give an excess of hydroxide ions and a pH greater than 7:

$$CH_3CO_2^-(aq) + H_2O(l) \rightarrow CH_3CO_2H(aq) + OH^-(aq)$$

Similarly, a salt derived from a weak base, e.g. ammonia, changes the pH because the ions react with water to give an excess of hydrogen ions and a pH less than 7:

$$NH_4^+(aq) + H_2O(l) \rightarrow NH_3(aq) + H_3O^+(aq)$$

These reactions are examples of what is referred to in many texts as the hydrolysis of salts. The pH of the resulting solutions may be determined directly if the equilibrium constants are known. However, an initial step is often necessary, in which the equilibrium constant is deduced from the listed value of the dissociation constant for the conjugate acid or base. We illustrate this in a Worked Example.

Worked Example Calculate the pH of a 0.10 M solution of sodium ethanoate, $CH_3CO_2^-Na^+(aq)$.

K_a (CH_3CO_2H) = 1.7 x 10^{-5} mol dm^{-3}

W

Solution

1. Calculate the equilibrium constant for the reaction:

$$CH_3CO_2^-(aq) + H_2O(l) \rightleftharpoons CH_3CO_2H(aq) + OH^-(aq)$$

In this reaction, $CH_3CO_2^-$ is behaving as a base and, as shown on page 131, the base dissociation constant is related to the dissociation constant for the conjugate acid, CH_3CO_2H. It is K_a, rather than K_b, which is likely to be found in data books.

$$K_b = \frac{K_w}{K_a} = \frac{1.0 \times 10^{-14}\ mol^2\ dm^{-6}}{1.7 \times 10^{-5}\ mol\ dm^{-3}} = 5.9 \times 10^{-10}\ mol\ dm^{-3}$$

2. Rewrite the equation leaving space above and below for initial and equilibrium concentrations. Fill in the initial concentrations and let the equilibrium concentration of hydroxide ions be x mol dm^{-3}.

	$CH_3CO_2^-(aq)$ + $H_2O(l)$	$\rightleftharpoons$	$CH_3CO_2H(aq)$	+ $OH^-(aq)$
Initial concn/mol dm^{-3}	0.10		0	0 (approx.)
Equilibrium concn/mol dm^{-3}	$0.10-x$		x	x

3. Write an expression for the equilibrium constant (hydrolysis constant):

$$K_b = \frac{[CH_3CO_2H(aq)][OH^-(aq)]}{[CH_3CO_2^-(aq)]}$$ (sometimes called K_h)

4. Substitute the values:

$$5.9 \times 10^{-10}\ mol\ dm^{-3} = \frac{(x\ mol\ dm^{-3})(x\ mol\ dm^{-3})}{(0.10-x)\ mol\ dm^{-3}}$$

5. Assume that $(0.10-x) \simeq 0.10$

$\therefore\ 5.9 \times 10^{-10} = \dfrac{x^2}{0.10}$ and

$x = \sqrt{5.9 \times 10^{-11}} = 7.7 \times 10^{-6}$

$\therefore\ [OH^-(aq)] = 7.7 \times 10^{-6}$ mol dm^{-3}

6. As before, calculate $[H^+(aq)]$ from $[OH^-(aq)]$ and pH from $[H^+(aq)]$:

$$[H^+(aq)] = \frac{K_w}{[OH^-(aq)]} = \frac{1.0 \times 10^{-14}\ mol^2\ dm^{-6}}{7.7 \times 10^{-6}\ mol\ dm^{-3}} = 1.3 \times 10^{-9}\ mol\ dm^{-3}$$

$$pH = -\log[H^+(aq)] = -\log(1.3 \times 10^{-9}) = \boxed{8.9}$$

Now try some similar problems yourself.

Exercise 262 Calculate the pH of the following solutions:

(a) 0.020 M KCN,

(b) 1.0 M NH_4Cl,

(c) 0.020 M NaF.

If an aqueous acid is partially neutralized, the resulting solution is a mixture of the weak acid and one of its salts. Such mixtures have special properties and are known as buffer solutions.

BUFFER SOLUTIONS

The composition of a buffer solution is such that small additions of a strong acid or a strong base have hardly any effect on the pH. To understand the way a buffer resists changes in pH, we can apply the equilibrium law.

Buffer solutions and the equilibrium law

Consider a buffer made from the weak acid, HA, in solution with its sodium salt, Na^+A^-. The acid dissociates in solution according to the equation:

$$HA(aq) \rightleftharpoons H^+(aq) + A^-(aq)$$

and $$K_a = \frac{[H^+(aq)][A^-(aq)]}{[HA(aq)]}$$

Rearranging this expression gives:

$$[H^+(aq)] = K_a \frac{[HA(aq)]}{[A^-(aq)]} \quad \text{- - - - - - - - - - - (1)}$$

Taking logarithms of each side gives:

$$\log[H^+(aq)] = \log K_a + \log\frac{[HA(aq)]}{[A^-(aq)]}$$

Multiplying through by minus 1 gives

$$-\log[H^+(aq)] = -\log K_a - \log\frac{[HA(aq)]}{[A^-(aq)]}$$

i.e. $$pH = pK_a - \log\frac{[HA(aq)]}{[A^-(aq)]} \quad \text{- - - - - - - - - - - (2)}$$

These expressions show that the pH of a buffer solution should remain constant as long as the ratio of concentrations of un-ionized acid molecules to anions stays constant. You can use either expression (1) or expression (2) to calculate the pH of a buffer solution, as we show in a Worked Example.

Worked Example Calculate the hydrogen ion concentration and pH of 1.0 dm^3 of a buffer solution made by dissolving 0.100 mol of ethanoic acid and 0.100 mol of sodium ethanoate in pure water.

W

$K_a(CH_3CO_2H) = 1.7 \times 10^{-5}$ mol dm^{-3}

Solution

1. Write the equation for the dissociation of ethanoic acid, leaving space above and below for initial and equilibrium amounts. Insert the initial amounts.

Initial amount/mol	0.100		0.100	1.0×10^{-7}
	$CH_3CO_2H(aq)$	$\rightleftharpoons$	$CH_3CO_2^-(aq)$ +	$H^+(aq)$
Equilibrium amount/mol				

2. Let the amount of hydrogen ions produced = x mol. Therefore, the amount of $CH_3CO_2^-$ at equilibrium = $(0.100 + x)$ and the amount of CH_3CO_2H at equilibrium = $(0.100 - x)$. Insert these values below the equation:

Initial amount/mol	0.100	0.100	1.0×10^{-7}
	$CH_3CO_2H(aq)$ ⇌	$CH_3CO_2^-(aq)$ +	$H^+(aq)$
Equilibrium amount/mol	$(0.100 - x)$	$(0.100 + x)$	$x + y$

3. Now you can make two important assumptions. Firstly, you can ignore the small amount, y mol, of hydrogen ions which arise from the ionization of water. It is possible to calculate y ($\sim 6 \times 10^{-10}$), but you can see that it must be very much smaller than 1.0×10^{-7} because the addition of acid suppresses the ionization of water. However, x must be greater than 1.0×10^{-7} because a substantial amount of acid has been added,

   i.e. $x >> y$ and $x + y \simeq x$

   Secondly, since ethanoic acid is a weak acid ($K_a = 1.7 \times 10^{-5}$), only a small proportion of the molecules are ionized, and the ionization is further suppressed by the addition of ethanoate ions (Le Chatelier's principle). This means that x must be so much smaller than 0.100 that it is insignificant by comparison. (Applying the simple rule introduced on page 130, $c_0/K_a = 0.100/1.7 \times 10^{-5} >> 1000$ and, therefore, $c_0 - x \simeq x$.)

   i.e. $0.100 >> x$ and $0.100 \pm x \simeq 0.100$

4. Write in these approximated equilibrium amounts.

Initial amount/mol	0.100	0.100	1.0×10^{-7}
	$CH_3CO_2H(aq)$ ⇌	$CH_3CO_2^-(aq)$ +	$H^+(aq)$
Equilibrium amount/mol	$(0.100 - x)$	$(0.100 + x)$	$x + y$
(approx)	0.100	0.100	x

5. Substitute these amounts in either of the expressions (1) or (2).

$$[H^+(aq)] = K_a \frac{[CH_3CO_2H(aq)]}{[CH_3CO_2^-(aq)]}$$

$$= 1.7 \times 10^{-5} \text{ mol dm}^{-3} \frac{0.100 \text{ mol}/1.00 \text{ dm}^3}{0.100 \text{ mol}/1.00 \text{ dm}^3} = \boxed{1.7 \times 10^{-5} \text{ mol dm}^{-3}}$$

$$pH = pK_a - \log \frac{[CH_3CO_2H(aq)]}{[CH_3CO_2^-(aq)]}$$

$$= pK_a - \log \frac{0.100 \text{ mol dm}^{-3}}{0.100 \text{ mol dm}^{-3}} = pK_a - \log 1 = pK_a = -\log(K_a) = \boxed{4.77}$$

In all the calculations you do which concern buffer solutions, you can make similar assumptions to the ones made above. The amounts of added weak acid and salt are always so much greater than the amount of hydrogen ion that they change hardly at all on mixing. This means that you can use the added amounts to calculate the equilibrium concentrations simply by dividing by the volume.

We suggest that you use the full method in the exercise which follows, so that you understand why you can use the simplified method in future calculations.

<u>Exercise 263</u> (a) Calculate the hydrogen ion concentration in a 0.10 M solution of propanoic acid, $C_2H_5CO_2H$. $K_a = 1.3 \times 10^{-5}$ mol dm^{-3}.

(b) Calculate the hydrogen ion concentration in a 0.10 M solution of propanoic acid which contains sodium propanoate to a concentration of 0.050 mol dm^{-3}. Assume that the total volume of the buffer is 1.0 dm^3. Explain why this answer differs from your answer to (a).

In explaining how a buffer system works, we have said that the pH stays constant as long as the ratio of concentrations of acid and conjugate base is unchanged. As you have just seen, in a buffer solution, the concentrations of acid and conjugate base are very large compared with the concentration of hydrogen ions. A change in either of these large quantities will have very little effect on their ratio; hence, the concentration of hydrogen ions in solution, and the pH, remain almost constant.

In the following exercises, make the simplifying assumptions straight away and use either expression (1) or expression (2) on page 135. Expression (2) is easier to use but harder to remember - make sure you know how to derive it.

Exercise 264 A certain buffer solution, made from a weak acid, HA, and its sodium salt, has a pH of 5.85. The solution was made by mixing 10 cm^3 of 0.090 M HA and 20 cm^3 of 0.15 M NaA. Calculate K_a for the acid.

Exercise 265 25.0 cm^3 of a solution of 0.10 M ethanoic acid is titrated with 0.10 M sodium hydroxide. When sufficient sodium hydroxide has been added to neutralize half the ethanoic acid (the 'half neutralization point'), calculate:

(a) the concentrations of ethanoic acid and ethanoate ions,

(b) the pH of the solution. $K_a(CH_3CO_2H) = 1.7 \times 10^{-5}$ mol dm^{-3}.

Exercise 266 Calculate the pH of the following buffer solutions containing butanoic acid and potassium butanoate.

	$[CH_3CH_2CH_2CO_2H]$/mol dm^{-3}	$[CH_3CH_2CH_2CO_2^-K^+]$/mol dm^{-3}
(a)	1.0	1.0
(b)	2.0	1.0
(c)	2.0	0.5
(d)	0.5	2.0

Exercise 267 A series of buffer solutions was prepared from methanoic acid, HCO_2H, and potassium methanoate, HCO_2K, using the following concentrations. Calculate the pH of each buffer solution.
K_a for methanoic acid = 1.6×10^{-4} mol dm^{-3}.

Solution	$[HCO_2H]$/mol dm^{-3}	$[HCO_2K]$/mol dm^{-3}
(a)	0.10	1.0
(b)	2.0	0.20
(c)	0.50	2.5

Exercise 268 A buffer solution is made by mixing 500 cm^3 of 2.0 M sodium dihydrogenphosphate(V), NaH_2PO_4, and 1000 cm^3 of 0.50 M disodium hydrogenphosphate(V), Na_2HPO_4.

K_a for the dihydrogenphosphate(V) ion is 6.2×10^{-8} mol dm^{-3}. Calculate the pH of the buffer.

Now we consider the small changes in pH which occur when a buffer solution is contaminated.

Contamination of buffer solutions

In this section, you learn how to calculate the change in pH that would be expected when a buffer is contaminated with either acid or alkali. Read through the Worked Example and then do the exercises which follow.

W

Worked Example 1.00 dm^3 of a buffer solution contains 0.500 mol of propanoic acid and 1.00 mol of sodium propanoate. It is contaminated by the addition of 100 cm^3 of 1.0 M hydrochloric acid. K_a for propanoic acid is 1.30×10^{-5} mol dm^{-3}. Calculate:

(a) the pH of the uncontaminated buffer solution,

(b) the change in pH caused by the addition of the acid.

Solution

(a) Calculate as in previous exercises.

$$[H^+(aq)] = K_a \frac{[HA(aq)]}{[A^-(aq)]} = 1.30 \times 10^{-5} \text{ mol dm}^{-3} \times \frac{0.500 \text{ mol dm}^{-3}}{1.00 \text{ mol dm}^{-3}}$$

$$= 6.50 \times 10^{-6} \text{ mol dm}^{-3}$$

$$pH = -\log[H^+(aq)] = \boxed{5.19}$$

(b) 1. Calculate the amount of hydrochloric acid added:

$$n = cV = 1.00 \text{ mol dm}^{-3} \times 0.100 \text{ dm}^3 = 0.100 \text{ mol}$$

2. Calculate the amounts of propanoate ion and propanoic acid after adding hydrochloric acid by reference to the equation:

$$H^+(aq) + CH_3CH_2CO_2^-(aq) \rightleftharpoons CH_3CH_2CO_2H(aq)$$

Assume that all of the added hydrogen ions react so that:

amount of propanoate ion after reaction =
initial amount - amount reacted = (1.00 - 0.100) mol = 0.900 mol

Amount of propanoic acid after reaction =
initial amount + amount formed = (0.500 + 0.100) mol = 0.600 mol

These amounts are the new equilibrium amounts.

3. Substitute these new values in the expression for $[H^+(aq)]$:

$$[H^+(aq)] = K_a \frac{[HA(aq)]}{[A^-(aq)]}$$

$$= 1.30 \times 10^{-5} \text{ mol dm}^{-3} \times \frac{0.600 \text{ mol}/V}{0.900 \text{ mol}/V}$$

$$= 8.67 \times 10^{-6} \text{ mol dm}^{-3}$$

4. Calculate the pH.

$$pH = -\log[H^+(aq)] = -\log(8.67 \times 10^{-6}) = (-6 + 0.938) = \boxed{5.06}$$

5. Compare this with the answer to part (a); note that the pH has changed by 5.06 - 5.19 = -0.13 units; i.e. the pH has decreased by 0.13 pH unit - very little for a substantial contamination.

Exercise 269 Calculate the change in pH of 100 cm^3 of a buffer solution containing 0.50 mol dm^{-3} of ethanoic acid and 0.50 mol dm^{-3} of potassium ethanoate when it is contaminated by:

(a) 1.0 cm^3 of 1.0 M hydrochloric acid,

(b) 2.0 cm^3 of 1.5 M potassium hydroxide.

$K_a(CH_3CO_2H) = 1.70 \times 10^{-5}$ mol dm^{-3}

Exercise 270 A buffer solution is prepared, containing 2.0 mol dm^{-3} of butanoic acid and 0.50 mol dm^{-3} of sodium butanoate. K_a for butanoic acid is 1.5×10^{-5} mol dm^{-3}.

(a) Calculate the pH of the uncontaminated buffer solution.

(b) Calculate the change in pH when 1.0 dm^3 of the buffer solution is contaminated by the addition of 20 cm^3 of 2.0 M sodium hydroxide solution.

(c) Calculate the change in pH when 1.0 dm^3 of the same buffer solution is contaminated by the addition of 100 cm^3 of 0.50 M hydrochloric acid.

Exercise 271 A buffer solution is made up by mixing 500 cm^3 of 1.5 M ethanoic acid and 1000 cm^3 of 0.50 M sodium ethanoate. 150 cm^3 of this solution are taken and placed in a beaker containing 50 cm^3 0.10 M sodium hydroxide. How much difference does this make to the pH of the original buffer solution? K_a for ethanoic acid is 1.7×10^{-5} mol dm^{-3}.

After doing these exercises, you should have some idea of the amounts of acid and alkali a buffer solution can absorb, without any significant change in its pH. To compare what happens when pure water is contaminated with acid or alkali, try the next exercise, in which the same amounts of acid and alkali as in Exercise 269 are added to the same volume of water.

Exercise 272 Calculate the change in pH of 100 cm^3 of pure water when

(a) 1.0 cm^3 of 1.0 M hydrochloric acid is added;

(b) 2.0 cm^3 of 1.5 M sodium hydroxide is added.

Most calculations and exercises concern acidic buffers, i.e. those made from a weak acid and one of its salts. However, you should also be able to discuss basic buffers, which we consider in the next section.

Basic buffers

As we mentioned earlier, a weak base and a salt of its conjugate acid can be used as a buffer system. In solution, a weak base such as phenylamine, $C_6H_5NH_2$, tends to accept a proton to become its conjugate acid:

$$C_6H_5NH_2(aq) + H_2O(l) \rightleftharpoons C_6H_5NH_3^+(aq) + OH^-(aq)$$

For this reaction, $K_c = \frac{[C_6H_5NH_3^+(aq)][OH^-(aq)]}{[C_6H_5NH_2(aq)][H_2O(l)]}$

Since water is present in excess, its concentration can be regarded as constant and we can combine it with K_c to give $K_c \times [H_2O(l)] = K_b$, the base dissociation constant. The expression then becomes:

$$K_b = \frac{[C_6H_5NH_3^+(aq)][OH^-(aq)]}{[C_6H_5NH_2(aq)]}$$

However, since any calculation of pH involves hydrogen ion concentration at some point, it is more convenient to treat the system instead as a weak acid dissociating and use K_a in the usual way. The above system becomes:

$$C_6H_5NH_3^+(aq) \rightleftharpoons C_6H_5NH_2(aq) + H^+(aq)$$

and $$K_a = \frac{[C_6H_5NH_2(aq)][H^+(aq)]}{[C_6H_5NH_3^+(aq)]}$$

Remembering that $K_w = [H^+(aq)][OH^-(aq)]$, you can see from the expressions for K_a and K_b that $K_b = K_w/K_a$ (see also page 131).

To make sure that you are confident with this way of treating basic buffers, read through the next Worked Example and try the exercise following it.

Worked Example A buffer solution is prepared by mixing 750 cm^3 of 0.500 M phenylmethylamine, $C_6H_5CH_2NH_2$, and 250 cm^3 of 1.00 M phenylmethylamine hydrochloride, $C_6H_5CH_2NH_3^+Cl^-$. Assuming that K_a for phenylmethylamine hydrochloride is 4.30×10^{-10} mol dm^{-3}, calculate the pH of the buffer.

W

Solution

1. Write an equation representing the equilibrium as the dissociation of an acid.

$$C_6H_5CH_2NH_3^+(aq) \rightleftharpoons C_6H_5CH_2NH_2(aq) + H^+(aq)$$

2. Write an expression for K_a and rearrange it to give an expression for $[H^+(aq)]$.

$$K_a = \frac{[C_6H_5CH_2NH_2(aq)][H^+(aq)]}{[C_6H_5CH_2NH_3^+(aq)]}$$

$$\therefore \quad [H^+(aq)] = K_a \times \frac{[C_6H_5CH_2NH_3^+(aq)]}{[C_6H_5CH_2NH_2(aq)]}$$

Note that this is simply another example of the general equation:

$$[H^+(aq)] = K_a \times \frac{[\text{acid}]}{[\text{conjugate base}]}$$

3. Calculate the amount of each substance added:

$n(C_6H_5CH_2NH_2) = cV = 0.500 \text{ mol dm}^{-3} \times 0.750 \text{ dm}^3 = 0.375 \text{ mol}$

$n(C_6H_5CH_2NH_3^+Cl^-) = cV = 1.00 \text{ mol dm}^{-3} \times 0.150 \text{ dm}^3 = 0.250 \text{ mol}$

4. Since K_a is so small, assume that these amounts are also the equilibrium amounts, and substitute them in the expression for $[H^+(aq)]$.

$$[H^+(aq)] = K_a \times \frac{[C_6H_5CH_2NH_3^+(aq)]}{[C_6H_5CH_2NH_2(aq)]}$$

$$= 4.30 \times 10^{-10} \text{ mol dm}^3 \times \frac{0.250 \text{ mol}/V}{0.375 \text{ mol}/V} = 2.87 \times 10^{-10} \text{ mol dm}^3$$

5. $pH = -\log[H^+(aq)] = -\log(2.87 \times 10^{-10}) = -10 + 0.457 = \boxed{9.54}$

Exercise 273 Calculate the pH of a buffer solution made by mixing 750 cm^3 of 0.20 M ammonium chloride and 750 cm^3 of 0.10 M ammonia solution. Assume that K_a for the ammonium ion is 6.00×10^{-10} mol dm^{-3}.

The equations derived on page 135 are useful for calculating the composition of a buffer from its pH. In other words, if you want to make up a buffer of a particular pH, knowing pK_a for the weak acid, the equations allow you to calculate the ratio of concentrations of acid to conjugate base that you would need.

Calculating the composition of a buffer from the pH

This calculation takes two main forms. We now go through these, giving a Worked Example followed by an exercise in each case.

Worked Example In what proportions must 0.020 M solutions of benzoic acid, $C_6H_5CO_2H$, and sodium benzoate, $C_6H_5CO_2Na$, be mixed to give a buffer solution of pH 4.5? K_a for benzoic acid is 6.4×10^{-5} mol dm^{-3}.

W

Solution

1. First calculate pK_a for benzoic acid.

$$pK_a = -\log K_a = -\log(6.4 \times 10^{-5}) = -(-5 + 0.8062) = 4.2$$

2. Write down the expression for the pH of this buffer solution:

$$pH = pK_a - \log \frac{[C_6H_5CO_2H(aq)]}{[C_6H_5CO_2^-(aq)]}$$

As we have shown earlier, this expression is derived from the equilibrium law expression, in this case for benzoic acid. If you cannot remember it, it takes only a few moments to derive it.

3. Substitute the known values of pH and pK_a into the expression for pH.

$$4.5 = 4.2 - \log \frac{[C_6H_5CO_2H(aq)]}{[C_6H_5CO_2^-(aq)]}$$

$$\therefore \log \frac{[C_6H_5CO_2H(aq)]}{[C_6H_5CO_2^-(aq)]} = 4.2 - 4.5 = -0.3 = -1 + 0.7$$

$$\therefore \frac{[C_6H_5CO_2H(aq)]}{[C_6H_5CO_2^-(aq)]} = \text{antilog } (-1 + 0.70) = 0.50$$

4. Since K_a is small, assume that the equilibrium amounts are the same as the added amounts.

Then the mixture should be made in the proportions of one volume of 0.020 M benzoic acid to two volumes of 0.020 M sodium benzoate.

Exercise 274 In what ratio must 0.50 M solutions of ethanoic acid and sodium ethanoate be mixed to produce buffer solutions of:

(a) pH 4.70, (b) pH 4.40?

$pK_a(CH_3CO_2H) = 4.74$

Worked Example A buffer of pH 3.50 is to be made by dissolving sodium iodoethanoate, $CH_2ICO_2^-Na^+$ in iodoethanoic acid, CH_2ICO_2H. What concentration of sodium iodoethanoate is needed? $pK_a(CH_2ICO_2H) = 3.17$.

W

Solution

In this Worked Example, we again make the assumption that because the dissociation of iodoethanoic acid is small, we can take initial concentrations as equal to equilibrium concentrations.

1. Write down the expression for the pH of this buffer solution:

$$pH = pK_a - \log \frac{[CH_2ICO_2H(aq)]}{[CH_2ICO_2^-(aq)]}$$

2. Let the concentration of sodium iodoethanoate be x mol dm^{-3} and substitute this and the values for pH and pK_a into the equation:

$$3.50 = 3.17 - \log \frac{0.50}{x}$$

3. Rearrange the equation; using the relation $\log \frac{a}{b} = \log a - \log b$

$$3.50 - 3.17 = -\log 0.50 - (-\log x)$$

$$0.33 = -\log 0.50 + \log x$$

$$\therefore \log x = 0.33 + (-1 + 0.6990) = 0.33 + (-0.301) = 0.029$$

$$\therefore x = \text{antilog } 0.029 = 1.069$$

$\therefore$ the concentration of sodium iodoethanoate needed is **1.07 mol dm^{-3}**

Exercise 275 What amount of sodium methanoate, $HCO_2^-Na^+$, must be added per dm^3 of 0.50 M solutions of methanoic acid, HCO_2H, to produce buffer solutions of

(a) pH = 3.8, (b) pH = 4.1?

$K_a(HCO_2H) = 1.58 \times 10^{-4}$ mol dm^{-3}

Exercise 276 A buffer solution of pH 4.0 is to be made from 1.0 M propanoic acid and 1.0 M sodium propanoate. What volume of each solution must be mixed to give 100 cm^3 of the buffer solution required? pK_a (propanoic acid) = 4.9.

Exercise 277 A buffer solution with a pH of 11.8 is needed. 0.50 M solutions of disodium hydrogenphosphate(V), Na_2HPO_4, and sodium phosphate(V), Na_3PO_4, are available. pK_a for the HPO_4^{2-} ion is 12.4.

$$HPO_4^{2-}(aq) \rightleftharpoons PO_4^{3-}(aq) + H^+(aq)$$

What volumes of each of the solutions would you have to mix together to make 200 cm^3 of a buffer solution of pH 11.8?

Exercise 278 What concentration of ammonium chloride solution should be mixed with an equal volume of 0.40 M ammonia solution to give a buffer solution of pH 9.5? pK_a for $NH_4^+(aq)$ is 9.3.

Exercise 279 In a buffered solution,

A

$$pH = -\log K_a + \log \frac{[\text{base}]\ \text{eqm}}{[\text{acid}]\ \text{eqm}}$$

Human plasma is buffered mainly by dissolved carbon dioxide which has reacted to form carbonic(IV) acid:

$$H_2CO_3(aq) \rightleftharpoons H^+(aq) + HCO_3^-(aq)$$

(a) Explain how carbonic(IV) acid can buffer human plasma. Give an example to illustrate your answer.

(b) When the concentrations of carbonic(IV) acid and hydrogencarbonate ion are equal, the concentration of hydrogen ion is 7.9×10^{-7} mol dm^{-3}. Calculate the value of $\log K_a$ for carbonic(IV) acid.

(c) Usually the pH of human plasma is about 7.4. Calculate the ratio of the concentrations of hydrogencarbonate ion and carbonic(IV) acid in plasma.

(d) If the total concentration of hydrogencarbonate ion and carbonic(IV) acid was equivalent to 2.52×10^{-2} mol dm^{-3} of carbon dioxide, calculate the separate concentrations of hydrogencarbonate ion and the carbonic acid in plasma.

An efficient buffer system, equally resistant to both acid and base contamination, would have equal concentrations of weak acid and conjugate base, i.e. $[HA(aq)] = [A^-(aq)]$. In these circumstances,

$$pH = pK_a - \log \frac{[HA(aq)]}{[A^-(aq)]} = pK_a - \log 1 = pK_a$$

Therefore, a buffer is at its most effective at the pH equal to the pK_a of the acid. In practice, the acid to be used should be chosen so that the required pH falls within 1 pH unit either side of the pK_a value, i.e. $pK_a \pm 1$.

The equation you have used to calculate the pH of a buffer solution can also be applied to the calculation of the pH range of an acid-base indicator. This is because most indicators are weak acids, and observed colour depends on the ratio of the concentrations of the weak acid and its conjugate base.

CALCULATING THE pH RANGE OF AN INDICATOR

An acid-base indicator may be regarded as a weak acid, represented by the general formula HIn. Its usefulness depends on the fact that the weak acid and its conjugate base have different colours:

$$\underset{\text{COLOUR A}}{HIn(aq)} \rightleftharpoons H^+(aq) + \underset{\text{COLOUR B}}{In^-(aq)}$$

This equilibrium will lie to the left or to the right according to the pH of the solution to which the indicator is added. If the intensities of the two colours are similar, then it is a good approximation to say that different colours will be observed as follows:

$[HIn(aq)] > 10 \times [In^-(aq)]$	colour A
$[In^-(aq)] > 10 \times [HIn(aq)]$	colour B
$[HIn(aq)] = [In^-]$	intermediate colour

Use these relationships to calculate the pH range of an indicator in the next exercise.

Exercise 280 (a) Bromocresol green is a weak acid for which $pK_a = 4.7$. The acid form is yellow and the conjugate base is blue.

(i) Write an expression relating the pH of a solution to the relative concentrations of the two forms of the added indicator.

(ii) If you used this indicator in an acid-base titration, at what pH values would you expect a colour change to begin and end? At what pH would the solution appear green?

(b) Use your data book to select an indicator for the titration of 0.20 M CH_3CO_2H with 0.20 M NaOH. (Refer to the Worked Example on page 134, if necessary.)

END-OF-CHAPTER QUESTIONS

Exercise 281 (a) Define the term pH of a solution and calculate the pH values of:

(i) 5×10^{-4} M hydrochloric acid,

(ii) 5×10^{-4} M sodium hydroxide solution.

$K_w = [H^+][OH^-] = 1.0 \times 10^{-14}$ mol^2 dm^{-6}

(b) Define the term acidity constant (acid ionization constant, dissociation constant) of an acid. The acidity constant of ethanoic acid (acetic acid) at room temperature is 1.8×10^{-5} mol dm^{-3}. Assuming that the extent of ionization is extremely small, calculate the pH of a 0.100 M solution of ethanoic acid at that temperature.

Exercise 282 The pH of 0.0400 M methanoic acid, HCO_2H, is 2.59 at 25 °C. Calculate the acid dissociation constant at this temperature.

Exercise 283 Explain the action of a buffer solution based on propanoic acid, $C_2H_5CO_2H$.

Calculate the pH of a buffer solution made by adding 6.50 g of sodium propanoate to 1.00 dm^3 of 0.100 M propanoic acid.

$K_a = 1.35 \times 10^{-5}$ mol dm^{-3} at 25 °C
[H = 1.0, C = 12.0, O = 16, Na = 23)

Exercise 284 (a) A solution of methanoic acid, HCO_2H, has a pH of 2.23. Calculate its concentration. $K_a = 1.78 \times 10^{-4}$ mol dm^{-3}.

(b) Calculate the pH of 0.0100 M dichloroethanoic acid, $CHCl_2CO_2H$. $K_a = 5.13 \times 10^{-2}$ mol dm^{-3}.

Exercise 285 (a) (i) The dissociation constant of propanoic acid ($C_2H_5CO_2H$) is 1.26 x 10^{-5} mol dm^{-3} (i.e. $10^{-4.9}$ mol dm^{-3}). Calculate the pH of 0.1 M propanoic acid.

A

$$C_2H_5CO_2H(aq) \rightleftharpoons C_2H_5CO_2^-(aq) + H^+(aq)$$

(ii) What is the hydroxide ion concentration of this solution?

(iii) What is meant by a buffer solution? How many moles of potassium propanoate must be added to 1 dm^3 of the 0.1 M propanoic acid in (i) to make a solution of pH 5? Would the resulting solution be a buffer solution?

(iv) Picric acid, dissociation constant 5.0 x 10^{-3} mol dm^{-3} (i.e. $10^{-2.3}$ mol dm^{-3}), can act as an acid-base indicator, the indicator anions being red, and the free acid yellow. What colour would it show in a solution of pH 5?

(b) My book of data lists the acid dissociation constant of the ammonium ion as 6.3 x 10^{-10} mol dm^{-3} (i.e. $10^{-9.2}$ mol dm^{-3}):

$$NH_4^+(aq) \rightleftharpoons H^+(aq) + NH_3(aq)$$

From this, calculate the equilibrium constant for the following equilibrium:

$$NH_3(aq) + H_2O(l) \rightleftharpoons NH_4^+(aq) + OH^-(aq)$$

(c) The sketch shows how the conductance varies as a solution of potassium hydroxide is added to 20 cm^3 of 0.1 M HCl.

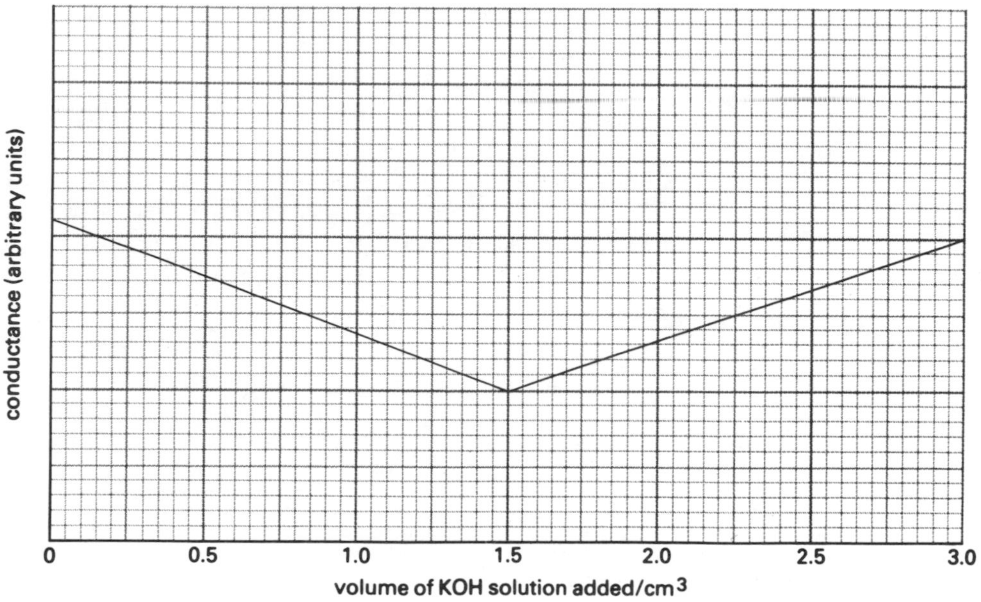

Fig.17.

What is the concentration, in mol dm^{-3}, of the KOH solution?

(K_w = 1.00 x 10^{-14} mol^2 dm^{-6})

Exercise 286 (a) The indicator phenolphthalein may be regarded as a weak monobasic acid, HIn. Explain briefly why the indicator is coloured when the solution in which it is present has an excess of alkali.

(b) The following is a curve showing how the pH changes when a solution of 0.1 M sodium hydroxide is added to 20 cm^3 of ethanoic acid during titration.

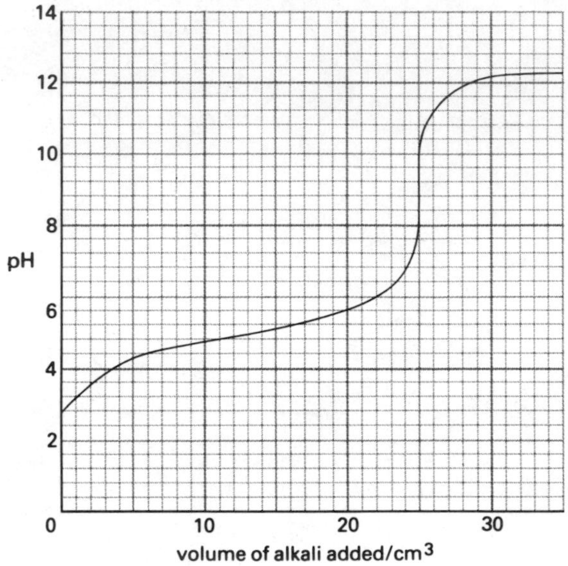

Fig.18.

(i) What is the molarity of the ethanoic acid?

(ii) What is the pH of the solution in the titration flask when 12.5 cm^3 of alkali has been added?

(iii) Calculate the dissociation constant, K_a, of the ethanoic acid.

(iv) Indicate by a sketch how the curve given differs from a possible curve for the titration of 0.1 M sodium hydroxide against an acid of the same concentration as the ethanoic acid, but whose dissociation constant is less than the ethanoic acid.

(c) 1000 cm^3 of a buffer solution has been prepared by dissolving potassium hydroxide and methanoic acid (HCO_2H) in water. What could be the effect on the pH of the solution of adding a further 100 cm^3 of water? Give your reasons.

Exercise 287 A buffer solution having a pH of 4.4 is needed. Methanoic acid, pK_a = 3.8, and sodium methanoate would be suitable. If 0.80 M methanoic acid is used, what should the concentration of sodium methanoate be?

Exercise 288 A buffer solution is made from 1.0 dm^3 of 1.0 M ethanoic acid and 1.0 dm^3 of 0.50 M sodium ethanoate. 200 cm^3 of this solution is placed in a beaker. Someone accidentally mixes into the buffer solution 10 cm^3 of 1.0 M hydrochloric acid. What difference will this make to the pH? K_a for ethanoic acid is 1.7×10^{-5} mol dm^{-3}.

EQUILIBRIUM III – REDOX REACTIONS

8

INTRODUCTION AND PRE-KNOWLEDGE

In this chapter we assume that you are familiar with the definitions of oxidation and reduction in terms of electron transfer (see also Chapter 2). We extend these ideas to account for the potential difference generated between two different metals when placed in an electrolyte and introduce the concept of standard electrode potentials, which are very useful in considering whether a particular chemical reaction is likely to occur in aqueous solution.

We also show you how to use standard electrode potentials in the determination of solubility products and other equilibrium constants by alternative methods to those already described in Chapter 6.

STANDARD ELECTRODE POTENTIALS

When a metal electrode is dipped into a solution of its ions, a potential difference is set up between the two. This can occur in two ways; either the metal atoms tend to form positive ions leaving the metal with a surplus of electrons, or the aqueous ions tend to gain electrons from the metal leaving the electrode with a net positive charge, as shown in Fig. 19.

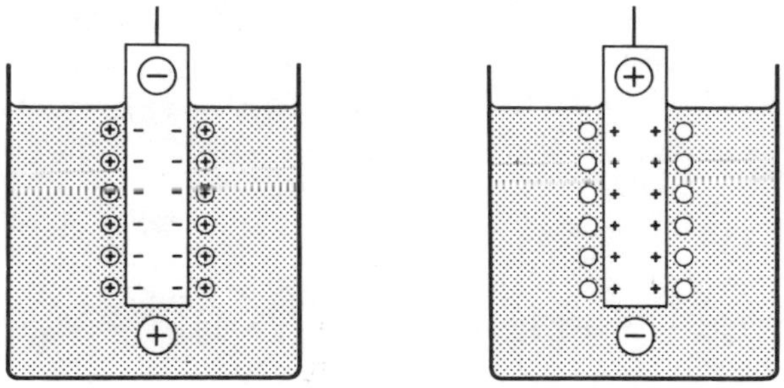

⊕ = additional ions released from electrode

○ = additional atoms deposited from solution

Fig.19. Absolute electrode potentials

However, we cannot be quite certain which of these processes occurs in a particular case because we cannot measure the potential difference between electrode and solution (known as the absolute electrode potential). Measurement would require an electrical connection to be made between a voltmeter and the solution by inserting a metal wire or strip, which would inevitably act as another electrode with its own electrode potential.

That is to say, we can only measure the sum of two electrode potentials, never a single electrode potential. To overcome this difficulty, we arbitrarily assign an electrode potential of zero to one particular half-cell and compare all other half-cells with this standard. By international agreement, the standard hydrogen electrode has been chosen as the reference electrode.

The standard hydrogen electrode

Fig. 20 represents a standard hydrogen electrode connected by a salt bridge to another electrode, or half-cell, to make a cell. This cell generates a potential difference, known as the e.m.f., which can be measured by a high resistance voltmeter (i.e. one which allows no current to pass).

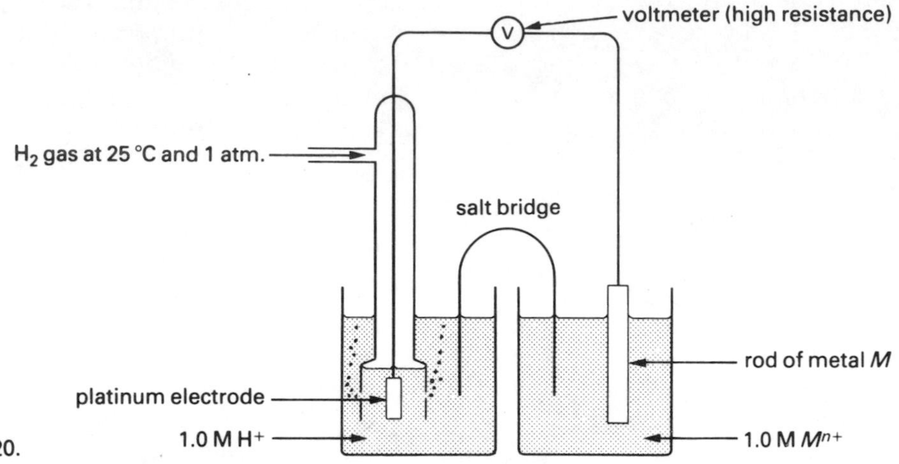

Fig.20.

In this book we use the conventions by which such a cell is represented in a cell diagram, as follows:

$$Pt(s),H_2(g) \mid H^+(aq,1.0\ M) \ \vdots\ M^{n+}(aq,1.0\ M) \mid M(s)$$

The standard electrode potential, $E^\ominus$, of the metal M is defined as equal to the e.m.f. of this cell, $\Delta E^\ominus$, since we assume that the hydrogen electrode has zero potential under standard conditions, i.e.

$$\Delta E^\ominus = E^\ominus_R - E^\ominus_H$$

where $E^\ominus_R$ = standard electrode potential of right-hand electrode $E^\ominus_H$ = standard electrode potential of the hydrogen electrode = 0.00 V

This is a particular case of the general statement for any cell under any conditions:

$$\boxed{\Delta E = E_R - E_L}$$

(R and L refer to the electrode potentials of the half-cells written on the right and left, respectively, of the cell diagram.)

Note that, by convention, the sign of the e.m.f. is the polarity of the right-hand electrode in the cell diagram, and the hydrogen electrode is always written on the left.

Because of these conventions, standard electrode potentials of half-cells are tabulated as reduction reactions since they always refer to the half-cell diagram which represents the half-equation written as reduction:

$M^{n+}(aq) \mid M(s)$ refers to $M^{n+}(aq) + ne^- \rightarrow M(s)$

This is why standard electrode potentials are sometimes called standard reduction potentials, or standard redox potentials.

To check that you understand the conventions, try the next exercise.

Exercise 289 Three cells were set up similar to the one shown in Fig. 20, using three different metals. The potential difference between the two electrodes, and the polarity of the metal, were as shown below.

Zinc negative, hydrogen positive	0.76 V
Copper positive, hydrogen negative	0.35 V
Silver positive, hydrogen negative	0.81 V

(a) For each cell, write a cell diagram and state the e.m.f.

(b) Write an equation for each metal half-cell together with the appropriate value for the standard electrode potential.

The electrode processes you have studied so far in this chapter have consisted of a metal in contact with a solution of its ions. We now extend this to cover other electrode systems involving ion/ion equilibria such as

$$Fe^{3+}(aq) + e^- \rightleftharpoons Fe^{2+}(aq)$$

and non-metal/non-metal-ion equilibra such as:

$$I_2(aq) + 2e^- \rightleftharpoons 2I^-(aq)$$

Such standard half-cells can be set up by dipping a platinum electrode into a mixture in which each reacting species has a concentration of 1.0 mol dm^{-3}. For instance, the standard $Fe^{3+}(aq)$, $Fe^{2+}(aq)$ half-cell would look like Fig. 21. The platinum electrode is inert and serves only to transfer electrons. It does not participate in the cell reaction except, perhaps, as a catalyst.

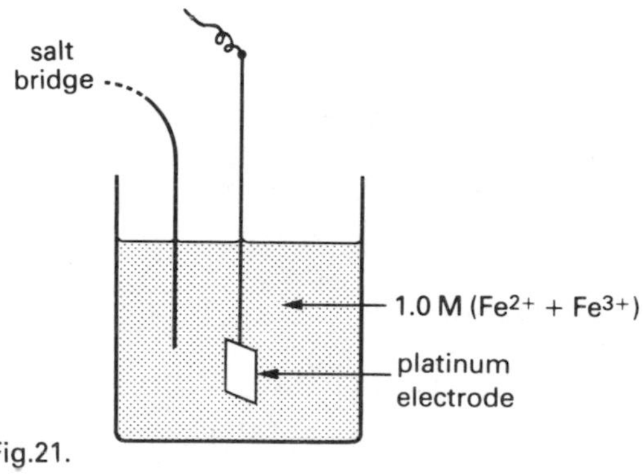

Fig.21.

The half-cell diagram for this electrode is:

$$Fe^{3+}(aq), Fe^{2+}(aq) \mid Pt$$

When a solution contains more than one type of ion, these are shown separated by commas, with the more reduced species nearer the electrode.

Sometimes, the electrode system contains more than two chemical species which take part in the cell reaction. For instance, the equilibrium between aqueous manganate(VII) and manganese(II) ions is established only in the presence of hydrogen ions, as shown below:

$$MnO_4^-(aq) + 8H^+(aq) + 5e^- \rightleftharpoons Mn^{2+}(aq) + 4H_2O(l)$$

The half-cell is shown in tables of electrode potentials as follows:

$$[MnO_4^-(aq) + 8H^+(aq)], [Mn^{2+}(aq) + 4H_2O(l)]$$

The square brackets are used to separate the oxidized and reduced forms of the equilibrium mixture. However, when writing a complete cell diagram, the half-cell may be abbreviated to:

$$MnO_4^-(aq), H^+(aq), Mn^{2+}(aq) \mid Pt$$

In the following Worked Example, we show you how to relate the information in a half-cell diagram to the equation for the half-reaction.

W

<u>Worked Example</u> For the half-cells shown below, write down the equation corresponding to the standard electrode potential.

$[MnO_4^-(aq) + 8H^+(aq)],[Mn^{2+}(aq) + 4H_2O(l) \mid Pt;$

$E^\ominus = +1.51$ V

Solution

The equation corresponding to a standard electrode potential is always written as reduction:

$$\text{oxidized form} + ne^- \rightarrow \text{reduced form}$$

The problem, therefore, reduces to calculating n in the equation:

$$MnO_4^-(aq) + 8H^+(aq) + ne^- \rightarrow Mn^{2+}(aq) + 4H_2O(l) \text{..........(1)}$$

You can either use the oxidation number method (Method 1) or the balancing of charge method (Method 2). We consider both methods in turn.

Method 1. The oxidation number method

1. Identify the atom which changes its oxidation number in the half-cell.

$$\overset{+7}{Mn}O_4^- \rightarrow \overset{+2}{Mn}{}^{2+}$$

2. Decide how many electrons are necessary to effect the change in oxidation number.

   To go from +7 to +2, five electrons are gained per Mn atom.

3. Substitute $n = 5$ in equation 1.

$$MnO_4\text{-}(aq) + 8H^+(aq) + 5e^- \rightarrow Mn^{2+}(aq) + 4H_2O(l)$$

Method 2. Balancing charge method

1. Count the charges on the left of equation 1 (excluding the electrons).

$$1(-1) + 8(+1) = +7$$

2. Count the charges on the right of equation 1.

$$1(+2) = +2$$

3. Work out the number of electrons which must be added so that the charge on the left equals the charge on the right:

$$+7 + 5e^- = +2$$

4. Write the complete half-equation including the number of electrons.

$$MnO_4^-(aq) + 8H^+(aq) + 5e^- \rightarrow Mn^{2+}(aq) + 4H_2O(l)$$

Decide which method you prefer and then apply it in the next exercise.

<u>Exercise 290</u> Write down the equations for the reactions which correspond to the standard electrode potentials for the following half-cells:

(a) $[2NO_3^-(aq) + 4H^+(aq)],[N_2O_4(g) + 2H_2O(l) \mid Pt;$ $E^\ominus = +0.80$ V

(b) $[2H_2SO_3(aq) + 2H^+(aq)],[S_2O_3^{2-}(aq) + 3H_2O(l)] \mid Pt;$ $E^\ominus = +0.40$ V

(c) $[IO^-(aq) + H_2O(l)],[I^-(aq) + 2OH^-(aq)] \mid Pt;$ $E^\ominus = +0.49$ V

In the following sections, we show you how useful standard electrode potentials are in making predictions about likely reactions.

Calculating cell e.m.f. from tabulated $E^\ominus$ values

Instead of measuring cell e.m.fs. experimentally, we can use tabulated $E^\ominus$ values to calculate them. We show you how to do this in a Worked Example.

W

Worked Example Calculate the magnitude and the sign of the e.m.f. of the cell represented below, and write an equation for the cell reaction.

$$Mn(s) \mid Mn^{2+}(aq) \,\vdots\, Co^{2+}(aq) \mid Co(s)$$

Solution

1. Look up the standard electrode potential for each half-cell reaction.

   $Co^{2+}(aq) \mid Co(s)$; $E^\ominus = -0.28$ V

   $Mn^{2+}(aq) \mid Mn(s)$; $E^\ominus = -1.19$ V

2. Write equations for the half-cell reactions.

   The right-hand side of the cell diagram corresponds to the electrode potential as tabulated in a data book, i.e. as a reduction:

   $Co^{2+}(aq) + 2e^- \rightleftharpoons Co(s)$; $E^\ominus = -0.28$ V

   The left-hand side of the cell diagram corresponds to the reverse of the electrode potential as tabulated, so the sign must be reversed:

   $Mn(s) \rightleftharpoons Mn^{2+}(aq) + 2e^-$; $E^\ominus = +1.19$ V

3. Add the two half-cell reactions together:

   $Co^{2+}(aq) + 2e^- \rightleftharpoons Co(s)$; $E^\ominus = -0.28$ V

   $Mn(s) \rightleftharpoons Mn^{2+}(aq) + 2e^-$; $E^\ominus = +1.19$ V

   $Mn(s) + Co^{2+}(aq) \rightarrow Co(s) + Mn^{2+}(aq)$; $\Delta E^\ominus = +0.91$ V

   The overall reaction occurs from left to right as shown, because $\Delta E^\ominus$ is positive. The right-hand electrode (cobalt) is positive, i.e. electron deficient, because it is supplying electrons to the cobalt ions in the solution. A negative value would show that the reaction proceeds in the opposite direction.

4. Complete the cell diagram by inserting $\Delta E^\ominus$.

   $Mn(s) \mid Mn^{2+}(aq) \,\vdots\, Co^{2+}(aq) \mid Co(s)$; $\boxed{\Delta E^\ominus = +0.91 \text{ V}}$

A short cut

The steps followed above can be summarised in the simple rule you have already met (page 148):

$$\Delta E^\ominus = E^\ominus_R - E^\ominus_L$$

($E^\ominus_R$ refers to the right-hand electrode in the cell diagram and $E^\ominus_L$ refers to the left-hand electrode in the cell diagram.)

So, a short cut through the calculation is achieved by substituting into this expression.

$$\Delta E^\ominus = E^\ominus_R - E^\ominus_L = -0.28 \text{ V} - (-1.19 \text{ V}) = +0.91 \text{ V}$$

Since $\Delta E^{\ominus}$ is positive, the reaction proceeds in the direction indicated by the cell diagram:

$$Mn(s) \mid Mn^{2+}(aq) \;\vdots\; Co^{2+}(aq) \mid Co(s); \quad \Delta E^{\ominus} = +0.91\ V$$

$$Mn(s) + Co^{2+}(aq) \rightarrow Mn^{2+}(aq) + Co(s)$$

You can apply this rule in the next exercise. Use the table of standard electrode potentials in your data book.

Exercise 291 For each of the following cells, calculate the cell e.m.f. and write the equation to represent the reaction which would occur if the cell were short-circuited. (Assume the temperature is 25 °C and all solutions are of concentration 1.0 mol dm^{-3}.)

(a) $Pt(s),H_2 \mid H^+(aq) \;\vdots\; Ag^+(aq) \mid Ag(s)$

(b) $Ni(s) \mid Ni^{2+}(aq) \;\vdots\; Zn^{2+}(aq) \mid Zn(s)$

(c) $Ni(s) \mid Ni^{2+}(aq) \;\vdots\; [NO_3^-(aq) + 3H^+(aq)],[HNO_2(aq) + H_2O(l)] \mid Pt$

(d) $Pt \mid 2H_2SO_3(aq),[4H^+(aq) + S_2O_6^{2-}(aq)] \;\vdots\; Cr^{3+}(aq) \mid Cr(s)$

In the next section we take these ideas one small step further.

Using $E^{\ominus}$ values to predict spontaneous reactions

We have shown that, given a cell diagram and the sign on the cell e.m.f., we are able to predict the cell reaction. We now show you a similar method for predicting whether the reaction described by a given equation is likely to occur or not.

Worked Example Use values of standard electrode potentials to decide whether or not the following reaction is likely to occur:

W

$$Pb(s) + Mn^{2+}(aq) \rightarrow Pb^{2+}(aq) + Mn(s)$$

Solution

1. Identify the two half-equations and write them down.

$$Pb(s) \rightleftharpoons Pb^{2+}(aq) + 2e^-$$

$$Mn^{2+}(aq) + 2e^- \rightarrow Mn(s)$$

2. Combine these in a cell diagram.

$$Pb(s) \mid Pb^{2+}(aq) \;\vdots\; Mn^{2+}(aq) \mid Mn(s)$$

3. Look up the standard electrode potentials and write them down.

$$E^{\ominus}(Pb^{2+}|Pb) = -0.13\ V \qquad E^{\ominus}(Mn^{2+}|Mn) = -1.18\ V$$

4. Determine the e.m.f. of this cell by substituting into the expression:

$$\Delta E^{\ominus} = E^{\ominus}_R - E^{\ominus}_L$$

$$\therefore \quad \Delta E^{\ominus} = -1.18\ V - (-0.13\ V) = -1.05\ V$$

The negative sign shows that the given reaction will <u>not</u> proceed spontaneously.

$$Pb(s) + Mn^{2+}(aq) \not\rightarrow Pb^{2+}(aq) + Mn(s)$$

(However, we would expect the reverse reaction to occur, because it has a positive value, +1.05 V, of $\Delta E^{\ominus}$.)

In the next exercise we ask you to predict whether or not certain reactions are likely to be spontaneous.

<u>Exercise 292</u> Use $E^{\ominus}$ values to predict whether the following reactions are spontaneous under standard conditions.

(a) $Br_2(aq) + 2I^-(aq) \rightarrow 2Br^-(aq) + I_2(aq)$

(b) $Br_2(aq) + 2Cl^-(aq) \rightarrow 2Br^-(aq) + Cl_2(aq)$

(c) $Zn(s) + 2Fe^{3+}(aq) \rightarrow Zn^{2+}(aq) + 2Fe^{2+}(aq)$

(d) $2MnO_4^-(aq) + 16H^+(aq) + 10Br^-(aq)$
$\rightarrow 8H_2O(l) + 2Mn^{2+}(aq) + 5Br_2(aq)$

(e) $2MnO_4^-(aq) + 16H^+(aq) + 5Cu(s)$
$\rightarrow 8H_2O(l) + 2Mn^{2+}(aq) + 5Cu^{2+}(aq)$

(f) $3S_2O_8^{2-}(aq) + 2Cr^{3+}(aq) + 7H_2O(l)$
$\rightarrow 6SO_4^{2-}(aq) + Cr_2O_7^{2-}(aq) + 14H^+(aq)$

Limitations of predictions made using $E^{\ominus}$ values

Consideration of $E^{\ominus}$ values in Exercise 292 led you to expect the following reaction to occur:

$$3S_2O_8^{2-}(aq) + 2Cr^{3+}(aq) + 7H_2O(l) \rightarrow Cr_2O_7^{2+}(aq) + 6SO_4^{2-}(aq) + 14H^+(aq)$$

However, the prediction is not borne out by experiment. This illustrates one of the limitations of using $E^{\ominus}$ values; they cannot predict reaction rates. Reactions like the one above are considered energetically favourable yet kinetically unfavourable. So we can only say that a reaction <u>could</u> occur, not that it definitely will.

A second possible reason for the non-occurrence of a reaction predicted from $E^{\ominus}$ values is that some alternative reaction may be more favoured.

Thirdly, you must remember that $E^{\ominus}$ values refer to standard conditions only; you will see later that changing the conditions can change the value of ΔE for a reaction, though not usually by very much.

We now describe another method for predicting spontaneous reactions. It simply applies in a different way the principle you have already learned.

The 'anti-clockwise rule'

We illustrate this rule by a Worked Example, followed by two exercises.

Worked Example Predict which reaction will occur on combining the following half-cells:

W

$I_2(aq) + 2e^- \rightleftharpoons 2I^-(aq)$; $E^\ominus = +0.53$ V

$Br_2(aq) + 2e^- \rightleftharpoons 2Br^-(aq)$; $E^\ominus = +1.06$ V

Solution

1. Write the half-cells with the more negative electrode potential written above the other. This is the order in which they are usually tabulated.

2. Draw in anticlockwise arrows, as shown below.

$$I_2(aq) + 2e^- \rightleftharpoons 2I^-(aq)$$
$$Br_2(aq) + 2e^- \rightleftharpoons 2Br^-(aq)$$

(This can easily be justified using the principles you have already learned. The direction of the arrows merely follows what you would expect from the values of $E^\ominus$.)

3. Re-write the half-equations following these arrows:

$$Br_2(aq) + 2e^- \rightarrow 2Br^-(aq)$$
$$2I^-(aq) \rightarrow I_2(aq) + 2e^-$$

and add them to obtain the predicted reaction:

$$Br_2(aq) + 2I^-(aq) \rightarrow 2Br^-(aq) + I_2(aq)$$

Exercise 293 Predict the reactions which might occur on combining the following half-cells:

(a) $Mn^{2+}(aq) + 2e^- \rightleftharpoons Mn(s)$; $E^\ominus = -1.19$ V

$Pb^{2+}(aq) + 2e^- \rightleftharpoons Pb(s)$; $E^\ominus = -0.13$ V

(b) $I_2(aq) + 2e^- \rightleftharpoons 2I^-(aq)$; $E^\ominus = +0.54$ V

$S(s) + 2e^- \rightleftharpoons S^{2-}(aq)$; $E^\ominus = -0.48$ V

(c) $2H^+(aq) + 2e^- \rightleftharpoons H_2(g)$; $E^\ominus = 0.00$ V

$Ag^+(aq) + e^- \rightleftharpoons Ag(s)$; $E^\ominus = +0.80$ V

Exercise 294 The following table gives a list of standard electrode potentials for a number of half-reactions.

Half-reaction	$E^\ominus$/V
$Al^{3+}(aq) + 3e^- \rightleftharpoons Al(s)$	-1.66
$I_2(aq) + 2e^- \rightleftharpoons 2I^-(aq)$	+0.54
$Fe^{3+}(aq) + e^- \rightleftharpoons Fe^{2+}(aq)$	+0.77
$H_2O_2(l) + 2H^+(aq) + 2e^- \rightleftharpoons 2H_2O(l)$	+1.77
$Co^{3+}(aq) + e^- \rightleftharpoons Co^{2+}(aq)$	+1.88

State the substances among those appearing in this table that can be used to convert Fe^{2+} into Fe^{3+}.

We have shown you several different methods for tackling problems involving cells and the use of $E^\ominus$ values. The next section helps you to choose which is most appropriate in a particular case.

Summary

Which method you use in predicting spontaneous reactions largely depends on the way the question is presented.

If you are given the cell diagram, then the direction of the spontaneous reaction can be predicted from the sign of the cell e.m.f. If you have to calculate the e.m.f., use $\Delta E^{\ominus} = E^{\ominus}_{R} - E^{\ominus}_{L}$.

If you are given a list of standard electrode potentials, then the direction of the spontaneous reaction can be predicted by writing your own cell diagram or using the anticlockwise rule.

Whichever method you choose, it is important that you understand that both use the same basic principles - oxidation occurs at the more negative electrode (i.e. the one with the more negative electrode potential), and reduction occurs at the more positive electrode (i.e. the electrode with the more positive electrode potential) in an electrochemical cell. All you need to know to be able to predict a cell reaction is the polarity of the electrodes.

So far, we have considered only cells operating under standard conditions, i.e. at 298 K and with all concentrations at 1.0 mol dm^{-3}. In the next section, we consider the effect of changing conditions, particularly the concentration.

THE EFFECT OF CONCENTRATION ON ELECTRODE POTENTIAL

Le Chatelier's principle allows us to predict the effect of diluting the solution in a metal/metal-ion half-cell, for example:

$$Ag^{+}(aq) + e^{-} \rightleftharpoons Ag(s)$$

If $[Ag^{+}(aq)]$ is decreased, the equilibrium position will shift to the left. This produces additional electrons which make the silver electrode more negative, or less positive, and so the electrode potential changes accordingly. The experimental results in the next exercise confirm this.

Exercise 295 A cell was set up in which one half-cell contained a standard copper electrode ($E^{\ominus}$ = +0.34 V) and the other a silver electrode dipping into a solution in which the concentration of silver ions could be varied. The e.m.f. was measured at different concentrations, as follows:

$[Ag^{+}(aq)]$/mol dm^{-3}	0.00010	0.00033	0.0010	0.0033	0.010	0.10
ΔE/V	+0.25	+0.28	+0.31	+0.34	+0.38	+0.43

(a) Calculate $E(Ag^{+}/Ag)$ in each case and plot against $\log[Ag^{+}(aq)]$. Use a scale on which E extends from -0.1 to +0.9 and $\log[Ag^{+}(aq)]$ extends from -15 to 0. You can make use of the extra area in later exercises.

(b) Use the graph to obtain:

(i) $E^{\ominus}(Ag^{+}/Ag)$ (i.e. when $[Ag^{+}(aq)]$ = 1.0 mol dm^{-3})

(ii) $[Ag^{+}(aq)]$ when $E(Ag^{+}/Ag)$ = 0.05 V

(c) What is the slope of the graph?

The graph you drew in the last exercise illustrates a general relationship between electrode potential and concentration which we explore in the next section.

The Nernst equation

The mathematical relationship between electrode potential and concentration of aqueous ions was worked out by Nernst and is called the Nernst equation:

$$E = E^{\ominus} + \frac{RT}{nF} \ln\frac{[\text{oxidized form}]}{[\text{reduced form}]} \qquad (\ln = \log_e)$$

or

$$E = E^{\ominus} + \frac{2.3RT}{nF} \log\frac{[\text{oxidized form}]}{[\text{reduced form}]} \qquad (\log = \log_{10})$$

It relates the electrode potential, E, the standard electrode potential, $E^{\ominus}$, and the concentrations of oxidized and reduced forms.

R = gas constant n = number of electrons transferred (some texts use z)
T = temperature F = the Faraday constant (9.65×10^4 C mol^{-1})
(9.65×10^4 J V^{-1} mol^{-1})

First we apply the Nernst equation to a metal/metal-ion system such as:

$$Ag^+(aq) + e^- \rightleftharpoons Ag(s)$$

$$E = E^{\ominus} + \frac{2.3RT}{1 \times F} \log\frac{[Ag^+(aq)]}{[Ag(s)]}$$

Since the 'concentration' of a solid is constant (taken as unity) the expression becomes:

$$E = E^{\ominus} + \frac{2.3RT}{F} \log[Ag^+(aq)] \qquad [\text{strictly } \log([Ag^+(aq)]/\text{mol dm}^{-3})]$$

For any metal/metal-ion system, ($M^{n+}|M$), the Nernst equation is often used in a simplified form:

$$E = E^{\ominus} + \frac{0.059\text{ V}}{n} \log[M^{n+}]$$

This equation has the form of a straight line ($y = c + mx$) where E and log[ion] are the variables, $E^{\ominus}$ is the intercept and 0.059 V/n is the slope. To confirm this, look at your answers to Exercise 295.

The next exercise links the Nernst equation in this form with some experimental results.

Exercise 296 The electrode potential of a metal was measured at various concentrations of the metal ions, all at 25 °C. The results were as follows:

c/mol dm^{-3}	0.50	0.20	0.10	0.050	0.020
E/V	-0.287	-0.299	-0.308	-0.317	-0.330

(a) Plot a graph of electrode potential, E, against log[ion].

(b) From the graph, calculate the standard electrode potential of the metal.

(c) What is the charge on the metal ion?

The following two exercises give you further practice in the graphical interpretation of the Nernst equation.

Exercise 297 Fig. 22 below summarises the results of an experiment in which the redox potential of a metal, X, was measured at different concentrations of its ions, $X^{n+}(aq)$. The temperature was 25 °C.

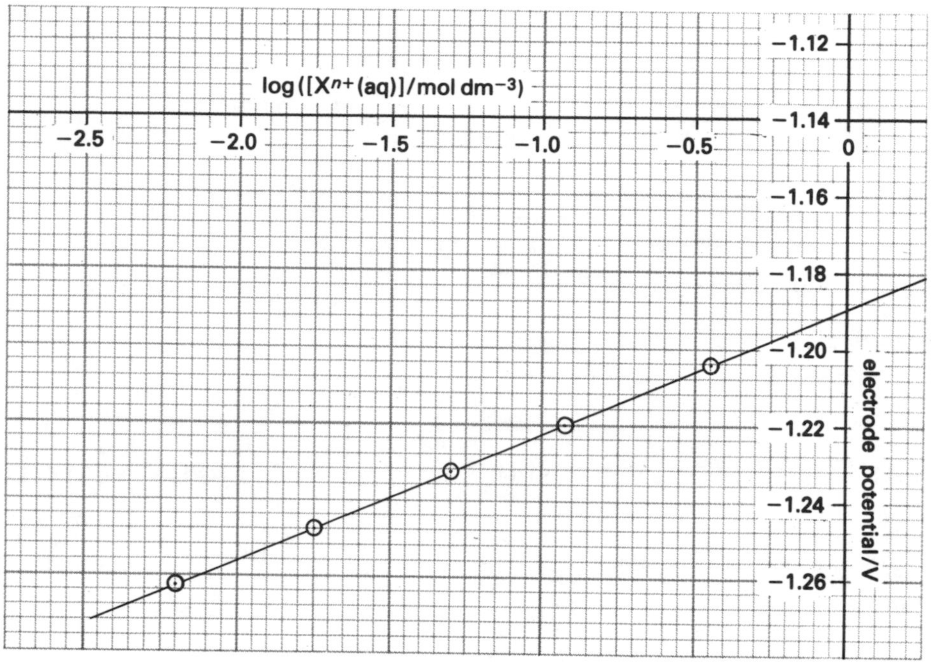

Fig.22.

(a) Calculate the charge on the metal ion by measuring the slope of the graph and using the Nernst equation.

(b) What is the standard redox potential of X?

(c) What is the e.m.f. of the cell shown below?

$X(s) \mid X^{n+}(aq, 1.0\ M) \,\vdots\, Zn^{2+}(aq), 1.0\ M) \mid Zn(s)$

$E^{\ominus}(Zn^{2+}|Zn) = -0.76\ V$

Exercise 298 The table below shows the redox potentials at 25 °C of some aqueous mixtures of two ions of the metal X in contact with a platinum electrode. The ions are $X^{2+}(aq)$ and $X^{n+}(aq)$, and n is greater than 2.

$[X^{2+}(aq) : [X^{n+}(aq)]$	E/V	$\log([X^{n+}(aq)]/[X^{2+}(aq)])$
8 : 1	0.113	-0.90
3 : 1	0.125	-0.48
1 : 2	0.149	0.30
1 : 8	0.167	0.90

(a) Plot a graph which will enable you to determine $E^{\ominus}(X^{2+}|X^{n+})$.

(b) What is the value of n?

Now that you know how the Nernst equation is constructed, you can easily understand some of its practical applications. We discuss these in the next section.

USES OF THE NERNST EQUATION

The great value of the Nernst equation is that it enables us to determine the concentration of ions in very dilute solutions. This provides us with methods for determining solubility product, the equilibrium constant for a redox reaction and the pH of a solution.

We have already described one method for determining the solubility product of calcium hydroxide in Chapter 6 (Equilibrium I: Principles). However, the titration method used there cannot be applied with any accuracy to salts which are less soluble than calcium hydroxide. We now show you another method using the Nernst equation.

Determining solubility products via the Nernst equation

We illustrate this method, and its advantages over titration, in the next exercise.

Exercise 299 A piece of silver foil was put into a solution of silver ions and connected to a standard hydrogen electrode as represented by the cell diagram:

$Pt,H_2(g) \mid H^+(aq) \,\vdots\, Ag^+(aq) \mid Ag(s)$; ΔE = +0.50 V

(a) Calculate the concentration of silver ions in the solution using the Nernst equation.

(b) Determine $[Ag^+(aq)]$ from the graph in the answer to Exercise 295. Compare your answers (a) and (b).

(c) Suppose you wanted to determine the concentration of silver ions by a titration method and you have a standard 0.010 M NaCl solution. What volume of the sodium chloride solution would be required in a titration against 50 cm^3 of 1.0×10^{-5} M Ag^+?

(d) Use the answer to (c) to comment on the accuracy of the titration versus electrochemical methods for determining small concentrations.

In order to determine a solubility product a half-cell is set up in which a metal electrode dips into a solution containing the metal ions in equilibrium with the solid salt. The concentration of the metal ion is found by measuring the electrode potential against a suitable standard, and the concentration of the anion by considering the conditions under which the salt was precipitated.

For example, if 20 cm^3 of 0.10 M KCl is added to 10 cm^3 of 0.10 M $AgNO_3$, a precipitate of AgCl is formed in a solution in which the concentration of chloride ions is found as follows:

10 cm^3 of the 0.10 M KCl is used up to form the precipitate, leaving an excess of 10 cm^3 in a total volume of 30 cm^3. The concentration of chloride ions is, therefore, $\frac{1}{3} \times 0.10$ mol dm^{-3}. (This assumes that the solubility of silver chloride in a solution containing excess chloride ions is negligibly small - for a justification of this see the discussion of the common ion effect in Chapter 6.)

The following exercises are based on the results of similar experiments.

Exercise 300 Four $Ag^+|Ag$ half-cells were set up as described above and their electrode potentials measured. In each case, the concentration of the anion was 0.033 mol dm^{-3}.

$AgCl(s) \rightleftharpoons Ag^+(aq) + Cl^-(aq)$ $E(Ag^+|Ag) = +0.340$

$AgBr(s) \rightleftharpoons Ag^+(aq) + Br^-(aq)$ $E(Ag^+|Ag) = +0.205$

$AgI(s) \rightleftharpoons Ag^+(aq) + I^-(aq)$ $E(Ag^+|Ag) = -0.020$

$AgIO_3(s) \rightleftharpoons Ag^+(aq) + IO_3^-(aq)$ $E(Ag^+|Ag) = +0.480$

Use the graph in Exercise 295 to determine $[Ag^+(aq)]$ in each case and hence determine the solubility product of each salt.

Exercise 301 Calculate the solubility products of the silver halides from the following information. 10 cm^3 of 0.10 M $AgNO_3$ solution was added to 15 cm^3 of a solution of each potassium halide, 0.10 M KX. A silver electrode was dipped into each mixture and the electrode potential measured, using a standard hydrogen electrode to complete the cell. The results were:

AgCl; $E = +0.34$ V AgBr; $E = +0.16$ V AgI; $E = +0.06$ V

Use the graph in Exercise 295.

Exercise 302 10 cm^3 of 1.0 M K_2CrO_4 solution was added to 10 cm^3 of 1.0 M $AgNO_3$ solution to give a precipitate of silver chromate, Ag_2CrO_4. The resulting mixture was used in the following cell and the e.m.f. measured.

$Cu(s) \mid Cu^{2+}(aq, 1.0\ M) \,\vdots\, Ag^+(aq) \mid Ag(s)$; $\Delta E = +0.14$ V

$E^{\ominus}(Ag^+|Ag) = +0.80$ V $E^{\ominus}(Cu^{2+}|Cu) = +0.34$ V

(a) Calculate the concentration of silver ions in the mixture, using the Nernst equation.

(b) What is the concentration of chromate ions?

(c) Calculate the solubility product of silver chromate.

A further application of the determination of small concentrations of ions by the Nernst equation is in the calculation of equilibrium constants.

Calculating equilibrium constants, K_c

We now show you, in a Worked Example, a method for determining the equilibrium constant, K_c, for a redox reaction, using the Nernst equation.

Worked Example Calculate the equilibrium constant from electrochemical data for the reaction:

W

$$Cu(s) + 2Ag^+(aq) \rightleftharpoons 2Ag(s) + Cu^{2+}(aq)$$

Solution

1. Draw a cell diagram for the reaction, showing one half-cell at standard conditions. (Here, assume it is the copper half-cell, since we have a plot of log[$Ag^+(aq)$] against $E(Ag^+|Ag)$ for the silver electrode but no plot for the copper electrode.)

2. At equilibrium there is no net reaction in either direction. This means that the electrode potential of the silver half-cell is equal to the standard electrode potential of the copper half-cell.

$$E(Ag^+|Ag) = E^\ominus(Cu^{2+}|Cu) = +0.34 \text{ V}$$

There are two alternative procedures now, 3A and 3B.

3A. Use the graph relating electrode potential to silver ion concentration (Exercise 295) to read the value of log[$Ag^+(aq)$] corresponding to an electrode potential of 0.34 V. Take the antilogarithm.

$$\log[Ag^+(aq)] = -8.1$$

$$\therefore \quad [Ag^+(aq)] = \text{antilog}(-8.1) = 7.9 \times 10^{-9} \text{ mol dm}^3$$

3B. Calculate [$Ag^+(aq)$] from the Nernst equation.

$$E = E^\ominus + \frac{0.059 \text{ V}}{n} \log[\text{ion}]$$

$$0.34 \text{ V} = 0.80 \text{ V} + 0.059 \text{ V} \times \log[Ag^+(aq)]$$

$$\therefore \quad \log[Ag^+(aq)] = \frac{0.34 \text{ V} - 0.80 \text{ V}}{0.059 \text{ V}} = -7.8$$

$$\therefore \quad [Ag^+(aq)] = \text{antilog}(-7.8) = 1.6 \times 10^{-8} \text{ mol dm}^{-3}$$

4. Substitute in the equilibrium law expression:

$$K_c = \frac{[Cu^{2+}(aq)]}{[Ag^+(aq)]^2} = \frac{1.0 \text{ mol dm}^{-3}}{(7.9 \times 10^{-9} \text{ mol dm}^{-3})^2} = \boxed{1.6 \times 10^{16} \text{ dm}^3 \text{ mol}^{-1}}$$

$$\text{or} \quad \frac{1.0 \text{ mol dm}^{-3}}{(1.6 \times 10^{-8} \text{ mol dm}^{-3})^2} = \boxed{3.9 \times 10^{15} \text{ dm}^3 \text{ mol}^{-1}}$$

The large value of K_c shows that the reaction goes virtually to completion.

So far, we have simplified the situation by assuming that if $\Delta E^\ominus$ is positive the reaction 'goes' and if $\Delta E^\ominus$ is negative the reaction does not 'go'. It is better to say that the more positive $\Delta E^\ominus$, the larger the value of K_c, and the more negative $\Delta E^\ominus$, the smaller the value of K_c. For values of $\Delta E^\ominus$ between about +0.3 V and -0.3 V, the values of K_c are such that the equilibrium mixtures contain significant amounts of products and reactants.

Calculate some equilibrium constants in the next exercise. Our answers show only the graphical method for (a) and (b), but you may wish to check them using the Nernst equation. However, do not expect precise agreement.

Exercise 303 Calculate the equilibrium constant K_c at 298 K for the following reactions.

(a) $Fe^{2+}(aq) + Ag^{+}(aq) \rightleftharpoons Fe^{3+}(aq) + Ag(s)$

(b) $Cu^{+}(aq) + Ag^{+}(aq) \rightleftharpoons Cu^{2+}(aq) + Ag(s)$

(c) $Co(s) + Pb^{2+}(aq) \rightleftharpoons Co^{2+}(aq) + Pb(s)$

Exercise 304 At room temperature the electrode potential for the system $Fe^{3+}(aq),Fe^{2+}(aq) \mid Pt$ is given by the equation:

$$E = E^{\ominus} + 0.06\ \text{V} \log_{10}\frac{[Fe^{3+}]}{[Fe^{2+}]} \quad \ldots\ldots\ldots\ldots\ldots\ldots\ldots (1)$$

For this system, $E^{\ominus} = 0.77$ V

Similarly, for the silver electrode, $Ag^{+}(aq) \mid Ag(s)$

$$E = E^{\ominus} + 0.06 \log_{10}[Ag^{+}] \quad \ldots\ldots\ldots\ldots\ldots\ldots\ldots (2)$$

and $E^{\ominus} = 0.80$ V

(a) Calculate the electrode potential of the half-cell $Fe^{3+}(aq),Fe^{2+}(aq) \mid Pt$ when the iron(III) ion concentration is 1.0 M and the iron(II) ion concentration is 0.6 M.

(b) What concentration of silver ions (Ag^{+}) in contact with metallic silver would give the same electrode potential as you have calculated in (a)?

(c) What would happen if a solution 1.0 M with respect to Fe^{3+}, 0.6 M with respect to Fe^{2+}, and containing silver ions of the concentration you have found in (b), was in contact with metallic silver? Explain.

(d) Use equations (1) and (2) to derive a value for the equilibrium constant, K_c, for the reaction

$$Fe^{3+}(aq) + Ag(s) \rightleftharpoons Fe^{2+}(aq) + Ag^{+}(aq)$$

The equilibrium constant, K_c, for a redox reaction may be found more directly by using a relationship derived from the Nernst equation.

The relationship between $\Delta E^{\ominus}$ and K_c

We now derive this relationship, using standard electrode potentials and the Nernst equation. Consider the cell:

$$Zn(s) \mid Zn^{2+}(aq) \,\vdots\, Cu^{2+}(aq) \mid Cu(s); \quad \Delta E^{\ominus} = +1.1\ \text{V}$$

Suppose the electrodes were connected with wire and the cell left to 'run down' until $\Delta E = 0$. In this condition, the cell is in equilibrium:

$$Zn(s) + Cu^{2+}(aq) \rightleftharpoons Zn^{2+}(aq) + Cu(s)$$

The equilibrium constant for this reaction is given by the expression:

$$K_c = \frac{[Zn^{2+}(aq)]}{[Cu^{2+}(aq)]}$$

The Nernst equation for the zinc electrode is:

$$E(Zn^{2+}|Zn) = E^{\ominus}(Zn^{2+}|Zn) + \frac{2.3\ RT}{nF}\log[Zn^{2+}(aq)] \quad\ldots\ldots\ldots\ldots\ldots\ldots\ldots (1)$$

Similarly, the Nernst equation for the copper electrode is:

$$E(Cu^{2+}|Cu) = E^{\ominus}(Cu^{2+}|Cu) + \frac{2.3\ RT}{nF}\log[Cu^{2+}(aq)] \quad\ldots\ldots\ldots\ldots\ldots\ldots\ldots (2)$$

At equilibrium, $E(Zn^{2+}|Zn) = E(Cu^{2+}|Cu)$

$\therefore$ expression (1) = expression (2), i.e.

$$E^{\ominus}(Cu^{2+}|Cu) + \frac{2.3\ RT}{nF}\log[Cu^{2+}(aq)] = E^{\ominus}(Zn^{2+}|Zn) + \frac{2.3\ RT}{nF}\log[Zn^{2+}(aq)]$$

Substituting this equation into the expression for the standard e.m.f. of this cell:

$$\Delta E^{\ominus} = E^{\ominus}[Cu^{2+}|Cu] - E^{\ominus}(Zn^{2+}|Zn)$$

$$= \frac{2.3\ RT}{nF}\log[Zn^{2+}(aq)] - \frac{2.3\ RT}{nF}\log[Cu^{2+}(aq)]$$

$$= \frac{2.3\ RT}{nF}\log\frac{[Zn^{2+}(aq)]}{[Cu^{2+}(aq)]}$$

$$\therefore\quad nF\ \Delta E^{\ominus} = 2.3\ RT\ \log\frac{[Zn^{2+}(aq)]}{[Cu^{2+}(aq)]}$$

i.e.

$$\boxed{nF\ \Delta E^{\ominus} = 2.3\ RT\ \log K_c}$$

Substitute values in this expression in the following exercises.

Exercise 305 The following cell is set up and short-circuited:

$$Cu(s)\ |\ Cu^{2+}(aq)\ \vdots\ Br_2(aq), 2Br^-(aq)\ |\ Pt$$

(a) Write an equation for the resulting cell equilibrium.

(b) Calculate K_c for this equilibrium and comment on the result.

Exercise 306 The oxidation of iron(II) ions by aqueous acidified dichromate(VI) ions proceeds according to the equation:

$$6Fe^{2+}(aq) + Cr_2O_7^{2-}(aq) + 14H^+(aq) \rightleftharpoons 6Fe^{3+}(aq) + 2Cr^{3+}(aq) + 7H_2O(l)$$

(a) Write half-equations for the two component half-reactions.

(b) Write a cell diagram for an electrochemical cell in which this reaction occurs.

(c) Calculate the standard e.m.f. of the cell at 298 K.

(d) Calculate the value of the equilibrium constant, K_c.

Exercise 307 For the following reactions at 25 °C, calculate the equilibrium constant, K_c, from values of $E^{\ominus}$.

(a) $Ag^+(aq) + Fe^{2+}(aq) \rightleftharpoons Ag(s) + Fe^{3+}(aq)$

(b) $3Cl_2(aq) + 2Cr^{3+}(aq) + 7H_2O(l) \rightleftharpoons 6Cl^-(aq) + Cr_2O_7^{2-}(aq) + 14H^+(aq)$

We now show you another application of the relationship between the concentration of an ion and the electrode potential, the measurement of pH.

Measuring the pH of a solution

In the exercise which follows you see how it is possible to determine the pH of a solution using hydrogen electrodes. The apparatus is shown in Fig. 23.

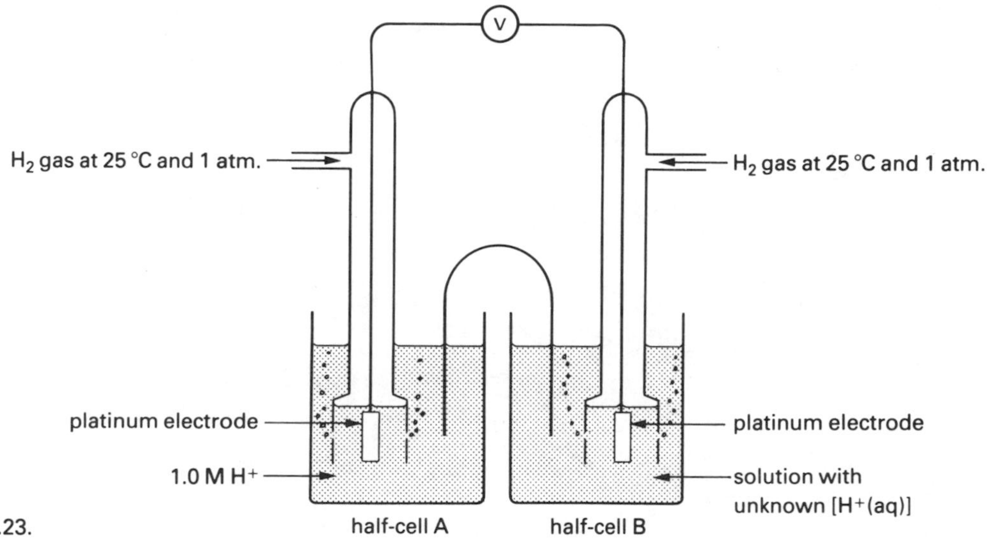

Fig.23.

Exercise 308 The apparatus in Fig. 23 was set up by a student to determine the pH of an unknown solution, X.

Half-cell A was kept as the standard hydrogen electrode throughout the experiment, but solutions of different hydrogen ion concentration were used in half-cell B. The e.m.f. of the cell was measured with each change in half-cell B; the following results were obtained:

$[H^+(aq)]/mol\ dm^{-3}$	0.10	0.050	0.020	0.0040	0.0010
$\Delta E/V$	+0.060	+0.078	+0.102	+0.144	+0.180

Solution X was placed in half-cell B and the e.m.f. of the cell was found to be +0.132 V.

(a) Using the results in the table, plot e.m.f. against $-\log[H^+(aq)]$.

(b) From the graph, determine the pH of solution X. ($pH = -\log[H^+(aq)]$.)

The procedure described in Exercise 308 above could, therefore, in principle, be used in the measurement of pH. However, you have already read about the difficulties of using the hydrogen electrode so, in practice, alternative electrode systems are used.

The theory of the glass electrode is not required for A-level, but you can read about it if you wish in a textbook of physical chemistry.

END-OF-CHAPTER QUESTIONS

Exercise 309 Some standard reduction potentials (in volts at 298 K) and the reactions to which they apply are given below:-

	$E^{\ominus}$/V	Electrode reaction
$Li^{+}(aq) \mid Li(s)$	-3.04	$Li^{+} + e^{-} \rightleftharpoons Li$
$Zn^{2+}(aq) \mid Zn(s)$	-0.76	$Zn^{2+} + 2e^{-} \rightleftharpoons Zn$
$Cu^{2+}(aq), Cu^{+}(aq) \mid Pt$	+0.17	$Cu^{2+} + e^{-} \rightleftharpoons Cu^{+}$
$Cu^{2+}(aq) \mid Cu(s)$	+0.34	$Cu^{2+} + 2e^{-} \rightleftharpoons Cu$
$I_2(aq), 2I^{-}(aq) \mid Pt$	+0.54	$I_2 + 2e^{-} \rightleftharpoons 2I^{-}$
$Fe^{3+}(aq), Fe^{2+}(aq) \mid Pt$	+0.74	$Fe^{3+} + e^{-} \rightleftharpoons Fe^{2+}$
$Ag^{+}(aq) \mid Ag(s)$	+0.80	$Ag^{+} + e^{-} \rightleftharpoons Ag$

(a) Define 'standard reduction potential' (standard electrode potential).

(b) (i) Which is the strongest oxidizing agent of all the species given?

(ii) Which species in the table can be oxidized by aqueous iodine?

(c) The following diagram represents an electrochemical system:

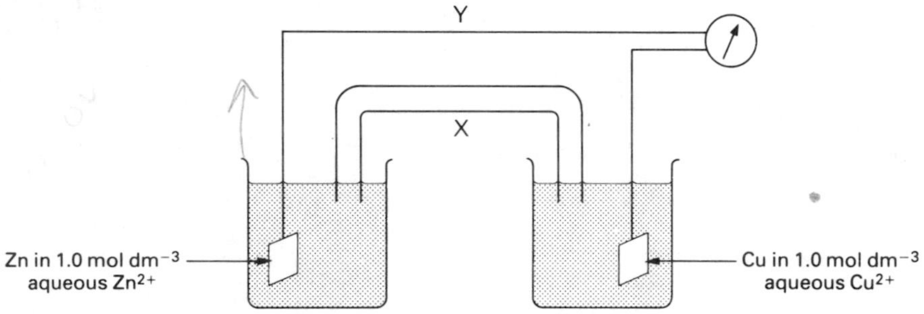

Fig.24.

(i) Explain why it is necessary to have two separate vessels joined by a bridge, X.

(ii) Calculate the e.m.f. of the cell from the data given.

(iii) Give the chemical equation for the overall reaction taking place in the cell.

(iv) State the direction of electron flow in the outer circuit Y, and explain why there is electron flow in this direction.

Exercise 310 The standard electrode potentials of four half-reactions are given below:

$Sn^{2+}(aq) + 2e^{-} \rightleftharpoons Sn(s)$; $E^{\ominus} = -0.14$ V

$Fe^{3+}(aq) + e^{-} \rightleftharpoons Fe^{2+}(aq)$; $E^{\ominus} = +0.77$ V

$2Hg^{2+}(aq) + 2e^{-} \rightleftharpoons Hg_2^{2+}(aq)$; $E^{\ominus} = +0.92$ V

$\frac{1}{2}Br_2(aq) + e^{-} \rightleftharpoons Br^{-}(aq)$; $E^{\ominus} = +1.07$ V

Based on this information, which of the following reactions will probably take place?

(a) $Sn(s) + 2Hg^{2+}(aq) \rightarrow Sn^{2+}(aq) + Hg_2^{2+}(aq)$

(b) $2Fe^{2+}(aq) + 2Hg^{2+}(aq) \rightarrow 2Fe^{3+}(aq) + Hg_2^{2+}(aq)$

(c) $2Br^-(aq) + Sn^{2+}(aq) \rightarrow Br_2(aq) + Sn(s)$

(d) $2Fe^{2+}(aq) + Br_2(aq) \rightarrow 2Fe^{3+}(aq) + 2Br^-(aq)$

Exercise 311

(a) Write down the equations for the half-cell reactions corresponding to these three standard electrode potentials:

$Cr^{3+}(aq),Cr^{2+}(aq) \mid Pt;\quad E^\ominus = -0.41\ V$

$Ce^{4+}(aq),Ce^{3+}(aq) \mid Pt;\quad E^\ominus = +1.70\ V$

$[Cr_2O_7^{2-}(aq) + H^+(aq)],[Cr^{3+}(aq) + H_2O(l)] \mid Pt;\quad E^\ominus = +1.33\ V$

(b) Find the e.m.fs. of the following cells (all concentrations are 1.0 mol dm^{-3}):

(i) $Pt \mid Cr^{2+}(aq),Cr^{3+}(aq) \;\vdots\; Ce^{4+}(aq),Ce^{3+}(aq) \mid Pt$

(ii) $Pt \mid Cr^{3+}(aq) + H_2O(l)],[Cr_2O_7^{2-}(aq) + H^+(aq)]$
$\vdots\; Ce^{4+}(aq),Ce^{3+}(aq) \mid Pt$

(c) Write down the cell reactions which would occur in cells (i) and (ii) in part (b) above if the cells were short-circuited.

Exercise 312

A

This question concerns the equilibrium between solid zinc carbonate and its ions in aqueous solution at 25 °C.

$ZnCO_3(s) \rightleftharpoons Zn^{2+}(aq) + CO_3^{2-}(aq)$

In order to obtain a value of the solubility product, K_{sp}, for zinc carbonate at 25 °C, a cell was set up using a zinc electrode in zinc carbonate suspension as one half-cell and a copper electrode in a solution of copper sulphate of concentration 1 mol dm^{-3} as the other half-cell. The e.m.f. obtained was 1.26 V.

(a) How would you check that equilibrium had been established?

(b) Give the expression for K_{sp} for zinc carbonate.

(c) (i) Write the cell diagram for the complete cell system.

(ii) Give the sign for the voltage corresponding to the cell diagram which you have written in (i).

(d) How would you make a suitable salt bridge for the system?

(e) At 25 °C, the value of the electrode potential, E, in a solution of known concentration is related to the standard electrode potential, $E^\ominus$, by the relation

$$E = E^\ominus + \frac{0.06\ V}{z}\ lg[ion]$$ (z = no. of electrons transferred.)

The standard redox potentials are:

$Zn^{2+}(aq) \mid Zn(s);\quad E^\ominus = -0.76\ V$

$Cu^{2+}(aq) \mid Cu(s);\quad E^\ominus = +0.34\ V$

(i) Calculate the concentration of zinc ions, $Zn^{2+}(aq)$, in the equilibrium mixture.

(ii) Hence, calculate K_{sp} for zinc carbonate.

Exercise 313 The graph below shows the variation of the redox potential of the $Cu^{2+}(aq)/Cu(s)$ electrode with the logarithm of the copper(II) ion concentration. Use the graph to determine the standard redox potential of the $Cu^{2+}(aq)/Cu(s)$ electrode.

A part

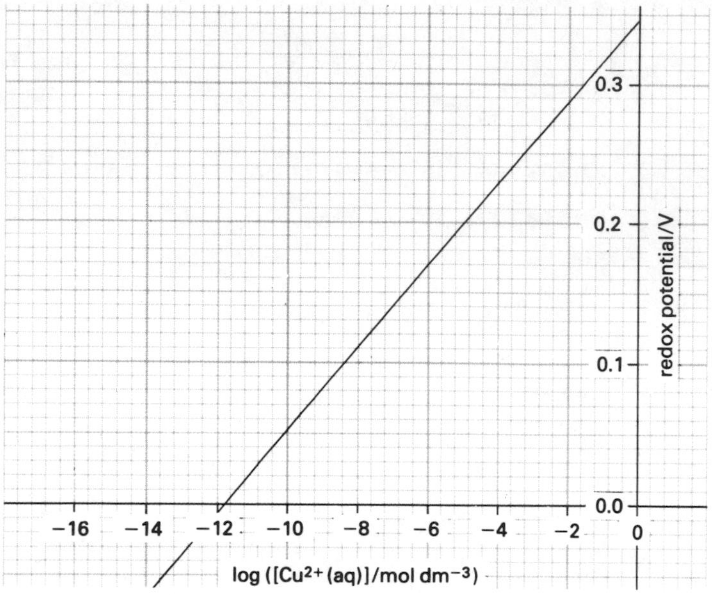

Fig.25.

Exercise 314 The table below gives some standard redox potentials.

A part

Half-cell reaction	$E^\ominus$/V
$Ce^{4+}(aq) + e^- \rightleftharpoons Ce^{3+}(aq)$	+1.45
$Ag^+(aq) + e^- \rightleftharpoons Ag(s)$	+0.80
$Fe^{3+}(aq) + e^- \rightleftharpoons Fe^{2+}(aq)$	+0.77
$Cu^{2+}(aq) + 2e^- \rightleftharpoons Cu(s)$	+0.34
$Fe^{2+}(aq) + 2e^- \rightleftharpoons Fe(s)$	-0.44
$Zn^{2+}(aq) + 2e^- \rightleftharpoons Zn(s)$	-0.76

(a) Using the data in the table:

(i) select two half-cell reactions that may be combined to produce a cell with a standard e.m.f. of +1.10 V. Write an overall equation for the cell reaction and indicate the direction of flow of electrons in a conductor connecting the cell terminals;

(ii) calculate the standard e.m.f. of a cell in which the following reaction occurs and explain the reaction in terms of oxidation and reduction;

$$Zn(s) + Fe^{2+}(aq) \rightarrow Zn^{2+}(aq) + Fe(s).$$

(b) The dependence of redox potential, E, on the ion concentration at 300 K, is given by

$$E = E^\ominus + \frac{0.06\ V}{n} \log\frac{[\text{oxidized form}]}{[\text{reduced form}]}$$

where $E^\ominus$ is the standard redox potential and n the number of moles of electrons in the half-cell reaction.

Using this equation and data from the table above, calculate the equilibrium constant, K_c, for the reaction at 300 K:

$$Ce^{4+}(aq) + Fe^{2+}(aq) \rightleftharpoons Ce^{3+}(aq) + Fe^{3+}(aq)$$

VAPOUR PRESSURE AND COLLIGATIVE PROPERTIES

9

INTRODUCTION AND PRE-KNOWLEDGE

If a liquid is introduced into a closed container, some of it evaporates until equilibrium is reached. Taking water as an example:

$$H_2O(l) \rightleftharpoons H_2O(g)$$

The equilibrium constant, K_c, for this system is equal to the partial pressure of water, p_{H_2O}, which is known as the saturated vapour pressure of water. (The word 'saturated' is often omitted.)

Vapour pressure varies from substance to substance and also with temperature, as shown in Fig. 26. The diagram also shows that the boiling-point of a liquid is the temperature at which the vapour pressure is equal to atmospheric pressure (760 mmHg). Liquids with lower vapour pressures have higher boiling points, and vice versa.

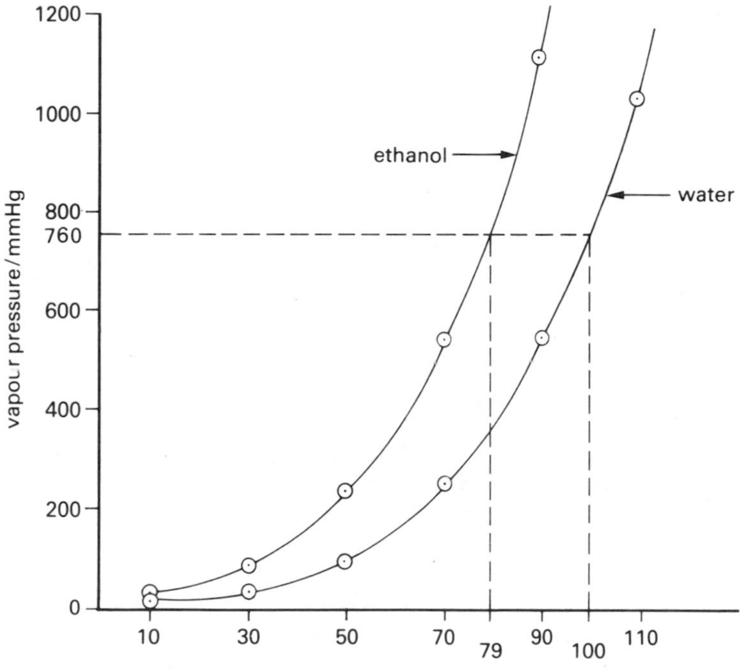

Fig. 26.

In the following sections, we assume that you are familiar with Dalton's law of partial pressures and the concept of mole fraction (see Chapter 5).

THE VAPOUR PRESSURE OF TWO-COMPONENT SYSTEMS

The vapour pressure of a mixture of two liquids is the sum of two partial vapour pressures, one for each liquid. If the liquids are similar to one another, for instance, hexane and heptane, the variation of total vapour pressure with mole fraction of each component is linear, as shown in Fig. 27.

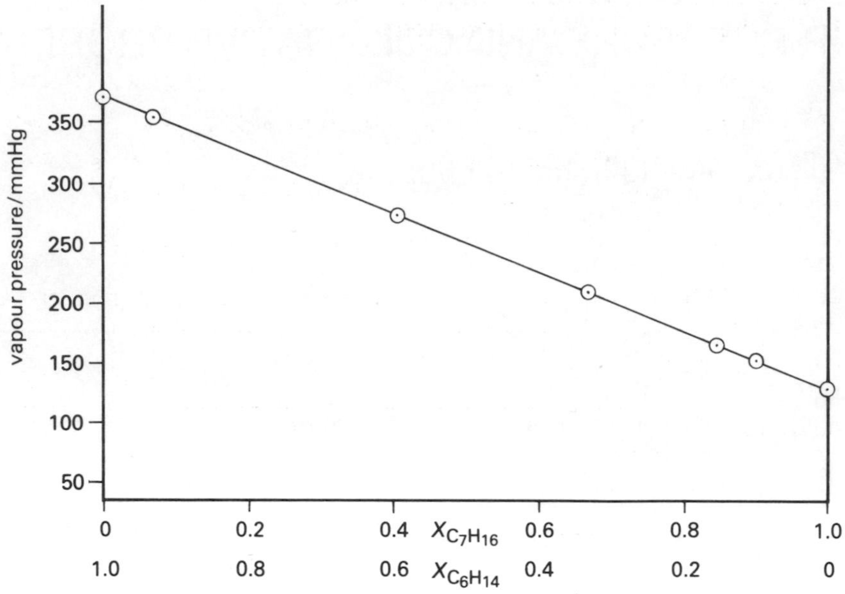

Fig.27.

Such a system is said to be an 'ideal mixture' or an 'ideal solution'. When the vapour pressure curve is not linear, which is more often the case, the mixture is 'non-ideal'. You will probably study non-ideal solutions from a qualitative point of view, but we shall deal with calculations only for ideal solutions.

The linear relationship between total vapour pressure and composition is often expressed as Raoult's law, which we now consider.

RAOULT'S LAW

Raoult's law states that the partial vapour pressure of any volatile component, A, of an ideal solution is equal to the vapour pressure of pure A multiplied by the mole fraction of A in the solution.

This law can be expressed mathematically, as follows:

$$p_A = p_A^o X_A$$

where p_A = partial vapour pressure of A in the solution
p_A^o = vapour pressure of pure A
X_A = mole fraction of A in the solution.

Similarly, for liquid B we have: $p_B = p_B^o X_B$

You met a formula very similar to this in Chapter 5 (Gases) when you considered methods of calculating partial pressures. Also in that chapter, you learned about Dalton's law of partial pressures. Since our ideal solution (A,B) consists of two volatile components, they both must contribute to the total vapour pressure, p_T.

Thus, $p_T = p_A + p_B$ (Dalton's law of partial pressures)

or $p_T = p_A^o X_A + p_B^o X_B$

In the next exercise you will use Raoult's law to construct a vapour-pressure/composition curve for the hexane/heptane mixture.

Exercise 315 (a) Use Raoult's law to determine the partial pressures of hexane, C_6H_{14}, and heptane, C_7H_{16}, at the mole fractions shown in the table, given that, at 47 °C,

$$p^o_{hexane} = 372 \text{ mmHg} \quad \text{and} \quad p^o_{heptane} = 127 \text{ mmHg}$$

Enter your answers in a copy of the table.

X_{hexane}	0.00	0.200	0.400	0.600	0.800	1.00
p_{hexane}						
$X_{heptane}$	0.00	0.200	0.400	0.600	0.800	1.00
$p_{heptane}$						

(b) Plot a graph of partial pressure of (i) hexane, and (ii) heptane, against mole fraction. Mark the x-axis 0→1 for $X_{heptane}$ and 1→0 for X_{hexane}.

(c) Now plot total vapour pressure, p_T, against mole fraction on the same graph. Use the equation

$$p_T = p_{hexane} + p_{heptane}$$

(d) What is the total vapour pressure of an equimolar mixture?

(e) What is the composition of a mixture which has a vapour pressure of 200 mmHg?

You can see that for two liquid mixtures which obey Raoult's law, the vapour pressure curve can be drawn simply by joining the vapour pressures of pure components with a straight line, as in Fig. 27 (page 168).

The calculations are similar in the following exercise, but you need not plot a graph.

Exercise 316 Two liquids, A and B, have vapour pressures of 75 mmHg and 130 mmHg respectively, at 25 °C. What is the total vapour pressure of the following ideal mixtures?

(a) 1 mol of A and 1 mol of B.

(b) 3 mol of A and 1 mol of B.

(c) 1 mol of A and 4 mol of B.

Exercise 317 Hexane and heptane are totally miscible and form an ideal two-component system. If the vapour pressures of the pure liquids are 56 000 and 24 000 N m^{-2} at 51 °C calculate (a) the total vapour pressure, and (b) the mole fraction of heptane in the vapour above an equimolar mixture of hexane and heptane.

A part

As you saw in the last exercise, the vapour above a mixture does not have the same composition as the liquid - it is always richer in the more volatile component. This is the basis of fractional distillation, as is shown in the next exercise.

Exercise 318 Two miscible liquids, A and B, which form an ideal solution, have vapour pressures of 3 kPa and 10 kPa respectively.

(a) Plot a graph of vapour pressure against mole fraction of A and B in the liquid. Label this line 'liquid composition'.

(b) In the following table, for each value of mole fraction of B in the liquid, calculate the mole fraction of B in the vapour:

X_B(liquid)	0.1	0.2	0.4	0.6	0.8
Total vapour pressure/kPa					
X_B(vapour)					

(c) Plot vapour pressure against X_B(vapour). Label this line 'vapour composition'.

(d) Suppose that an equimolar mixture is boiled and the vapour condensed. Use the graph to determine the composition of the condensed liquid.

(e) Suppose the procedure in (d) is carried out twice more. What would be the composition of the condensed liquid now?

Repeated evaporation and condensation, as described in the last exercise, occurs in a fractionating column and results in the complete separation of two miscible liquids in an ideal solution. Two immiscible liquids may be separated by steam distillation, which also depends on differences in vapour pressure, as we show in the next section.

STEAM DISTILLATION

Consider a mixture of immiscible liquids such as nitrobenzene and water. Fig. 28 shows the vapour pressure curves for this mixture. Study the diagram and use it in the exercise which follows.

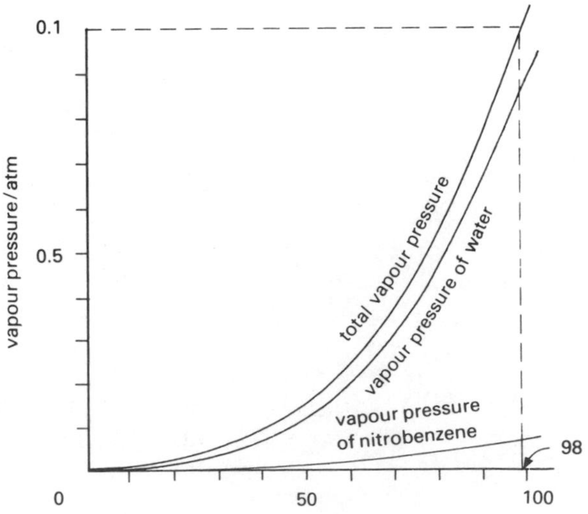

Fig.28.

Exercise 319 (a) In Fig. 28, can you see a relationship between the total vapour pressure of the mixture and the separate vapour pressures of water and nitrobenzene?

(b) What is the boiling-point of the mixture? Explain how you arrive at your answer.

(c) Is the boiling-point of the mixture higher or lower than the boiling-point of the individual pure liquids? Explain.

(d) At what temperature will the mixture steam-distil?

Water and another liquid, Y, undergoing steam distillation, each exert a vapour pressure proportional to the mole fraction present in the vapour. The distillate has the same composition as the vapour. It follows that:

$$\frac{p_{\text{water}}}{p_{\text{Y}}} = \frac{\text{amount of water}}{\text{amount of Y}} = \frac{\text{mass of water}}{18} \times \frac{\text{relative molecular mass of Y}}{\text{mass of Y}}$$

Use a rearranged form of this equation in the following exercises.

Exercise 320 In the extraction of bromobenzene by steam distillation at 101 kPa pressure, the mixture boils at 95.5 °C. At that temperature, the vapour pressure of pure water is 85.8 kPa. Calculate the percentage by mass of bromobenzene in the distillate.

A part

Exercise 321 (a) Explain the process of steam distillation, and state why it is used in the preparation of phenylamine (aniline) from nitrobenzene.

A

(b) The vapour pressures of pure water and pure phenylamine (anilino) at the stated temperatures are:

Temperature/°C	85	90	95	100	105
Vapour pressure/kPa of water /mmHg	57.9 434	70.1 526	84.5 634	101.3 760	120.5 906
Vapour pressure/kPa of water /mmHg	3.0 22.9	3.9 29.2	4.9 36.5	6.1 45.7	7.3 55.0

With the aid of suitable graphs

(i) find to the nearest °C the temperature at which a mixture of phenylamine (aniline) and water will steam-distil under standard atmospheric pressure.

(ii) Calculate the percentage by mass of phenylamine (aniline) which would be expected to be present in the distillate.

In the next section, we consider two-component systems consisting of a volatile solvent and a non-volatile solute.

RAOULT'S LAW FOR SOLUTIONS OF A NON-VOLATILE SOLUTE

The addition of a non-volatile solute (usually a solid) to a volatile solvent results in the lowering of the vapour pressure. Since the solute is non-volatile, it makes practically no contribution to the vapour pressure and may be thought of as decreasing the concentration of the solvent. We can apply Raoult's law to such solutions provided they are dilute enough for us to ignore the overall effect of interaction between solvent and solute. Dilute solutions can be regarded as ideal solutions.

In 1886, Raoult investigated the effect of non-volatile solutes on the vapour pressures of solvents quantitatively and discovered a very important relationship. You discover this for yourself by doing the next exercise.

Exercise 322 The table below gives values of the vapour pressure, p, of sugar solutions at various concentrations (i.e. at the stated mole fractions of water and sugar). There is also a column for the relative lowering of vapour pressure, which is given by the expression:

$$\frac{p^o - p}{p^o} \quad \begin{pmatrix} p^o = \text{vapour pressure of pure water} \\ p = \text{vapour pressure of sugar solution} \end{pmatrix}$$

X_{water}	X_{sugar}	p/mmHg	$\frac{p^o - p}{p^o}$
1.00	0.00	17.5	
0.950	0.050	16.6	
0.900	0.100	15.8	
0.850	0.150	14.9	

(a) Calculate values for the relative lowering of vapour pressure. What strikes you as significant about this?

(b) Your answer to (a) above should now enable you to write a general relationship between the lowering of the vapour pressure and the composition of a solution

(i) in words, (ii) mathematically.

Your answer to Exercise 322(b)(i) is a statement of Raoult's law. On initial inspection, this may not immediately seem related to the general expression of Raoult's law, which you met on page 168, i.e.

$$p_T = p_A^o X_A + p_B^o X_B$$

In fact, it is a specific case of the more general expression, and we show this in the next two exercises.

Exercise 323 (a) Use the values in Exercise 322 to plot a graph of vapour pressure of sugar solution (y-axis) against mole fraction of water (x-axis).

(b) What conclusion can you draw from this graph?

You have shown in your graph (Exercise 323) that the vapour pressure of the solution is proportional to the mole fraction of water:

$$p_{soln} \propto X_{H_2O} \text{ ....(1)}$$

The vapour above the sugar solution consists entirely of water vapour because sugar is non-volatile. The vapour pressure of the solution is, therefore, the partial pressure of the water vapour.

Since $p_{soln} = p_{H_2O}$, expression (1) now becomes:

$p_{H_2O} \propto X_{H_2O}$ or $p_{H_2O} = kX_{H_2O}$..........(2) where k is a constant.

In the next exercise, you will determine a value for the constant, k.

Exercise 324

(a) Measure the slope of the graph obtained from Exercise 323. This will give you a value for the constant, k, in equation (2) above.

(b) Look at the table in Exercise 322; to which quantity does the constant, k, in equation (2) correspond?

(c) Using your answer to (b), rewrite equation (2).

(d) Remembering that sugar is non-volatile, show that your equation in (c) is a special case of the equation:

$$p_T = p_A^oX_A + p_B^oX_B$$

(e) Raoult's law for solutions of non-volatile solutes is more usually expressed as:

$$\frac{p^o - p}{p^o} = X_{solute}$$

Show that the equation in (c) can be obtained by rearranging this expression.

Since the lowering of the vapour pressure is proportional to the mole fraction of the solute, it is clear that the effect depends on the number of solute particles, not on their nature. It follows that if a solute dissociates (e.g. an ionic salt) the effect is increased and if a solute associates (e.g. the dimerization of certain acids in non-polar solvents) the effect is decreased. Take this into account in the next exercise.

Exercise 325

The vapour pressure of water at 70 °C is 31.2 kPa. Calculate the vapour pressures of the following solutions at the same temperature, assuming that they obey Raoult's law.

(a) 9.0 g of glucose (M_r = 180) in 50 g of water.

(b) 1.0 M NaCl (assume complete dissociation)

(c) 1.0 M $MgCl_2$ (assume complete dissociation)

As you would expect from your previous study in this chapter, this lowering of vapour pressure is accompanied by a rise in boiling-point. You will see in the next section that, in addition, the freezing-point is lowered.

Elevation of boiling-point and depression of freezing-point

Several different (but closely related) expressions appear in textbooks to describe the relationship between the elevation of boiling-point and the concentration of non-volatile solutes. We use the following:

$$\Delta T_b = K_b \times \frac{m}{M}$$

where ΔT_b = elevation of boiling temperature (unit: K)

K_b = boiling-point (ebullioscopic) constant (unit: K kg mol^{-1})

m = mass of solute per kg of solvent (unit: g kg^{-1})

M = molar mass of solute (unit: g mol^{-1})

You can see the link between the elevation of boiling-point and Raoult's law by doing the following exercises.

Exercise 326 **A**

(a) State Raoult's law as applied to dilute solutions.

(b) In the same diagram, sketch the two curves which show the effect of temperature on the vapour pressures of (i) a pure solvent, and (ii) a solution of a non-volatile solute in the same solvent.

Account for the difference between these curves and use them to explain the difference in boiling-points between the solvent and the solution.

(c) Explain why the vapour pressure above a solvent rises with an increase in temperature.

(d) A solution was made by dissolving 8.4 g of a non-volatile organic compound X (relative molecular mass = 168) in 69 g of ethanol. Calculate:

(i) the mole fraction of X in the solution,

(ii) the vapour pressure of the solution at 20 °C given that the vapour pressure of pure ethanol at this temperature is 26.6 kPa,

(iii) the boiling-point of the solution, given that ethanol boils at 78 °C and the boiling-point constant for ethanol is 1.15 K kg mol^{-1}.

The next exercise concerns the vapour pressure curves for pure water and two dilute solutions which are shown in Fig. 29.

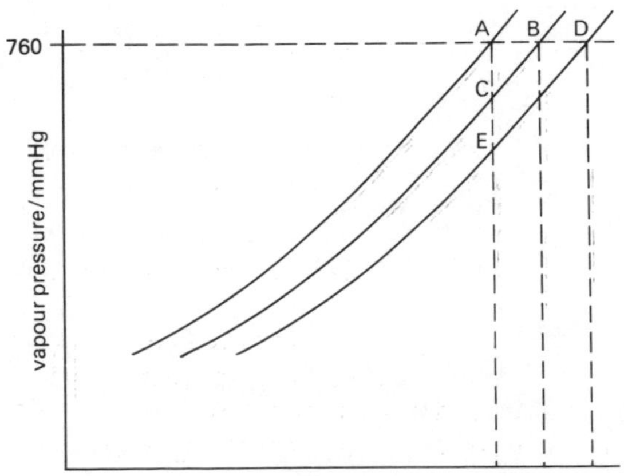

Fig.29.

Exercise 327 Because the solutions are dilute, lines BC and DE in Fig. 29 can be assumed to be straight and parallel. Use this assumption to show how the expression for the elevation of boiling-point (page 174) can be derived from Raoult's law.

The lowering of the vapour pressure of a solvent and the elevation of its boiling-point depend only on the number of solute particles in a near-ideal solution and not on the identity of the solute. Thus, the change in boiling-point of a solvent for the addition of any solute is unique to that solvent.

For example, the boiling-point constant of water is 0.52 K kg mol^{-1}, which means that if 1 mol of any solute is added to 1 kg of water, its boiling-point is raised by 0.52 K. From such constants it is, therefore, possible to calculate the molar mass of a solute. Do this in the following exercises.

Exercise 328 Calculate the molar mass of a solute given that 4.6 g of it raises the boiling-point of 56 g of water by 0.71 °C. (The boiling-point constant of water is 0.52 K kg mol^{-1}.)

Exercise 329 Calculate the molar mass of the solute in each case described below.

	Mass of solute/g	Solvent	Mass of solvent/g	K_b /K kg mol^{-1}	ΔT /K
(a)	2.11	Water	100	0.52	0.15
(b)	1.82	Ethanol	58.7	1.15	0.55
(c)	1.07	Benzene	49.1	1.53	0.67
(d)	4.13	Propanone	71.6	1.71	1.07

The next exercise points out some additional features of the use of boiling-point elevation to determine molar mass.

Exercise 330 (a) Calculations from the elevation of boiling-point of solutions of sodium chloride suggest that the molar mass is 29 g mol^{-1} and not 58.5 g mol^{-1}. Why do you think this is?

(b) What would be the apparent molar mass of iron(III) chloride in aqueous solution, measured in the same way?

(c) Why is water not used as the solvent if another can be found? (Look at the data in Exercise 329.)

(d) What practical difficulty is encountered when investigating solutes of high molar mass?

In Exercise 326, you sketched two curves showing the effect of temperature on the vapour pressures of (a) a pure solvent, and (b) a solution of a non-volatile solute in the same solvent. In the next exercise, we extend the curves to indicate the freezing-point of the solvent and its solutions.

Exercise 331 Study Fig. 30 and answer the questions which follow.

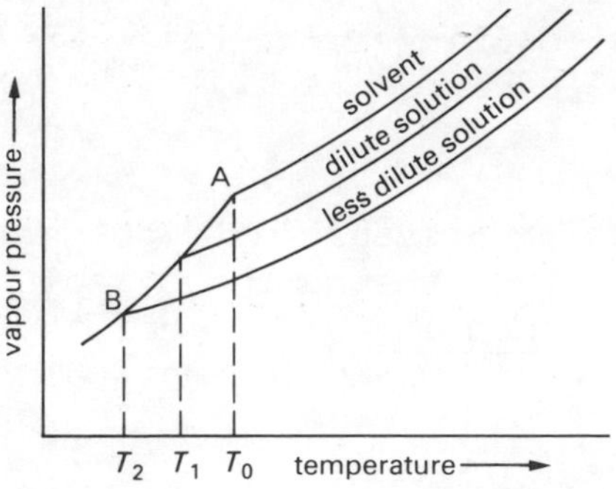

Fig. 30.

(a) What do the temperatures T_0, T_1 and T_2 represent?

(b) Which of the three liquids shown has the lowest freezing-point?

(c) What does the line AB represent?

From curves such as those shown in Fig. 30 above, it can be shown that the depression of freezing-point is directly proportional to the lowering of the vapour pressure. This is another statement of Raoult's law and leads to an equation similar to the one for boiling-point elevation:

$$\Delta T_f = K_f \times \frac{m}{M}$$

where ΔT_f = depression of freezing temperature (unit: K)

K_f = freezing-point (cryoscopic) constant (unit: K kg mol^{-1})

m = mass of solute per 1 kg of solvent (unit: g kg^{-1})

M = molar mass of solute (unit: g mol^{-1})

Substitute values in this expression in the following exercises.

Exercise 332 (a) The depression of freezing-point constant of concentrated sulphuric acid is 6.1 K kg mol^{-1}. State what is meant by this.

(b) Calculate the depression of freezing-point of concentrated sulphuric acid caused by dissolving 1 mol of concentrated nitric acid in 1000 g of concentrated sulphuric acid.

(Write a balanced equation first.)

Exercise 333 (a) 5.63 g of naphthalene, $C_{10}H_8$, dissolved in 100 g of benzene, depresses the freezing-point of benzene by 2.2 °C. Calculate the cryoscopic constant of benzene.

(b) 5.83 g of aluminium chloride dissolved in 100 g of benzene, depresses the freezing-point by 1.1 °C. Calculate the molar mass of aluminium chloride and comment on the result.

(C = 12.0, H = 1.0, Al = 17.0, Cl = 35.5)

Exercise 334 1.54 g of a compound, A, was mixed with 17.84 g of naphthalene, $C_{10}H_8$, and melted. On cooling, the mixture solidified at 48.2 °C. Pure naphthalene has a melting-point of 53.0 °C and a cryoscopic constant of 7.10 K kg mol^{-1}. Calculate the molar mass of A.

We end our discussion of colligative properties with a section on osmosis and osmotic pressure, which are of great interest to biologists in connection with the structure and function of cell membranes. For chemists, the measurement of osmotic pressure affords a method for determining large molar masses.

OSMOTIC PRESSURE

Osmosis is the process which occurs when a solvent and a solution in the same solvent are separated by a semi-permeable membrane, i.e. one which allows the passage of solvent particles but not solute particles. Solvent flows continuously into the solution until the concentration of the solution is vanishingly small, or until a pressure is built up to prevent the flow.

Osmotic pressure is the excess hydrostatic pressure which must be applied across a semi-permeable membrane in order to prevent osmosis occurring.

There is a close parallel between the osmotic pressure of an ideal solution and the gaseous pressure of an ideal gas. The osmotic pressure, Π, of a solution containing n mol of solute in a total volume V is given by the expression:

$$\Pi V = nRT$$

Substitute values in this expression in the following exercises.

Exercise 335 Calculate the osmotic pressure of each of the following aqueous solutions. (R = 0.0821 atm dm^3 K^{-1} mol^{-1}.)

(a) A 0.100 M solution of any non-electrolyte at 25 °C.

(b) 36.0 g of glucose, $C_6H_{12}O_6$, in 250 cm^3 of solution at 20 °C.

(c) 7.80 g of ethanamide, CH_3CONH_2, in 100 cm^3 of solution at 35 °C.

(d) 2.15 g of urea, $(NH_2)_2CO$, in 1500 cm^3 of solution at 20 °C.

Exercise 336 (a) An aqueous solution at 15 °C has an osmotic pressure of 1.12 atm. What would be the new osmotic pressure if the volume is doubled by adding water and the temperature raised to 25 °C?

(b) Calculate the molar mass of a non-electrolyte which has an osmotic pressure of 3.45 atm at 35 °C in a solution containing 42.1 g per dm^3.

(c) 72.0 g of a substance was dissolved in 525 cm^3 of water at 30 °C. The solution had an osmotic pressure of 2.89 atm. What is the molar mass of the substance?

Exercise 337 The following table gives values of osmotic pressure at 25 °C for various concentrations of naphthalene in methylbenzene (toluene).

Plot the osmotic pressure of these solutions against concentration, and use the graph to calculate the relative molar mass of naphthalene.

Concentration/g dm^{-3}	Osmotic pressure/kPa
0.5	10.0
1.0	20.0
1.5	28.0
2.0	37.0
2.5	46.0
3.0	56.0
4.0	74.0
5.0	92.0
6.0	110.0

Osmotic pressure is a colligative property, like the elevation of boiling-point and depression of freezing-point. Consequently, the effect is increased if the solute dissociates or ionizes. Remember this in the following exercises.

Exercise 338 What would be the osmotic pressure of the following solutions at 25 °C?

(a) 0.001 M $C_6H_{12}O_6$

(b) 0.001 M NaCl (assume complete ionization)

(c) 0.001 M $Fe_2(SO_4)_3$ (assume complete ionization).

Osmotic pressure is a much larger effect than elevation of boiling-point or depression of freezing-point. Consequently, a very small amount (in moles) of solute gives a measurable osmotic pressure. This makes it possible to determine much higher molar masses by osmotic pressure measurements than by means of the other colligative effects.

Even so, when the molar mass is very high, say greater than 10 000 g mol^{-1}, solutions which give measurable osmotic pressures cannot really be described as ideal, because they are not dilute, and the expression $\Pi V = nRT$ does not hold. Rearranging this expression gives:

$$\Pi V = \frac{m}{M}RT \quad \text{or} \quad \frac{\Pi V}{m} = \frac{RT}{M} \quad \text{or} \quad \frac{\Pi}{c} = \frac{RT}{M} \quad (c = \text{concentration of solute})$$

Π/c can be measured for a series of non-ideal solutions of different concentrations, c. An 'ideal' value of Π/c can be obtained by plotting Π/c against c and extrapolating to $c = 0$. By equating this value of Π/c to RT/M, a value of M can be obtained, as in the next exercise.

Exercise 339 The osmotic pressures (Π) of a series of solutions of different concentration (c) of a sample of polystyrene in butanone are measured at 27 °C. The height of butanone records the pressure (cm) for each concentration (g cm^{-3}), the density (d) of butanone at 27 °C being 0.80 g cm^{-3}. If the intercept at c = 0 of a plot of Π/c against c is 110 cm^4 g^{-1}, calculate the average molar mass of the sample of polystyrene. Care should be exercised over units.

R = 8.31 J mol^{-1} K^{-1} J = kg m^2 s^{-2} g = 981 cm s^{-2}

END-OF-CHAPTER QUESTIONS

Exercise 340 Mixtures of water and methanol obey Raoult's vapour pressure law.

(a) State Raoult's law.

(b) If the vapour pressure of pure water at 298 K is 24 mmHg, calculate the partial vapour pressure of water in a mixture of 36 g water and 32 g methanol at this temperature.

(H = 1, C = 12, O = 16)

Exercise 341 (a) The vapour pressure of water is 7.38 kPa at 40 °C. What is the new vapour pressure after dissolving:

(i) 5 g of glucose, $C_6H_{12}O_6$, in 100 g of water,

(ii) 5 g of calcium chloride, $CaCl_2$, in 200 g of water?

(b) Calculate the freezing-points and boiling-points of each solution. (For water, K_b = 0.52 K kg mol^{-1}, K_f = 1.86 K kg mol^{-1}.)

Exercise 342 Describe how you would measure the relative molecular mass of a compound either by the method of boiling-point elevation or by the method of freezing-point depression.

These two properties, among others, are described as colligative properties. Explain the meaning of the term colligative property and give two other examples of colligative properties.

A solution of mercury(II) nitrate containing 3.270 g in 600 g of water has a freezing-point of -0.093 °C and a solution of mercury(II) chloride containing 8.131 g in 750 g of water has a freezing-point of -0.075 °C. Calculate the apparent relative molecular mass of each of the salts and discuss the significance of your results.

(Freezing-point depression constant = 1.86 K kg mol^{-1})

Exercise 343 The boiling-point of pure benzene is 80.20 °C at atmospheric pressure. When 1.59 g of phosphorus is dissolved in 50.0 g of benzene, the boiling-point is raised to 80.85 °C. What is the molecular formula of phosphorus in benzene? (K_b for benzene = 2.54 K kg mol^{-1})

A

Exercise 344

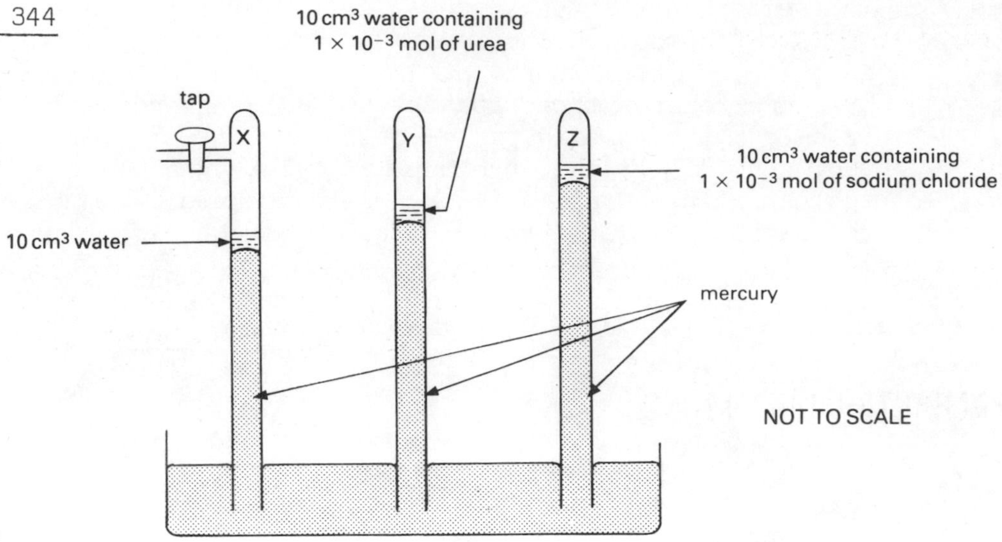

Fig.31.

(a) (i) What is to be found in each of the spaces marked X, Y and Z?

(ii) Explain the differences in mercury levels in the three tubes in terms of the properties of dilute solutions.

(iii) What changes, if any, would occur to the mercury levels if the apparatus was placed in a thermostatic bath maintained at a higher temperature? Explain your answer.

(iv) If the pressure in X was reduced by means of a vacuum pump, explain what would happen.

(b) Calculate the boiling-point of an aqueous solution of urea, $CO(NH_2)_2$, of concentration 12.0 g dm^{-3} at a pressure of 101.3 kPa. Assume that the volume of the solute is negligible compared to that of the solution, and the boiling-point elevation constant for water is 0.52 K mol^{-1} kg.

(c) (i) Write an expression for the mole fraction of a solute in solution.

(ii) Calculate the mole fraction of sodium chloride in an aqueous solution containing 10 g of sodium chloride per 100 g of water.

Exercise 345 Calculate the osmotic pressures of the following:

(a) 2.10 g of sucrose, $C_{12}H_{22}O_{11}$ in 50.0 cm^3 of aqueous solution at 36 °C.

(b) 100 cm^3 of an aqueous solution at 20 °C containing 1.00 g of a protein of molar mass 6,000 g mol^{-1}.

Exercise 346 The osmotic pressure at 35 °C of 500 cm^3 of a solution containing 7.50 g of a certain protein is 2.15 kPa. What is the approximate molar mass of the protein? (R = 82.1 dm^3 kPa K^{-1} mol^{-1}.)

How could you obtain a more accurate value?

REACTION KINETICS

10

INTRODUCTION AND PRE-KNOWLEDGE

The rate at which a reaction proceeds may be defined in a number of ways and it is important to be clear which definition you use in a particular case. The most useful definition is the rate of increase in concentration of one specified product of the reaction. However, others are sometimes used because they are related to the means of measurement, e.g. the rate at which the mass of a reactant decreases, or the rate at which the volume of a gaseous product increases.

We assume that you have already investigated qualitatively the effect on the rate of reaction of such factors as concentration of reactant, surface area of solid reactant and the presence of a catalyst. In this chapter we consider some of these factors quantitatively. In particular, we introduce an expression known as the rate equation, which relates the rate of reaction at a constant temperature to the concentrations of the reactants.

We also assume that you can plot graphs accurately and obtain from them values for the slope at any point and the intercept.

THE RATE OF A REACTION

The study of reaction kinetics relies on experimental work. We begin by presenting you with a set of experimental results for a reaction in which a single substance (the reactant) breaks down. We will take you through a series of exercises on the data, to show you how to find the rate equation for the reaction. Our purpose is to illustrate the key ideas of reaction rate, order of reaction and rate constant.

2,4,6-trinitrobenzoic acid in solution loses carbon dioxide when heated, as shown by the following equation:

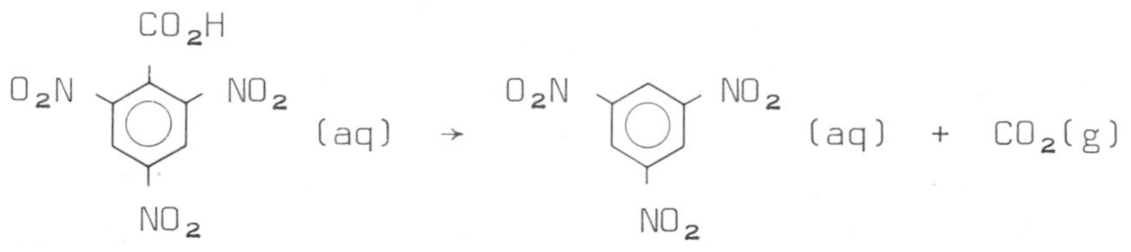

Since one of the products is gaseous, the rate of reaction could be studied by measuring the volume of carbon dioxide produced. Alternatively, you could follow the decrease in the amount of 2,4,6-trinitrobenzoic acid itself.

The scientists who did the experiment in 1931 (E.A. Moelwyn-Hughes and O.N. Hinshelwood) chose the second method. They set up several mixtures at 90 °C. After various reaction times they withdrew a sample and added a large volume of iced water to quench (i.e. stop) the reaction. They then titrated each mixture with 5.0×10^{-3} M barium hydroxide solution using bromothymol blue as indicator. Some of their results appear in the next exercise.

Exercise 347 The following table shows how the concentration of 2,4,6-trinitrobenzoic acid varied with time in a particular experiment.

Time/min	Concentration of 2,4,6-trinitrobenzoic acid/mol dm^{-3}
0	2.77×10^{-4}
18	2.32×10^{-4}
31	2.05×10^{-4}
55	1.59×10^{-4}
79	1.26×10^{-4}
157	0.58×10^{-4}
Infinity	0.00

(a) Plot a graph of concentration of 2,4,6-trinitrobenzoic acid (vertical axis) against time (horizontal axis). Use a large piece of graph paper to make it easier to draw tangents to the curve at various points.

(b) Draw tangents to the curve at 10, 50, 100 and 150 minutes and calculate their slopes. (Note that the unit is mol dm^{-3} min^{-1}, a unit of rate of reaction. However, the slope is negative in this case because the concentration of a reactant decreases, but the rate is always positive.) Complete a copy of the table below.

Time /min	Concentration /mol dm^{-3}	Slope /mol dm^{-3} min^{-1}	Rate /mol dm^{-3} min^{-1}
10			
50			
100			
150			

(c) Plot a graph of rate of reaction (vertical axis) against concentration of 2,4,6-trinitrobenzoic acid (horizontal axis):

(i) Does your graph go through the origin? Explain why it should.

(ii) Use your graph to state the relationship between the rate of reaction and concentration of reactant both in words and mathematically.

The expression you have just worked out in Exercise 347 is called the rate equation for the reaction. We now consider rate equations in more detail.

RATE EQUATIONS, RATE CONSTANTS AND ORDERS OF REACTION

A rate equation (rate expression) relates the rate of a reaction at a fixed temperature to the concentrations of the reacting species by means of constants called (a) the rate constant (velocity constant), and (b) the order(s) of reaction.

The general form of a rate equation is:

$$\text{rate} = k\,[\text{A}]^a\,[\text{B}]^b\,[\text{C}]^c$$

k = rate constant (velocity constant) at the particular temperature
a = order of reaction with respect to reactant A
b = order of reaction with respect to reactant B
c = order of reaction with respect to reactant C
$a + b + c$ = overall order of reaction

In the next exercise, compare this equation with the one you derived in Exercise 347.

Exercise 348 Use the rate equation which you worked out in Exercise 347 to state:

(a) the order of reaction with respect to 2,4,6-trinitrobenzoic acid;

(b) the overall order of reaction.

You should be able to work out a value for the rate constant for a first-order reaction like this one, both graphically and by substituting into the rate equation. To make sure you can use both methods, try the next two exercises.

Exercise 349 Using both your graph and rate equation from Exercise 347, work out a value for the rate constant (k) of the decarboxylation of 2,4,6-trinitrobenzoic acid at 90 °C.

What is the unit of the rate constant for this reaction?

Another method to calculate the rate constant is to substitute values (from your graph) into the rate equation, rearranged to give:

k = rate/concentration

and average the results. Try this in the next exercise.

Exercise 350 (a) Use your graph from Exercise 347 to calculate a value for the rate of reaction at zero time (the initial rate).

(b) Calculate the rate constant (k) for each of the times (10, 50, 100 and 150) minutes) listed and for zero time.

(c) Average your results.

(d) Is the value of k you have just calculated likely to be more or less accurate than the one obtained in Exercise 349?

All the exercises so far in this chapter have been based on data from the decarboxylation of 2,4,6-trinitrobenzoic acid. The following summary will help you to identify the main points in these exercises so that you can apply them to other reactions.

Summary

So far in this chapter, we have shown you how to:

(a) plot a concentration/time curve from experimental data;

(b) draw tangents to this curve and hence calculate values of reaction rate;

(c) use rate/concentration data obtained from this curve to identify a first-order reaction by plotting rate versus concentration;

(d) express the relationship between rate and concentration of a reactant mathematically (a rate equation or rate expression);

(e) calculate a value for the rate constant in two ways.

Now try the next exercise, where you are given rate/concentration data for a different reaction and asked to deduce the rate expression, the order of reaction and the rate constant.

Exercise 351 Below are some data for the decomposition of dinitrogen pentoxide in tetrachloromethane solution (tcm):

$2N_2O_5(\text{tcm}) \rightarrow 4NO_2(\text{tcm}) + O_2(\text{g})$

Concentration N_2O_5/mol dm^{-3}	Rate of reaction as decrease in concentration of N_2O_5 per second /10^{-5} mol dm^{-3} s^{-1}
2.21	2.26
2.00	2.10
1.79	1.93
1.51	1.57
1.23	1.20
0.92	0.95

Plot a graph of the rate of the reaction against the concentration of N_2O_5, and try to answer these questions:

(a) What is the rate expression for the reaction?

(b) What order is this reaction with respect to N_2O_5?

(c) What is the value of the constant in the rate expression? Include the correct unit.

In the next section we aim to extend your knowledge of rate equations and orders of reaction to more complicated reactions.

Using rate equations

So far you have studied reactions in which one compound decomposes. We now take a brief look at reactions involving more than one reactant. To see that you have understood the difference between overall order and order of reaction with respect to a single reactant, try the following three short exercises.

<u>Exercise 352</u> The reaction between mercury(II) chloride and ethanedioate ions takes place according to the equation:

$$2HgCl_2(aq) + C_2O_4^{2-}(aq) \rightarrow 2Cl^-(aq) + 2CO_2(g) + Hg_2Cl_2(s)$$

The rate equation for the reaction is

$$\text{rate} = k\ [HgCl_2(aq)][C_2O_4^{2-}(aq)]^2$$

(a) What is the order of reaction with respect to each reactant?

(b) What is the overall order of reaction?

In the next exercise you may be surprised to find a fractional order in the rate equation. You will not come across fractional orders very often at A-level, but you should be aware that they exist and are a sign of a more complex reaction mechanism.

<u>Exercise 353</u> For the following reaction:

$$C_2H_4(g) + I_2(g) \rightarrow C_2H_4I_2(g)$$

the rate equation is:

$$\text{rate} = k\ [C_2H_4(g)][I_2(g)]^{3/2}$$

(a) What is the order of reaction with respect to each reactant?

(b) What is the overall order of reaction?

The next exercise deals with the overall order of a hypothetical reaction.

<u>Exercise 354</u> Substances P and Q react together to form products and the overall order of reaction is three. Which of the following rate equations could not be correct?

A rate = $k\ [P]^2[Q]$

B rate = $k\ [P]^0[Q]^3$

C rate = $k\ [P][Q]^2$

D rate = $k\ [P][Q]^3$

E rate = $k\ [P][Q]^2[H^+]^0$

Now do the next exercise, to check that you can work out units for the rate constant, k, for a reaction which is not first-order.

Exercise 355 The equation for the reaction between peroxodisulphate ions and iodide ions is:

$$S_2O_8^{2-}(aq) + 2I^-(aq) \rightarrow 2SO_4^{2-}(aq) + I_2(aq)$$

The rate equation is:

$$\text{rate} = k\,[S_2O_8^{2-}(aq)][I^-(aq)]$$

If concentrations are measured in mol dm^{-3} and rate in mol dm^{-3} s^{-1}, what is the unit of k? Show your working clearly.

Half-life of first-order reactions

One of the features of a first-order reaction is that it has a constant half-life, i.e. the time taken for half of any given amount of substance to react is constant at constant temperature. This is easily derived from the rate equation using calculus but, since integrated rate equations are not required for most A-level syllabuses, we have left the derivation till the end of the chapter (pages 200 - 201).

Half-life is a term most often associated with the rate of decay of radio-isotopes. We now consider the rate of radioactive decay before applying the idea of constant half-life to chemical reactions.

The rate of decay of radioisotopes

Radioactive isotopes decay in various ways (see Chapter 3). However, the rate of decay for any particular isotope is directly proportional to the amount present, i.e. the decay is a first-order process:

$$\text{rate} = k\,[\text{reactant}]$$

Furthermore, the rate is unaffected by external factors such as temperature and pressure. Each radioisotope has its own characteristic half-life; this gives us a simple method of comparison for different isotopes and also a means of identification.

Constant half-life is illustrated by the tabulated data for the decay of a transuranium element, americium-239.

Mass of sample/mg	time/hr
0.512	0
0.256	12
0.128	24
0.064	36
0.032	48
0.016	60
0.008	72
0.004	84
0.002	96
0.001	108
0.0005	120

You can see from the table on the previous page that for each time interval of 12 hr the mass of the sample is halved. The half-life, therefore, is constant at 12 hr. Use the same data in the following exercise.

Exercise 356 Using the data on the previous page, plot the mass of the sample of americium-239 (vertical axis) against time.

(a) Confirm that the half-life is constant by reading from the graph the time taken for the mass to fall

(i) from 0.100 mg to 0.050 mg,

(ii) from 0.050 mg to 0.025 mg.

(b) How long would it take for the mass to reach zero? Explain.

(c) What percentage of the original radioactivity remains after ten half-lives?

Exercise 357 (a) State what is meant by the term 'half-life'.

(b) The following is a decay curve for a radioactive element X.

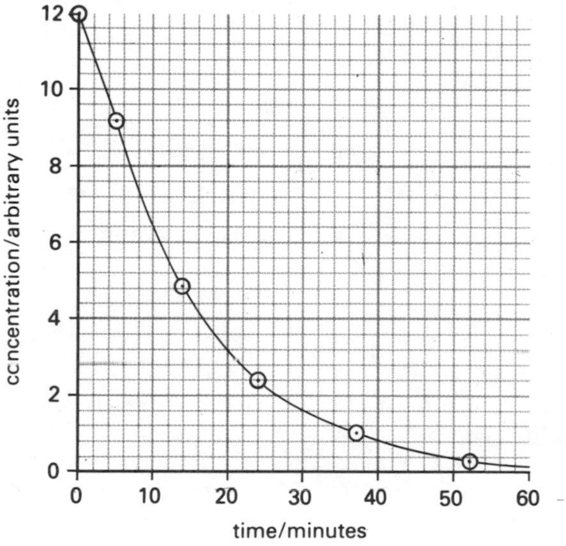

Fig. 32.

(i) Determine the half-life of X.

(ii) What is the order of the decay reaction?

(c) On a copy of the axes shown, sketch the approximate relationship between rate of decay and concentration of X.

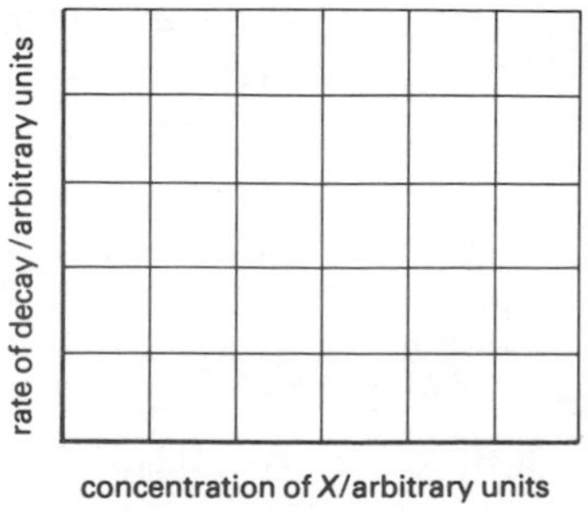

Fig. 33.

An important application of constant half-life is in radioactive dating.

Radioactive dating

Half-lives of certain radioactive elements have been used to calculate the age of rocks and estimate the age of the earth. In this section we examine the use of carbon-14 in the dating of archaeological remains.

Cosmic radiation is constantly converting small amounts of atmospheric nitrogen into carbon-14, which is radioactive. This carbon-14 is converted into carbon dioxide and removed from the atmosphere by photosynthesis. An equilibrium is established between these two opposing processes so that the concentration of $^{14}_{6}C$ remains constant in the atmosphere and, as a result, in all living organisms that depend, directly or indirectly, on photosynthesis.

When an organism dies, its uptake of $^{14}_{6}C$ ceases and its radioactivity decreases. Comparison of an old specimen with the radioactivity of a living organism gives us a method of determining its age, as in the next exercise.

Exercise 358

(a) Complete the following nuclear equation which shows how carbon-14 is formed from nitrogen-14 by the action of cosmic radiation:

$$^{14}_{7}N + ^{1}_{0}n \rightarrow ^{14}_{6}C + ^{A}_{Z}X$$

(b) Carbon from a piece of wood from a beam found in an ancient tomb gave a reading of 7.5 counts per minute per gram. New wood gives a reading of 15 counts per minute per gram. Estimate the year in which the tomb was built given that the half-life of $^{14}_{6}C$ is 5730 years.

(c) What is assumed to be constant over the period?

Most A-level syllabuses require only simple calculations involving whole numbers of half-lives. A more general application involves the use of integrated rate equations, which we deal with at the end of this chapter. Now we consider some chemical reactions, which also have constant half-lives.

Constant half-life in first-order chemical reactions

In the following exercises you apply the half-life concept to some chemical reactions which follow first-order kinetics.

Exercise 359

A part

The decomposition of benzenediazonium chloride in aqueous solution is a reaction of the first order which proceeds according to the equation:

$$C_6H_5N_2Cl(aq) \rightarrow C_6H_5Cl(aq) + N_2(g)$$

A certain solution of benzenediazonium chloride contains initially an amount of this compound which gives 80 cm^3 of nitrogen on complete decomposition. It is found that, at 30 °C, 40 cm^3 of nitrogen are evolved in 40 minutes. How long after the start of the decomposition will 70 cm^3 of nitrogen have been evolved? (All volumes of nitrogen refer to the same temperature and pressure.)

Exercise 360 The following chemical reaction, under certain conditions, proceeds with the order shown.

A part

$2N_2O_5(g) \rightarrow 4NO_2(g) + O_2(g)$ First order

(a) In one experiment on the reaction, the initial concentration of N_2O_5 was 32 units and the half-life was 5 minutes. On a copy of the graph below, plot the relation between the concentration of N_2O_5 and time for this experiment.

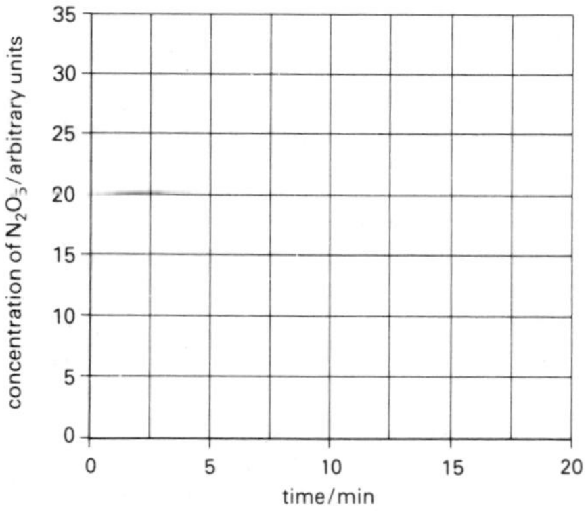

Fig. 34

(b) When the reaction was carried out in tetrachloromethane, it was found that an initial concentration of N_2O_5 of 2.00 mol dm^{-3} gave an initial rate of reaction of 2.10×10^{-5} mol dm^{-3} s^{-1}. Calculate the rate constant for the reaction and state the units.

In the next section, we show you how to determine the order of reaction with respect to one reactant at a time, from experimental data, in a reaction which involves two or more reactants.

The order of reaction with respect to individual reactants

We show you how to determine the order of reaction with respect to individual reactants in the following Worked Example.

Worked Example The data below were obtained for the reaction between nitrogen monoxide and hydrogen at 700 °C. The stoichiometric equation for the reaction is:

W

$2H_2(g) + 2NO(g) \rightarrow 2H_2O(g) + N_2(g)$

Experiment number	Initial concentration /mol dm^{-3}		Initial rate /mol dm^{-3} s^{-1}
	H_2	NO	
1	0.01	0.025	2.4×10^{-6}
2	0.005	0.025	1.2×10^{-6}
3	0.01	0.0125	0.6×10^{-6}

Determine the order of reaction with respect to NO and to H_2. Write the rate equation for the reaction.

Solution

1. Consider Experiments 1 and 2 in the table on page 189.

   Initial $[NO(g)]$ is constant. Therefore, any variation in initial rate is due to the variation in initial $[H_2(g)]$.

2. Deduce the order of reaction with respect to hydrogen.

   Halving initial $[H_2(g)]$ halves the rate of reaction.

   $\therefore$ rate $\propto [H_2]^1$, i.e. the order with respect to H_2 is 1

3. Consider Experiments 1 and 3 in the table.

   Initial $[H_2(g)]$ is constant. Therefore, any variation in initial rate is due to the variation in initial $[NO(g)]$.

4. Deduce the order of reaction with respect to nitrogen monoxide.

   Halving initial $[NO(g)]$ gives one quarter the rate, i.e. $(\frac{1}{2})^2$.

   $\therefore$ rate $\propto [NO(g)]^2$, i.e. the order with respect to NO is 2

5. Write the rate equation:

   $$\text{rate} = k\,[H_2(g)][NO(g)]^2$$

Use the same method for the following exercises.

Exercise 361 Two gases, A and B, react according to the stoichiometric equation:

A part

$$A(g) + 3B(g) \rightarrow AB_3(g)$$

A series of experiments carried out at 298 K in order to determine the order of this reaction gave the following results.

Expt.	Initial concentration of A c/mol dm^{-3}	Initial concentration of B c/mol dm^{-3}	Initial rate of formation of AB_3 /mol dm^{-3} min^{-1}
1	0.100	0.100	0.00200
2	0.100	0.200	0.00798
3	0.100	0.300	0.01805
4	0.200	0.100	0.00399
5	0.300	0.100	0.00601

(a) What is the order of the reaction between A and B with respect to

(i) substance A (ii) substance B?

(b) Write down a rate equation for the reaction between A and B.

(c) Using the experimental data given for Experiment 1 in the table, calculate the rate constant, k, for the reaction. Give the appropriate units of k.

Exercise 362 A series of experiments was carried out on the reaction:

A part

$$2H_2(g) + 2NO(g) \rightarrow 2H_2O(g) + N_2(g)$$

The initial rate of reaction at 750 °C was determined by noting the rate of formation of nitrogen, and the following data recorded:

Expt.	Initial concentration of nitrogen monoxide /mol dm^{-3}	Initial concentration of hydrogen /mol dm^{-3} s^{-1}	Rate of formation of nitrogen /mol dm^{-3} s^{-1}
1	6.0×10^{-3}	1.0×10^{-3}	2.88×10^{-3}
2	6.0×10^{-3}	2.0×10^{-3}	5.77×10^{-3}
3	6.0×10^{-3}	3.0×10^{-3}	8.62×10^{-3}
4	1.0×10^{-3}	6.0×10^{-3}	0.48×10^{-3}
5	2.0×10^{-3}	6.0×10^{-3}	1.92×10^{-3}
6	3.0×10^{-3}	6.0×10^{-3}	4.30×10^{-3}

The rate equation for the reaction is:

$$\text{rate} = k\,[H_2]^m[NO]^n$$

Deduce the order of the reaction with respect to

(a) hydrogen,

(b) nitrogen monoxide.

The kinetics of the iodination of propanone are investigated in the next exercise.

Exercise 363 Under conditions of acid catalysis, propanone reacts with iodine, as follows:

A

$$(CH_3)_2CO(aq) + I_2(aq) \rightarrow CH_2ICOCH_3(aq) + HI(aq)$$

50 cm^3 of 0.02 M I_2(aq) and 50 cm^3 of acidified 0.25 M propanone were mixed together. 10 cm^3 portions of the reaction mixture were removed at 5-minute intervals and rapidly added to an excess of 0.5 M $NaHCO_3$(aq). The iodine remaining was titrated against aqueous sodium thiosulphate.

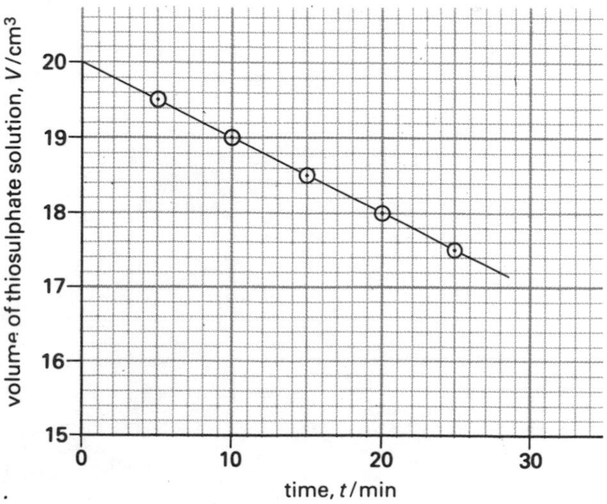

Fig. 35.

Exercise 363 (continued) The graph (Fig. 35) records the volume of aqueous sodium thiosulphate required to react with the iodine remaining at different times after mixing the reactants.

(a) Why are the 10 cm^3 portions of the reaction mixture added to aqueous sodium hydrogencarbonate before titration with sodium thiosulphate?

(b) How does the rate of change of iodine concentration vary during the experiment?

(c) What is the rate of reaction in terms of cm^3 of sodium thiosulphate per minute?

(d) What is the order of the reaction with respect to iodine?

(e) Suppose the reaction is first order with respect to propanone. What would be the rate of reaction (in cm^3 of sodium thiosulphate min^{-1}) if 0.50 M propanone were used instead of 0.25 M propanone?

(f) Indicate by means of a sketch how the volume of aqueous sodium thiosulphate used would vary with time if no catalytic acid was present in the reacting mixture of propanone and iodine.

(g) Explain your answer to (f).

Finally, in this section, we present an exercise where you are given the rate equation and the rate constant for a reaction, at a particular temperature, and asked to calculate the rates at different initial concentrations of reactants.

Exercise 364 Hydrogen and iodine react together to produce hydrogen iodide:

$$H_2(g) + I_2(g) \rightleftharpoons 2HI(g)$$

The rate equation for the reaction is:

$$\text{rate} = k\,[H_2(g)][I_2(g)]$$

and the rate constant at 374 °C is 8.58×10^{-5} mol^{-1} dm^3 s^{-1}.

(a) Work out the rate of reaction at each of the following initial concentrations of hydrogen and iodine, in experiments A, B and C.

Expt.	Initial $[H_2]$ /mol dm^{-3}	Initial $[I_2]$ /mol dm^{-3}
A	0.010	0.050
B	0.020	0.050
C	0.020	0.10

(b) From your answers to (a), what can you say about the effect of concentration of reactants on reaction rate?

We now go on to consider the effect of temperature on the rate of reaction, and introduce the important concept of activation energy.

ACTIVATION ENERGY AND THE EFFECT OF TEMPERATURE ON REACTION RATE

The rate of many gaseous and aqueous reactions often doubles for a temperature rise of only 10 K. Why is it that such a small rise in temperature can cause a large percentage increase in the reaction rate? We consider the question in the following section, which also helps to explain why some reactions, which we might expect to proceed on the basis of values of $\Delta H^{\ominus}$ or $\Delta G^{\ominus}$, do not appear to occur.

Consider the following reaction:

$$C_8H_{18}(l) + 12\tfrac{1}{2}O_2(g) \rightarrow 8CO_2(g) + 9H_2O(g); \quad \Delta H^{\ominus} = -5498 \text{ kJ mol}^{-1}$$

This reaction is highly exothermic and can occur with explosive force under the right conditions. However, it does not take place at room temperature!

There are many other examples like this, where there is an 'energy barrier' to be overcome before reaction can take place. This barrier is called the activation energy for the reaction. Since collisions between reactant particles are always occurring, it seems reasonable to assume that particles do not always react when they collide. A reaction occurs as a result of collisions between particles which possess more than a certain minimum amount of energy - the activation energy, E_a (sometimes E_A).

An energy-profile diagram is a way of showing how the energy of reacting molecules changes. To ensure that you can distinguish between the activation energy and the enthalpy change for a reaction, try the next exercise.

Exercise 365 (a) Draw an energy profile to represent the reaction:

$$2N_2O(g) \rightleftharpoons 2N_2(g) + O_2(g); \quad \Delta H^{\ominus} = -164.0 \text{ kJ mol}^{-1}$$

The activation energy, E_a, for the forward reaction is 250 kJ mol^{-1}. Use graph paper and let 1 cm represent 50 kJ mol^{-1}.

(b) Use your energy-profile diagram to calculate the activation energy for the back reaction.

Next, we go on to consider what proportion of particles in a gaseous reaction mixture reach the activation energy and how the proportion is affected by temperature.

The fraction of particles with energy greater than E_a

In Chapter 5 (Gases), we dealt with the calculation of the average speed of the molecules in a gas and showed that this is proportional to the absolute temperature. The distribution of speeds about this average is known as the Maxwell-Boltzmann distribution. Since the energy of a particle is proportional to the square of its speed, curves showing the distribution of energy have a rather similar shape.

Consider Fig. 36 below, which shows the distribution of kinetic energies of particles in the gas phase and the activation energy, E_a.

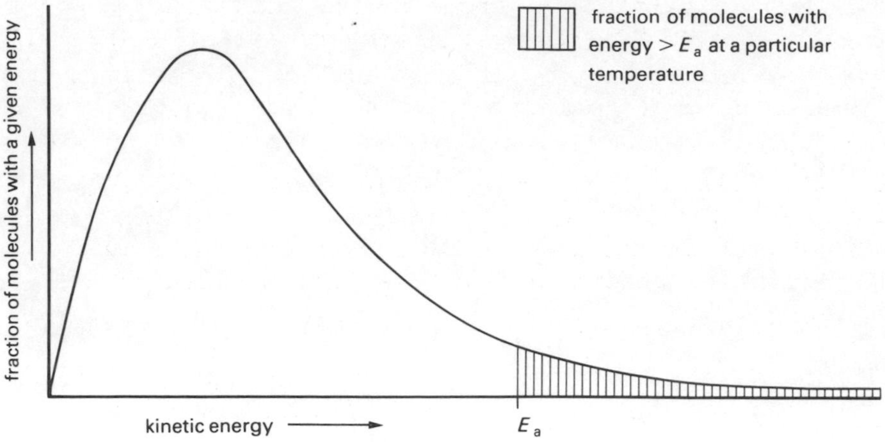

Fig. 36.

The area beneath the curve is proportional to the total number of particles involved and the shaded area under the curve is proportional to the number of particles with energy greater than E_a. Hence, the fraction of particles with energy greater than E_a is given by the ratio:

$$\frac{\text{shaded area under curve}}{\text{total area under curve}} = \frac{\text{number of particles with } E > E_a}{\text{total number of particles}}$$

Maxwell and Boltzmann derived a useful mathematical expression for this fraction in terms of the activation energy, E_a, the gas constant, R, and the absolute temperature, T:

$$\text{Fraction of particles with energy} > E_a = e^{-E_a/RT}$$

The following Worked Example shows you how to use this expression to calculate the number of molecules with energy greater than a given energy.

<u>Worked Example</u> Calculate the number of molecules in 1.0 mol of gas at 25 °C with energy greater than 55.0 kJ mol^{-1}.

W

<u>Solution</u>

1. Calculate the fraction of molecules with energy greater than 55.0 kJ mol^{-1}.

   Fraction of molecules with $E > 55.0$ kJ mol^{-1} = $e^{-E/RT}$ where R = 8.31 J K^{-1} mol, T = 298 K and E = 55000 J mol^{-1}.

   $\therefore$ Fraction of molecules $= e^{-(55000 \text{ J mol}^{-1})/(8.31 \text{ J K}^{-1} \text{ mol}^{-1})(298 \text{ K})}$

   $= e^{-22 \cdot 2} = 2.28 \times 10^{-10}$

2. Calculate the total number of molecules with energy greater than 55.0 kJ mol^{-1}.

   The total number of molecules present in 1.0 mol of gas is given by:

   $L = 6.02 \times 10^{23}$ mol^{-1} where L is the Avogadro constant.

   Since fraction of molecules with $E > 55.0$ kJ mol^{-1} $= \dfrac{\text{number of molecules with } E > 55.0 \text{ kJ mol}^{-1}}{\text{total number of molecules}}$

   Then number with $E > 55.0$ kJ mol^{-1} $= L\, e^{-E/RT}$

   $= 6.02 \times 10^{23} \text{ mol}^{-1} \times 2.28 \times 10^{-10} = \boxed{1.37 \times 10^{14} \text{ mol}^{-1}}$

Now you do a similar calculation for a different temperature.

Exercise 366 (a) Calculate the number of molecules in 1.0 mol of gas at 35 °C with energy greater than 55.0 kJ mol^{-1}

(b) Compare this value to that worked out at 298 K.

You have just calculated that for a rise in temperature of only 10 K, twice as many molecules exceed the energy barrier of 55.0 kJ mol^{-1}. Thus, on the Maxwell distribution curves, the shaded area under the (T + 10) K curve will be twice that on the T K curve (where T is the absolute temperature) as shown in Fig. 37.

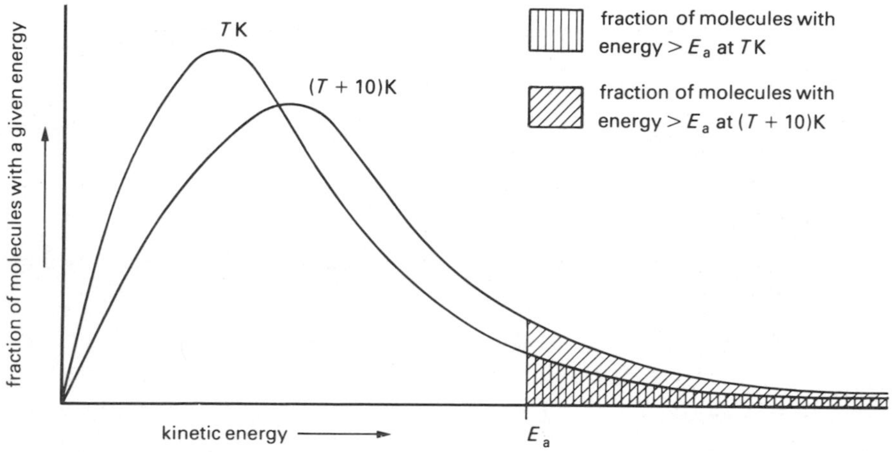

Fig. 37.

We now go on to consider the relationship between the rate constant, k, for a reaction and the fraction of particles with energy greater than E_a.

THE ARRHENIUS EQUATION

For particles to react, their energy must exceed E_a for that reaction. This suggests that at a given temperature

$$\text{Rate} \propto e^{-E_a/RT} \quad \text{....................(1)}$$

Since the rate changes during the progress of a reaction, it is more useful to use k (rate constant).

Consider the general reaction

$$A(g) + B(g) \rightarrow \text{products}$$

$$\text{Rate} = k[A]^x[B]^y \quad \text{at a room temperature } T_1$$

If the experiment is repeated using the same concentrations of A and B at a higher temperature, T_2, the rate increases. Since $[A]^x$ and $[B]^x$ have the same initial value, then k must increase. Therefore, we can say that the rate constant is a general measure of the reaction rate at a particular temperature. Equation (1) becomes:

$$k \propto e^{-E_a/RT} \quad \text{or} \quad k = Ae^{-E_a/RT} \text{..........(2)}$$

A is a constant, sometimes called the Arrhenius constant or pre-exponential constant. Equation (2) is often called the Arrhenius equation, named after the Swedish chemist, Svante Arrhenius, who formulated it in 1880.

Using the Arrhenius equation

The equation provides an extremely useful means of getting values for the activation energy and the pre-exponential factor for a reaction. It is usually changed to a logarithmic form to make it more manageable:

$$k = Ae^{-E_a/RT}$$

$$\therefore \quad \ln k = \ln Ae^{-E_a/RT} \qquad (\ln = \log_e)$$

$$= \ln A + \ln e^{-E_a/RT}$$

$$= \ln A - E_a/RT$$

Changing this to the $\log_{10}$ form:

$$\log_{10} k = \log_{10} A - \frac{E_a}{2.3\,RT} \quad \ldots\ldots\ldots\ldots\ldots (3)$$

Equation (3) is often quoted for you in A-level questions. The next exercise helps you to understand it and use it.

Exercise 367 (a) Which of the terms in equation (3) are variables for a particular reaction?

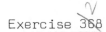

(b) Compare equation (3) to the equation for a straight line, i.e. $y = mx + c$. Which of the terms in equation (3) are analogous to y, m, x and c?

(c) Sketch the shape of the graph you would expect to obtain by plotting $\log_{10} k$ against $1/T$.

(d) Show how you could calculate A and E_a from your graph.

Use the method you have just devised in the next exercise.

Exercise 368 When gaseous hydrogen iodide decomposes in accordance with the equation:

A

$$2HI(g) \rightleftharpoons H_2(g) + I_2(g)$$

the reaction is found to be second order with respect to hydrogen iodide. In a series of experiments, the rate constant for this reaction was determined at several different temperatures. The results obtained are shown in the table below.

Temperature (T)/K	Rate constant (k) /dm^3 mol^{-1} s^{-1}
633	1.78×10^{-5}
666	1.07×10^{-4}
697	5.01×10^{-4}
715	1.05×10^{-3}
781	1.51×10^{-2}

Exercise 368 (continued)

Arrhenius deduced that for a reaction:

$$\log k = \frac{-E}{2.3\,RT} + \log A$$

where R is the gas constant and A is a constant.

(a) Use the data given to determine the activation energy, E, for the reaction.

(b) Determine by what factor the rate increases when the temperature rises from 300 K to 310 K.
(You may wish to use the alternative form of the Arrhenius equation

$$k = A\ 10^{-E/2.3RT})$$

(c) The kinetic energy of a fixed mass of gas is directly proportional to its temperature measured on the Kelvin scale. Calculate the ratio of the kinetic energies of a fixed mass of gas at 310 K and at 300 K.

(d) Compare this ratio with the increase in the rate of reaction determined from experimental data. Discuss the significance of this result.

In the next exercise, we present data for a reaction between peroxodisulphate(VI) ions and cobalt metal. In this case, you take the units of k as loss in mass of metal per minute and work out a value for the activation energy for the reaction.

Exercise 369

A

When strips of cobalt foil are rotated rapidly at a constant rate in sodium peroxodisulphate(VI) solution, there is a slow reaction and the cobalt dissolves. The reaction can be followed by removing and weighing the foil at intervals.

Experiment 1 at 0.5 °C		Experiment 2 at 13.5 °C		Experiment 3 at 25 °C	
Time /min	Mass /mg	Time /min	Mass /mg	Time /min	Mass /mg
0	130	0	130	0	130
20	120	10	115	4	116
60	98	15	106	8	103
80	86	30	86	12	87

Determine a rate constant, k, at each temperature as a loss in mass per minute (mg min^{-1}), and then determine the activation energy, E, of the reaction using the relationship

$$k \propto 10^{-E/2.3RT}$$

What difference in the results would you predict if the cobalt foil were NOT rotated?

In the next section, we show how catalytic activity can be explained in terms of activation energy.

Catalysts and activation energy

Most catalysts are thought to operate by providing an alternative reaction pathway with a lower activation energy. The catalysed reaction proceeds by a different mechanism, which may involve more reaction steps than the uncatalysed one, but it takes place more rapidly because each step has a lower activation energy than that of the uncatalysed reaction.

In the next exercise, you interpret an energy-profile diagram for the decomposition of hydrogen iodide, both catalysed and uncatalysed.

Exercise 3/0 The decomposition of hydrogen iodide is catalysed by platinum metal. The following diagram, Fig. 38, represents the energy changes that take place during the course of the uncatalysed reaction, with the changes during the catalysed reaction superimposed.

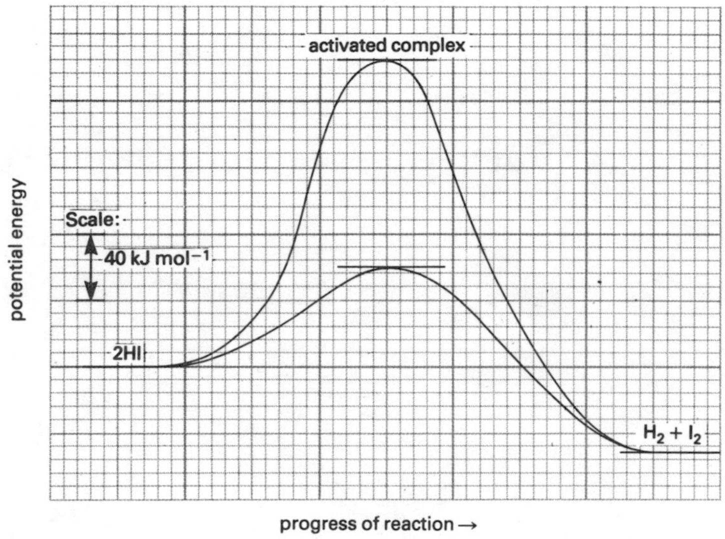

Fig. 38.

Use the scale of Fig. 38 to work out the following for both the forward and back reactions:

(a) the activation energies (uncatalysed),

(b) the activation energies (catalysed),

(c) the enthalpy changes.

Autocatalysis

Reactions in which a product acts as a catalyst are called 'autocatalytic'. Once the reaction starts, more catalyst is produced, so that the rate of reaction increases with time for at least part of its duration.

The next exercise is about the autocatalytic oxidation of manganate(VII) ions, MnO_4^-, by ethanedioate ions, $C_2O_4^{2-}$, catalysed by manganese(II) ions, Mn^{2+}.

$$2MnO_4^-(aq) + 16H^+(aq) + 2C_2O_4^{2-}(aq) \rightarrow 4CO_2(g) + 2Mn^{2+}(aq) + 8H_2O(l)$$

Exercise 371 The graph in Fig. 39 represents the reaction between manganate(VII) ions and ethanedioate ions.

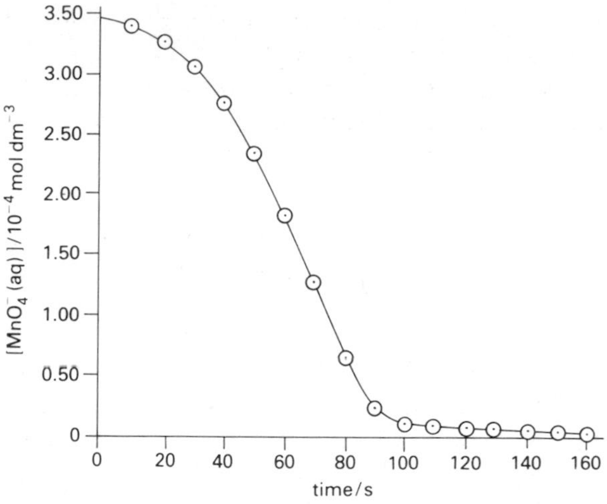

Fig. 39.

(a) Estimate from the graph the rates of reaction at the beginning of the process and after 50 seconds. How do you explain the difference in these values?

(b) The rate of reaction suddenly drops again after about 90 seconds. Give a reason for this.

The final section in this chapter is about the use of calculus in problems on reaction kinetics. It is not required by many A-level syllabuses.

USING DIFFERENTIAL NOTATION IN REACTION KINETICS

In your reading you may have seen the symbol d/dt written in front of a concentration term. For example, the rate equation for the decomposition of nitrogen pentoxide could be written as

$$-\frac{d[N_2O_5]}{dt} = k[N_2O_5]$$

This is a mathematical notation meaning 'the rate of change of $[N_2O_5]$ with time'. Thus, the equation says that the rate of change of concentration of N_2O_5 is proportional to the concentration of N_2O_5. The negative sign indicates that the concentration is decreasing.

In practical terms, $\frac{d[N_2O_5]}{dt}$ is the slope of the concentration/time curve.

When rates are expressed in this shorthand notation, much information can be expressed with brevity. For example, for the reaction

$$2N_2O_5(g) \rightarrow 4NO_2(g) + O_2(g)$$

the rate of formation of nitrogen dioxide is four times the rate of formation of oxygen and twice as fast as the disappearance of dinitrogen pentoxide, i.e.

$$\frac{d[NO_2]}{dt} = \frac{4d[O_2]}{dt} \quad \text{and} \quad \frac{d[NO_2]}{dt} = -\frac{2d[N_2O_5]}{dt}$$

The negative sign is necessary because $[NO_2]$ is increasing while $[N_2O_5]$ is decreasing. To see that you have understood this, try the following exercise.

Exercise 372 Phosphine decomposes when heated, as follows:

$$4PH_3(g) \rightarrow P_4(g) + 6H_2(g)$$

At a given instant, the rate at which phosphine decomposes is 2.4×10^{-3} mol dm^{-3} s^{-1}.

(a) Express the rate in three different ways, using differential notation, and show the relationships between them.

(b) What is the rate of formation of

(i) H_2, (ii) P_4?

The great value of expressing rates of reaction in differential notation is that it enables us to express rate equations in another very useful form by means of the mathematical technique called integration.

First-order integrated rate equations

The rate equation for the decomposition of dinitrogen pentoxide is:

$$\text{rate} = k[N_2O_5]$$

Using differential notation, and writing c for $[N_2O_5]$, this becomes

$$-\frac{dc}{dt} = kc$$

Integrating gives $\ln c = -kt + I$ (I = integration constant)

When $t = 0$, $c = c_0$ (the initial concentration) and $I = \ln c_0$. The equation thus becomes

$$\ln c = -kt + \ln c_0 \quad \text{or} \quad \log c = -kt/2.3 + \log c_0$$

This is an equation for a straight line and enables us to obtain the rate constant directly from concentration/time data by plotting $\log c$ against t and measuring the slope. In the following exercises, you can see how much easier this is than the method you learned earlier in the chapter, which involved two graphs and several intermediate calculations.

Exercise 373 Use the data in the table below to show that the hydrolysis of 2,4,6-trinitrobenzoic acid is a first-order reaction. Calculate the rate constant.

Time/min	Concentration of 2,4,6-trinitrobenzoic acid/mol dm^{-3}
0	2.77×10^{-4}
18	2.32×10^{-4}
31	2.05×10^{-4}
55	1.59×10^{-4}
79	1.26×10^{-4}
157	0.58×10^{-4}
Infinity	0.00

Exercise 374 The decomposition of dinitrogen pentoxide (N_2O_5) dissolved in tetrachloromethane (carbon tetrachloride) at 45 °C is first order. Using the concentrations of dinitrogen pentoxide and the times given, estimate by a graphical method the rate constant for the decomposition, stating the unit in which it is expressed.

Time, t/s	0	250	500	750	1000	1500	2000	2500
c/mol dm^{-3}	2.33	1.95	1.68	1.42	1.25	0.95	0.70	0.50

Earlier in this chapter we stated that first-order reactions have a constant half-life. In the next exercise, you confirm this mathematically by using the integrated rate equation.

Exercise 375 From the integrated rate equation for a first-order reaction, obtain an expression for the time taken for the concentration to fall to half its initial value. How does this show that the half-life is constant?

Finally, we take a brief look at an integrated rate equation for second-order reactions.

Second-order integrated rate equations

We limit our discussion to two simple types:

$$A + A \rightarrow \text{products} \qquad [A] = c$$

$$A + B \rightarrow \text{products} \qquad [A] = [B] = c$$

Each of these gives the same rate expression:

$$-\frac{dc}{dt} = kc^2$$

Integrating gives $kt = 1/c - I$

When $t = 0$, $c = c_0$ $\therefore$ $I = 1/c_0$ and the equation becomes

$$kt = \frac{1}{c} - \frac{1}{c_0} \quad \text{or} \quad t = \frac{1}{kc} - \frac{1}{kc_0}$$

This is an equation for a straight line and, once again, we can obtain the rate constant directly from concentration/time data. Try this in the next exercise - you should be able to see from the equation what graph to draw.

Exercise 376 Methyl ethanoate reacts with aqueous alkali according to the following equation:

$$CH_3CO_2CH_3(l) + OH^-(aq) \rightarrow CH_3CO_2^-(aq) + CH_3OH(aq)$$

The reaction may be followed by withdrawing samples at intervals and titrating with standard acid solution. The following results were obtained.

Exercise 376 (continued)

Time/min	$[OH^-(aq)]$ mol/dm^{-3}
3	7.40×10^{-3}
5	6.34×10^{-3}
7	5.50×10^{-3}
10	4.64×10^{-3}
15	3.63×10^{-3}
21	2.88×10^{-3}
25	2.54×10^{-3}

Determine a value for k, the rate constant for this reaction.

(Assume that the initial concentrations of methyl ethanoate and hydroxide ions were equal.)

END-OF-CHAPTER QUESTIONS

Exercise 377 In the Journal of the Chemical Society for 1950, Hughes, Ingold and Reed report some kinetic studies on aromatic nitration. In one experiment, ethanoic acid containing 0.2% water was used as solvent and pure nitric acid was added to make a 7 M solution.

A part

The kinetics of the nitration of ethylbenzene, $C_6H_5C_2H_5$, were studied with the following result at 20 °C.

Time /min	Concentration of ethylbenzene/mol dm^{-3}
0	0.090
8.0	0.063
11.0	0.053
13.0	0.049
16.0	0.037
21.0	0.024
25.0	0.009

(a) Determine an order of reaction from these results.

(b) What does this suggest about the mechanism of the reaction?

(c) What products are likely?

Exercise 378 The rate of decomposition of a compound X in aqueous solution is given by: rate = $k[X]^2$. What are the units of k?

A

Exercise 379 For the reaction $X + Y \rightarrow Z$, the rate expression is

$$\text{Rate} = k[X]^2[Y]^{\frac{1}{2}}$$

A

If the concentrations of X and Y are both increased by a factor of 4, by what factor will the rate increase?

Exercise 380 The rate of the reaction between peroxodisulphate ions and iodide ions in aqueous solution may be studied by measuring the amount of iodine formed after different reaction times. The stoichiometric equation is:

A part

$$S_2O_8^{2-}(aq) + 2I^-(aq) \rightarrow 2SO_4^{2-}(aq) + I_2(aq)$$

The graphs show the results of a kinetic investigation of this reaction.

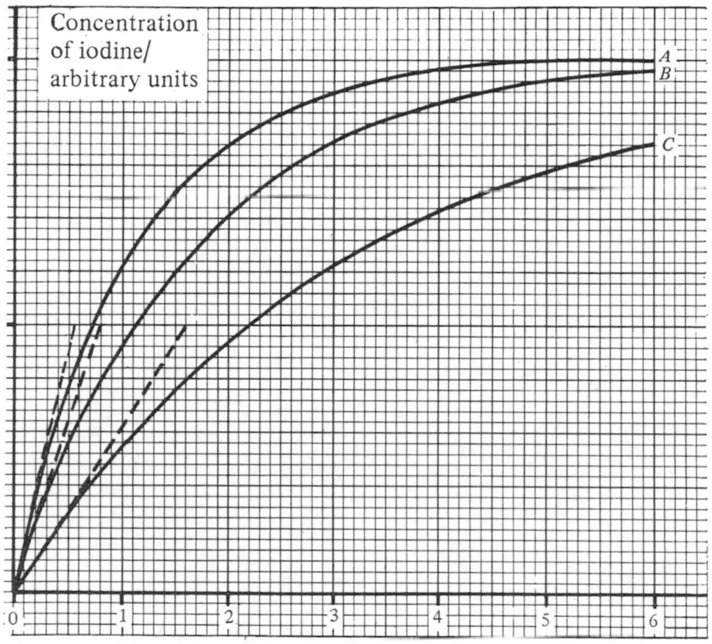

Fig. 40.

The experimental conditions were at a fixed temperature:

	Initial concentration of $S_2O_8^{2-}(aq)$/mol dm^{-3}	Initial concentration of $I^-(aq)$/mol dm^{-3}
Curve A	0.01	0.3
Curve B	0.01	0.2
Curve C	0.01	0.1

(a) Evaluate the initial reaction rates (given as $\Delta[I_2]/\Delta t$) for the three curves.

(b) (i) With respect to which reactant concentration do these reaction rates vary?

(ii) What is the order of the reaction with respect to this reactant?

(c) With reference to experimental curve A, find the times required for the completion of:

(i) one-half of the reaction,

(ii) three-quarters of the reaction.

In the light of these results, what is the order of the reaction with respect to the second reactant?

(d) From your answers to parts (b) and (c), suggest an overall rate equation for the reaction.

Exercise 381 Dinitrogen pentoxide (N_2O_5) decomposes according to the equation

A

$$2N_2O_5(g) \rightarrow 2N_2O_4(g) + O_2(g)$$

The reaction proceeds at a rate which is conveniently measurable at a temperature of 45 °C; at this temperature, the following results were obtained.

$[N_2O_5]$/mol dm^{-3}	Rate of disappearance of N_2O_5/mol dm^{-3} s^{-1}
22.3×10^{-3}	11.65×10^{-6}
17.4×10^{-3}	8.67×10^{-6}
13.2×10^{-3}	6.63×10^{-6}
9.5×10^{-3}	4.65×10^{-6}
4.7×10^{-3}	2.35×10^{-6}

(a) Plot these results on a suitable graph.

(b) The rate law for the reaction can be expressed in the form

$$\text{rate} = k[N_2O_5]^x$$

(i) What is the value of x as deduced from the graph?

(ii) What is the value of k at 45 °C? In what units is it expressed?

(c) It can be calculated from the figures given that the concentration would fall from 2.0×10^{-2} mol dm^{-3} to 1.0×10^{-2} mol dm^{3} in 1386 seconds. How much longer would it take to fall to 2.5×10^{-3} mol dm^{-3}?

(d) It will be seen from the equation that two N_2O_5 molecules give rise to three molecules of products. (You may neglect dissociation of N_2O_4 at this temperature.)

(i) Indicate how, in principle, you would use this fact to follow the extent of decomposition of N_2O_5 with time.

(ii) Sketch a simple apparatus that could be used for this purpose.

Exercise 382 The Arrhenius equation which involves the activation energy of a reaction, E, may be given in the forms

A part

either $k = Ae^{-E/RT}$ or $\ln k = \ln A - E/RT$

For the reaction

$$H_2(g) + I_2(g) \rightleftharpoons 2HI(g),$$

the values of k at 590 K and 700 K are 1.4×10^{-3} dm^3 mol^{-1} s^{-1} and 6.4×10^{-2} dm^3 mol^{-1} s^{-1} respectively. (R = 8.3 J K^{-1} mol^{-1})

(a) What do the symbols k, A, R, T represent?

(b) Explain the meaning of the term activation energy.

(c) Use an Arrhenius equation and the above data to calculate the activation energy for the reaction

$$H_2(g) + I_2(g) \rightarrow 2HI(g)$$

Exercise 383 An experiment was carried out to investigate the rate of reaction of an organic chloride of molecular formula C_4H_9Cl with hydroxide ions. The reaction was carried out in solution in a mixture of propanone and water with the following result:

A part

Time elapsed /sec	Concentration of C_4H_9Cl/mol dm^{-3}	Concentration of hydroxide ions/mol dm^{-3}
0	0.0100	0.0300
294	0.0050	0.0250
595	0.0025	0.0225

(a) Write a balanced equation for the reaction of an organic chloride of formula C_4H_9Cl with hydroxide ions.

(b) Suggest a reason why water alone is not used as the solvent in the experiment.

(c) (i) From the result given, deduce an order of reaction. Explain your answer.

(ii) Write a rate expression for the overall reaction.

Exercise 384 The reaction between potassium iodide and potassium peroxodisulphate(VI), $K_2S_2O_8$, in aqueous solution proceeds according to the overall equation

A

$$2KI(aq) + K_2S_2O_8(aq) \rightleftharpoons 2K_2SO_4(aq) + I_2(aq)$$

The rate of this reaction is found by experiment to be directly proportional to the concentration of the potassium iodide, and directly proportional to that of the potassium peroxodisulphate(VI).

(a) Write the above equation in ionic form.

(b) What is the overall order of the reaction?

(c) Show the meaning of the term velocity constant by writing an equation for the rate of this reaction, such that the concentration of the peroxodisulphate(VI) ions decreases with time.

(d) With an initial concentration of potassium iodide of 1.0×10^{-2} mol dm^{-3} and of potassium peroxodisulphate(VI) of 5.0×10^{-4} mol dm^{-3}, it is found that at 298 K, the initial rate of disappearance of the peroxodisulphate(VI) ions is 1.02×10^{-8} mol dm^{-3} s^{-1}. What is the velocity constant of the reaction at this temperature?

Exercise 385 Archaeologists can determine the age of organic matter by measuring the proportion of $^{14}_{6}C$ present. Assuming that carbon-14 has a half-life of 5600 years, what is the age of a piece of wood found to contain $\frac{1}{8}$ as much carbon-14 as living material?

A

APPENDIX

SIGNIFICANT FIGURES AND SCIENTIFIC MEASUREMENTS

These notes on the use of significant figures are not rigorous. We merely give some useful rules which you can apply to ordinary A-level calculations. For a detailed treatment, read a textbook on the theory of measurements.

The numerical value of any physical measurement is an approximation which is limited by the accuracy of the measuring instrument. Generally, the last digit in a measured quantity has an uncertainty associated with it. For example, in reading a thermometer, part of which is shown in Fig. 41, some may read the temperature as 21.1 °C and some may read it as 21.2 °C or 21.3 °C.

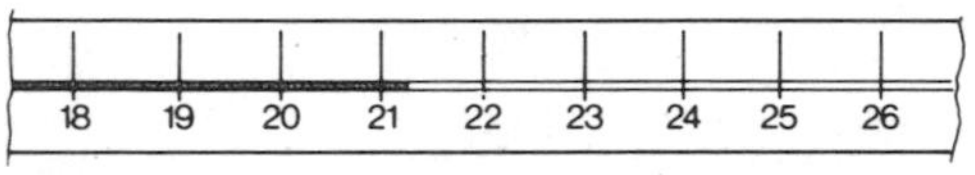

Fig. 41.

That is, there is no doubt that the temperature is between 21 °C and 22 °C, but there is some uncertainty in the decimal place. It is for this reason that we must consider the use of significant figures. Furthermore, it is important to consider the use of significant figures when so many calculations are made using electronic calculators giving as many as ten digits on their displays. You are very rarely justified in using all of them.

Zeros

A measured mass of 23 g has two significant figures, 2 and 3. If this same mass were written as 0.023 kg, it still contains two significant figures because zeros appearing as the first figures of a number are not significant - they merely locate the decimal point. However, the mass 0.0230 kg is expressed to three significant figures (2, 3 and the last 0).

The expression 'the length is 4700 m' does not necessarily show the accuracy of the measurement. To do this, the number should be written in standard form. If the measurement is made only to the nearest 1000 m, we use only one significant figure, i.e. $\ell = 5 \times 10^3$ m.

A more precise measurement, to the nearest 100 m, merits two significant figures, i.e. $\ell = 4.7 \times 10^3$ m, and so on, as summarised below.

Distance ℓ/m	Significant figures	Range of uncertainty	Precision of measurement
4700	unspecified	unspecified	unspecified
5×10^3	1	4.5 to 5.5	nearest 1000 m
4.7×10^3	2	4.65 to 4.75	nearest 100 m
4.70×10^3	3	4.695 to 4.705	nearest 10 m
4.700×10^3	4	4.6995 to 4.7005	nearest 1 m

Here is an exercise to see if you can recognise the number of significant figures in a measured quantity.

<u>Exercise 386</u> How many significant figures are in the following quantities?

(a)	2.54 g	(d)	14.0 cm^3	(g)	9.993 g cm^{-3}
(b)	2.205 g	(e)	1.86 x 10^5 s	(h)	5070 m s^{-1}
(c)	1.1 g	(f)	2.0070 g	(i)	127 000 kg

Now we look at significant figures in the results of combining uncertain values in calculations.

Addition and subtraction

After addition or subtraction, the answer should be rounded off to keep only the same number of decimal places as the <u>least</u> precise item. Here are some Worked Examples.

<u>Worked Examples</u> Add the following quantities:

W

(a)

46.247 cm^3
3.219 cm^3
0.224 cm^3
49.690 cm^3

Answer: 49.690 cm^3

Each volume to be added is expressed to the nearest 0.001 cm^3, so we can express the answer also to the nearest 0.001 cm^3.

(b)

26.6 cm^3
0.0028 cm^3
0.00002 cm^3
26.60282 cm^3

Answer: 26.6 cm^3

The number 26.6 is expressed to one place past the decimal point so you cannot have the answer quoted to a greater accuracy than one place past the decimal point.

(c)

2.40 cm^3
3.6584 cm^3
0.029 cm^3
6.0874 cm^3

Answer: 6.09 cm^3

The reasoning here is the same as in part (b). The least accurate measurement is 2.40 cm^3 so, in the answer, the volume cannot be quoted to more than two places past the decimal point. In this case, however, we round up rather than round down.

The reasoning is the same for subtraction, as we show in the following Worked Examples and Exercises.

Worked Examples Perform the following subtractions:

(a)

$$\begin{array}{r} 7.26\ \text{g} \\ -0.2\ \ \ \text{g} \\ \hline 7.06\ \text{g} \end{array} \qquad \text{Answer: } \boxed{7.1\ \text{g}}$$

(b)

$$\begin{array}{r} 539.27\ \text{g} \\ -\ 12.8\ \ \ \text{g} \\ \hline 526.47\ \text{g} \end{array} \qquad \text{Answer: } \boxed{526.5\ \text{g}}$$

Exercise 387 Add the following, expressing your answer to the correct number of significant figures:

(a)

$$\begin{array}{r} 203\ \ \ \ \ \text{g} \\ 4\ \ \ \ \ \text{g} \\ 0.77\ \text{g} \\ \hline \end{array}$$

(b)

$$\begin{array}{l} 0.0034\ \text{dm}^3 \\ 0.094\ \ \ \text{dm}^3 \\ 0.552\ \ \ \text{dm}^3 \\ \hline \end{array}$$

Exercise 388 Perform the following subtractions:

(a)

$$\begin{array}{r} 4.0\ \ \ \ \ \text{m} \\ -0.623\ \text{m} \\ \hline \end{array}$$

(b)

$$\begin{array}{r} 76\ \ \ \ \text{cm}^3 \\ -\ 0.3\ \text{cm}^3 \\ \hline \end{array}$$

The rules for multiplication and division are even easier.

Multiplication and division

The result of multiplying or dividing can contain only as many significant figures as are contained in the factor with the least number of significant figures.

Worked Example Calculate the density of an object which weighs 17.32 g and has a volume of 2.4 cm^3.

Solution

$$\text{density} = \frac{\text{mass}}{\text{volume}} = \frac{17.32\ \text{g}}{2.4\ \text{cm}^3}$$

A calculator gives the result as 7.2166667 g cm^{-3}. But the volume has the fewer significant figures - two. So the result is rounded off to 7.2 g cm^{-3}.

Exercise 389 Multiply the following, expressing your answer to the correct number of significant figures:

(a) 0.11 mol dm^{-3} x 0.0272 dm^3

(b) 2.43 mol x 27.9 g mol^{-1}

Exercise 390 Divide the following, expressing your answer to the correct number of significant figures:

(a) $\dfrac{9.2\ \text{g}}{19.00\ \text{g mol}^{-1}}$ (b) $\dfrac{0.20\ \text{g}}{0.1\ \text{cm}^3}$

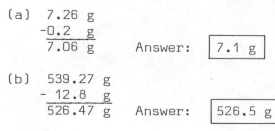

ANSWERS TO EXERCISES

Exercise 1

(a) $\text{time} = \dfrac{\text{number of molecules}}{\text{rate of removal}} = \dfrac{1.67 \times 10^{23}}{1\ \text{s}^{-1}}$

$= \boxed{1.67 \times 10^{23}\ \text{s}}$

(b) $1\ \text{y} = 365\ \text{dy} \times \dfrac{24\ \text{hr}}{\text{dy}} \times \dfrac{60\ \text{min}}{\text{hr}} \times \dfrac{60\ \text{s}}{\text{min}} = 3.15 \times 10^{7}\ \text{s}$

$\therefore\ 1\ \text{s} = \dfrac{1\ \text{y}}{3.15 \times 10^{7}}$

$\therefore$ time taken $= 1.67 \times 10^{23}\ \text{s}$

$= 1.67 \times 10^{23} \times \dfrac{1\ \text{y}}{3.15 \times 10^{7}} = \boxed{5.30 \times 10^{15}\ \text{y}}$

Exercise 2

(a) $\boxed{17.0\ \text{g mol}^{-1}}$ $14.0 + 3(1.01) = 17.0$

(b) $\boxed{199.9\ \text{g mol}^{-1}}$ $40.1 + 2(79.9) = 199.9$

(c) $\boxed{98.0\ \text{g mol}^{-1}}$ $3(1.01) + 31.0 + 4(16.0) = 98.0$

(d) $\boxed{322.3\ \text{g mol}^{-1}}$ $2(23.0) + 32.1 + 4(16.0) + 20(1.01) + 10(16.0) = 322.3$

Exercise 3

(a) 35.5 g

(b) 71.0 g

(c) 31.0 g

(d) 124.0 g

(e) 127 g (ignore the mass of the extra electrons)

Exercise 4

(a) Substituting into the expression

$n = \dfrac{m}{M}$

where $m = 30.0$ g and $M = 32.0\ \text{g mol}^{-1}$

gives $n = \dfrac{m}{M} = \dfrac{30.0\ \text{g}}{32.0\ \text{g mol}^{-1}} = \boxed{0.938\ \text{mol}}$

(b) $n = \dfrac{m}{M} = \dfrac{31.0\ \text{g}}{124.0\ \text{g mol}^{-1}} = \boxed{0.250\ \text{mol}}$

(c) $n = \dfrac{m}{M} = \dfrac{50.0\ \text{g}}{100.0\ \text{g mol}^{-1}} = \boxed{0.500\ \text{mol}}$

Exercise 5

(a) Substituting into the expression

$n = \dfrac{m}{M}$ in the form $m = nM$

where $n = 1.00$ mol and $M = 2.02\ \text{g mol}^{-1}$

gives $m = nM = 1.00\ \text{mol} \times 2.02\ \text{g mol}^{-1} = \boxed{2.02\ \text{g}}$

(b) $m = nM = 0.500\ \text{mol} \times 58.5\ \text{g mol}^{-1} = \boxed{29.3\ \text{g}}$

(c) $m = nM = 0.250\ \text{mol} \times 44.0\ \text{g mol}^{-1} = \boxed{11.0\ \text{g}}$

Exercise 6

(a) Substituting into the expression

$n = \dfrac{m}{M}$

where $m = 1.00$ g, $M = 17.0\ \text{g mol}^{-1}$

gives $n = \dfrac{m}{M} = \dfrac{1.00\ \text{g}}{17.0\ \text{g mol}^{-1}} = \boxed{0.0588\ \text{mol}}$

(b) If the number of molecules is to be the same, then the amount must be the same. For SO_2, $M = 64.1\ \text{g mol}^{-1}$ and from (a), $n = 0.0588$ mol

$\therefore$ substituting these values in the expression

$n = \dfrac{m}{M}$ in the form $m = nM$

gives $m = nM = 0.0588\ \text{mol} \times 64.1\ \text{g mol}^{-1} = \boxed{3.77\ \text{g}}$

Exercise 7

(a) Substituting into the expression

$$n = \frac{m}{M}$$

where m = 18.0 g, M = 12.0 g mol^{-1}

gives $n = \frac{m}{M} = \frac{18.0\ \text{g}}{12.0\ \text{g mol}^{-1}} = 1.50$ mol

Substituting into the expression

$$N = nL$$

where n = 1.50 mol, L = 6.02 x 10^{23} mol^{-1}

gives $N = nL$ = 1.50 mol x 6.02 x 10^{23} mol^{-1} = 9.03 x 10^{23}

Or, substituting $n = \frac{m}{M}$ into the expression $N = nL$

gives $N = \frac{mL}{M} = \frac{18.0\ \text{g} \times 6.02 \times 10^{23}\ \text{mol}^{-1}}{12.0\ \text{g mol}^{-1}} = \boxed{9.03 \times 10^{23}}$

(b) $N = \frac{mL}{M} = \frac{18.0\ \text{g} \times 6.02 \times 10^{23}\ \text{mol}^{-1}}{63.5\ \text{g mol}^{-1}} = \boxed{1.71 \times 10^{23}}$

(c) $N = \frac{mL}{M} = \frac{7.20\ \text{g} \times 6.02 \times 10^{23}\ \text{mol}^{-1}}{32.1\ \text{g mol}^{-1}} = \boxed{1.35 \times 10^{23}}$

Note that in (c), the number of atoms is the same whatever the molecular formula.

Exercise 8

(a) Substituting into the expression

$$n = \frac{m}{M}$$

where m = 1.00 g, M = 17.0 g mol^{-1}

gives $n = \frac{1.00\ \text{g}}{17.0\ \text{g mol}^{-1}} = 0.0588$ mol

Substituting into the expression

$$N = nL$$

gives N = 0.0588 mol x 6.02 x 10^{23} mol^{-1} = $\boxed{3.54 \times 10^{22}}$

Or, combining $n = \frac{m}{M}$ with $N = nL$ and substituting gives

$N = \frac{mL}{M} = \frac{1.00\ \text{g} \times 6.02 \times 10^{23}\ \text{mol}^{-1}}{17.0\ \text{g mol}^{-1}} = \boxed{3.54 \times 10^{22}}$

(b) $N = \frac{mL}{M} = \frac{3.28\ \text{g} \times 6.02 \times 10^{23}\ \text{mol}^{-1}}{64.1\ \text{g mol}^{-1}} = \boxed{3.08 \times 10^{22}}$

(c) $N = \frac{mL}{M} = \frac{7.20\ \text{g} \times 6.02 \times 10^{23}\ \text{mol}^{-1}}{8 \times 32.1\ \text{g mol}^{-1}} = \boxed{1.69 \times 10^{22}}$

Exercise 9

(a) 0.500 mol of NaCl contains 0.500 mol of Na^+ and 0.500 mol of Cl^-

∴ Total amount of ions, n = 1.00 mol

Substituting into the expression

$$N = nL$$

gives N = 1.00 mol x 6.02 x 10^{23} mol^{-1} = $\boxed{6.02 \times 10^{23}}$

(b) The amount of NaCl is calculated by substituting into the expression

$$n = \frac{m}{M}$$

where m = 14.6 g and M = 58.5 g mol^{-1}

∴ $n = \frac{14.6\ \text{g}}{58.5\ \text{g mol}^{-1}} = 0.250$ mol

0.250 mol of NaCl contains 0.250 mol of Na^+ and 0.250 of Cl^-

∴ total amount of ions, n = 0.500 mol

Substituting into the expression

$$N = nL$$

gives N = 0.500 mol x 6.02 x 10^{23} mol^{-1} = $\boxed{3.01 \times 10^{23}}$

(c) The amount of $CaCl_2$ is given by substituting in the expression

$$n = \frac{m}{M}$$

where m = 18.5 g and M = 111.0 g mol^{-1}

∴ $n = \frac{18.5\ \text{g}}{111.0\ \text{g mol}^{-1}} = 0.167$ mol

Since each mole of $CaCl_2$ contains 3 mol of ions (Ca^{2+}, Cl^-, Cl^-), the amount of ions, n = 3 x 0.167 mol = 0.501 mol

Substituting into the expression

$$N = nL$$

gives N = 0.501 mol x 6.02 x 10^{23} mol^{-1} = $\boxed{3.02 \times 10^{23}}$

Exercise 10

(a) From the equation for the reaction we know that

amount of Mg = amount of S

The amount of S is found by using the expression

$$n = \frac{m}{M}$$

where m = 16.0 g and M = 32.1 g mol^{-1}

$$\therefore\ n = \frac{m}{M} = \frac{16.0\ \text{g}}{32.1\ \text{g mol}^{-1}} = 0.498\ \text{mol}$$

∴ the amount of Mg also = 0.498 mol

The mass of Mg is found by using the expression

$n = \frac{m}{M}$ in the form $m = nM$

where n = 0.498 mol and M = 24.3 g mol^{-1}

$$\therefore\ m = nM = 0.498\ \text{mol} \times 24.3\ \text{g mol}^{-1} = \boxed{12.1\ \text{g}}$$

(b) For $NaNO_3$, $n = \frac{m}{M} = \frac{4.25\ \text{g}}{85.0\ \text{g mol}^{-1}} = 0.0500$ mol

But amount of O_2 = ½ x amount of $NaNO_3$

= ½ x 0.0500 mol = 0.0250 mol

For O_2, $m = nM$ = 0.0250 mol x 32.0 g mol^{-1} = $\boxed{0.800\ \text{g}}$

Exercise 11

For P, substituting into the expression

$$n = \frac{m}{M}$$

where m = 4.00 g and M = 31.0 g mol^{-1}

gives $n = \frac{m}{M} = \frac{4.00\ \text{g}}{31.0\ \text{g mol}^{-1}} = 0.129$ mol

From the equation,

$$\frac{\text{amount of } P_2O_5}{\text{amount of P}} = \frac{2}{4} = \frac{1}{2}$$

∴ amount of P_2O_5 = ½ x amount of P

= ½ x 0.129 mol = 0.0645 mol

For P_2O_5, substituting into the expression

$n = \frac{m}{M}$ in the form $m = nM$

where n = 0.0645 mol and M = 142 g mol^{-1}

gives m = 0.0645 mol x 142 g mol^{-1} = $\boxed{9.16\ \text{g}}$

Exercise 12

The reacting amount of Al is given by substituting into the expression

$$n = \frac{m}{M}$$

where m = 0.27 g and M = 27.0 g mol^{-1}

$$\therefore\ n = \frac{0.27\ \text{g}}{27.0\ \text{g mol}^{-1}} = 0.010\ \text{mol}$$

The amount of Cu formed is given by substituting into the expression

$$n = \frac{m}{M}$$

where m = 0.96 g and M = 63.8 g mol^{-1}

$$\therefore\ n = \frac{0.96\ \text{g}}{63.5\ \text{g mol}^{-1}} = 0.015\ \text{mol}$$

$$\therefore\ \frac{\text{amount of Al}}{\text{amount of Cu}} = \frac{0.010\ \text{mol}}{0.015\ \text{mol}} = \frac{2}{3}$$

We can build up the equation from this ratio

$2Al(s) + ?\ CuSO_4(aq) \rightarrow 3Cu(s) + ?\ Al_2(SO_4)_3(aq)$

To equalise Cu atoms, the stoichiometric coefficient for $CuSO_4$ must be 3.
To equalise Al atoms, the stoichiometric coefficient for $Al_2(SO_4)_3$ must be 1.

∴ $\boxed{2Al(s) + 3CuSO_4(aq) \rightarrow 3Cu(s) + Al_2(SO_4)_3(aq)}$

Exercise 13

We must calculate the amount of each reagent to determine which limits the reaction.

For Fe, substituting into the expression

$$n = \frac{m}{M}$$

where m = 2.8 g and M = 55.8 g mol^{-1}

gives $n = \dfrac{2.8\ \text{g}}{55.8\ \text{g mol}^{-1}} = 0.050$ mol

For S, substituting into the expression

$$n = \frac{m}{M}$$

where m = 2.0 g and M = 32.1 g mol^{-1}

gives $n = \dfrac{2.0\ \text{g}}{32.1\ \text{g mol}^{-1}} = 0.062$ mol

From the equation, one mole of iron reacts with one mole of sulphur, so the amount of iron limits the amount of iron(II) sulphide formed.

∴ amount of FeS = amount of Fe = 0.050 mol

Substituting into the expression

$n = \frac{m}{M}$ in the form $m = nM$

where n = 0.050 mol and M = 87.9 g mol^{-1}

gives m = 0.050 mol x 87.9 g mol^{-1} = **4.4 g**

Exercise 14

	Co	S	O	H_2O
Mass/g	2.10	1.14	2.28	4.50
Molar mass/ g mol^{-1}	58.9	32.1	16.0	18.0
Amount/mol	$\frac{2.10}{58.9}$ = 0.0357	$\frac{1.14}{32.1}$ = 0.0355	$\frac{2.28}{16.0}$ = 0.143	$\frac{4.50}{18.0}$ = 0.250
Amount/ smallest amount = relative amount	$\frac{0.0357}{0.0355}$ = 1.01	$\frac{0.0355}{0.0355}$ = 1.00	$\frac{0.143}{0.0355}$ = 4.03	$\frac{0.250}{0.0355}$ = 7.04
Simplest ratio	1	1	4	7

The formula is **$CoSO_4 \cdot 7H_2O$**

Exercise 15

	$BaCl_2$	H_2O
Mass/g	8.53	1.47
Molar mass/g mol^{-1}	208.2	18.0
Amount/mol	$\frac{8.53}{208.2}$ = 0.0410	$\frac{1.47}{18.0}$ = 0.0817
Amount/smallest amount = relative amount	$\frac{0.0410}{0.0410}$ = 1.00	$\frac{0.0817}{0.0410}$ = 1.99
Simplest ratio	1	2

The formula is $BaCl_2 \cdot 2H_2O$, i.e. **$x = 2$**

Exercise 16

The mass of water removed = 0.585 g - 0.535 g = 0.050 g

	$UO(C_2O_4)\cdot 6H_2O$	H_2O(removed)
Mass/g	0.585	0.050
Molar mass/g mol^{-1}	450	18.0
Amount/mol	$\frac{0.585}{450}$ = 0.00130	$\frac{0.050}{18.0}$ = 0.0028
Amount/smallest amount = relative amount	$\frac{0.00130}{0.00130}$ = 1.00	$\frac{0.0028}{0.0013}$ = 2.2
Simplest ratio	1	2

Thus, the ratio of the amount of original compound to the removed water is 1:2. This means that for every 1 mol of compound, 2 mol of water were removed. The resulting substance would therefore have the formula $UO(C_2O_4)\cdot 4H_2O$.

Exercise 17

	C	H	O
Mass/g	40.0 g	6.6 g	53.4 g
Molar mass/g mol^{-1}	12.0	1.0	16.0
Amount/mol	$\frac{40.0}{12.0}$ = 3.33	$\frac{6.6}{1.0}$ = 6.6	$\frac{53.4}{16.0}$ = 3.34
Amount/smallest amount = relative amount	$\frac{3.33}{3.33}$ = 1.00	$\frac{6.6}{3.33}$ = 2.0	$\frac{3.34}{3.33}$ = 1.00
Simplest ratio	1	2	1

The empirical formula is **CH_2O**

Exercise 18

	Na	Al	Si	O	H_2O
Mass/g	12.1	14.2	22.1	42.1	9.48
Molar mass/ g mol^{-1}	23.0	27.0	28.1	16.0	18.0
Amount/ mol	$\frac{12.1}{23.0} = 0.526$	$\frac{14.2}{27.0} = 0.526$	$\frac{22.1}{28.1} = 0.786$	$\frac{42.1}{16.0} = 2.63$	$\frac{9.48}{18.0} = 0.527$
Amount/ smallest amount = relative amount	$\frac{0.526}{0.526} = 1.00$	$\frac{0.526}{0.526} = 1.00$	$\frac{0.786}{0.526} = 1.49$	$\frac{2.63}{0.526} = 5.00$	$\frac{0.527}{0.526} = 1.00$
Simplest ratio	2	2	3	10	2

Note: Since 1.49 is very close to 1.5, we are justified in rounding up.
The empirical formula is $\boxed{Na_2Al_2Si_3O_{10}\cdot 2H_2O}$.

Exercise 19

	C	H	O
Mass/g	3.91	0.87	5.22
Molar mass/g mol^{-1}	12.0	1.0	16.0
Amount/mol	0.326	0.87	0.326
Amount/smallest amount = relative amount	$\frac{0.326}{0.326} = 1.00$	$\frac{0.87}{0.326} = 2.67$	$\frac{0.326}{0.326} = 1.00$
Simplest ratio	3	8	3

Note: We are not justified in rounding off 2.67 to 3. 2.67 is the decimal equivalent of 8/3, so we treble the relative amounts, thus converting 2.67 to 8. The resulting empirical formula is $\boxed{C_3H_8O_3}$.

Exercise 20

Amount of CO_2, $n = \frac{m}{M} = \frac{0.66\ g}{44.0\ g\ mol^{-1}} = 0.015\ mol$

∴ amount of C = 0.015 mol

amount of H_2O, $n = \frac{m}{M} = \frac{0.33\ g}{18.0\ g\ mol^{-1}} = 0.020\ mol$

∴ amount of H = 0.020 mol x 2 = 0.040 mol

	C	H
Amount/mol	0.015	0.040
$\frac{\text{Amount}}{\text{Smaller amount}}$	$\frac{0.015}{0.015} = 1.0$	$\frac{0.040}{0.015} = \frac{8}{3}$
Simplest ratio	3	8

Empirical formula = $\boxed{C_3H_8}$

Exercise 21

Amount of CO_2, $n = \frac{m}{M} = \frac{0.374\ g}{44.0\ g\ mol^{-1}} = 8.50 \times 10^{-3}\ mol$

∴ amount of C = 8.50×10^{-3} mol

Amount of H_2O, $n = \frac{m}{M} = \frac{0.154\ g}{18.0\ g\ mol^{-1}} = 8.56 \times 10^{-3}\ mol$

∴ amount of H = 2 x amount of water = 0.0171 mol

Mass of carbon, $m = nM = 8.50 \times 10^{-3}\ mol \times 12.0\ g\ mol^{-1} = 0.102\ g$

Mass of hydrogen, $m = nM = 0.0172\ mol \times 1.0\ g\ mol^{-1} = 0.0171\ g$

Mass of oxygen = mass of sample - (mass of C + mass of H)

= 0.146 g - (0.102 + 0.0171) g = 0.027 g

	C	H	O
Mass/g			0.027
Molar mass/g mol^{-1}			16.0
Amount/mol	8.50×10^{-3}	0.0171	1.7×10^{-3}
$\frac{\text{Amount}}{\text{Smallest amount}}$	$\frac{8.50 \times 10^{-3}}{1.7 \times 10^{-3}} = 5.0$	$\frac{0.0171}{1.7 \times 10^{-3}} = 10$	$\frac{1.7 \times 10^{-3}}{1.7 \times 10^{-3}} = 1.0$
Simplest ratio of relative amounts	5	10	1

Empirical formula = $\boxed{C_5H_{10}O}$

Exercise 22

Amount of CO_2, $n = \frac{m}{M} = \frac{0.3771\ g}{44.0\ g\ mol^{-1}} = 8.57 \times 10^{-3}$ mol

∴ amount of C = 8.57×10^{-3} mol

Amount of H_2O, $n = \frac{m}{M} = \frac{0.0643\ g}{18.0\ g\ mol^{-1}} = 3.57 \times 10^{-3}$ mol

∴ amount of H = $2 \times 3.57 \times 10^{-3}$ mol = 7.14×10^{-3} mol

Amount of AgBr, $n = \frac{m}{M} = \frac{0.2685\ g}{187.8\ g\ mol^{-1}} = 1.43 \times 10^{-3}$ mol

∴ amount of Br = 1.43×10^{-3} mol.

Total mass of C, H and Br

$= \{(8.57 \times 12.0) + (7.14 \times 1.01) + (1.43 \times 79.9)\} \times 10^{-3}$ g

= 0.2243 g = mass of sample

∴ no other element is present.

	C	H	Br
Amount/mol	8.57×10^{-3}	7.14×10^{-3}	1.43×10^{-3}
$\frac{\text{Amount}}{\text{Smallest amount}}$	$\frac{8.57}{1.43} = 5.99$	$\frac{7.14}{1.43} = 4.99$	$\frac{1.43}{1.43} = 1.00$
Simplest ratio of relative amounts	6	5	1

Empirical formula = $\boxed{C_6H_5Br}$

Exercise 23

(a) Substituting into the ideal gas equation:

$pV = (m/M)\ RT$ in the form $M = \frac{mRT}{pV}$

gives $M = \frac{0.24\ g \times 0.0821\ atm\ dm^3\ K^{-1}\ mol^{-1} \times 373\ K}{0.98\ atm \times 0.134\ dm^3} = \boxed{56\ g\ mol^{-1}}$

(b) Substituting into the expression:

molecular formula = (empirical formula)$_n$

where $n = \frac{\text{relative molecular mass}}{\text{relative empirical-formula-mass}} = \frac{56}{14} = 4$

gives molecular formula = $(CH_2)_4$ = $\boxed{C_4H_8}$

Exercise 24

(a) Amount of CO_2, $n = \frac{m}{M} = \frac{1.10\ g}{44\ g\ mol^{-1}} = 0.025$ mol

∴ amount of C = 0.025 mol

Amount of H_2O, $n = \frac{m}{M} = \frac{0.45\ g}{18\ g\ mol^{-1}} = 0.025$ mol

∴ amount of H = 2 × 0.025 mol = 0.050 mol

Mass of C, $m = nM$ = 0.025 mol × 12 g mol^{-1} = 0.30 g

Mass of H, $m = nM$ = 0.050 mol × 1.0 g mol^{-1} = 0.050 g

Mass of oxygen = mass of sample − (mass of C + mass of H)

= 0.43 g − (0.30 g + 0.050 g) = 0.08 g

	C	H	O
Mass/g			0.08 g
Molar mass/g mol^{-1}			16
Amount/mol	0.025	0.050	0.005
$\frac{\text{Amount}}{\text{Smallest amount}}$	$\frac{0.025}{0.005} = 5$	$\frac{0.050}{0.005} = 10$	$\frac{0.005}{0.005} = 1$
Simplest ratio of relative amounts	5	10	1

Empirical formula = $\boxed{C_5H_{10}O}$

(b) $M_r(C_5H_{10}O) = (5 \times 12) + (10 \times 1) + (1 \times 16) = 86$

∴ molecular formula = empirical formula = $\boxed{C_5H_{10}O}$

(c) $C_5H_{10}O + 7O_2 \rightarrow 5CO_2 + 5H_2O$

Exercise 25

(a) Substituting into the expression

$c = \frac{n}{V}$

where n = 0.100 mol and V = (2000/1000) dm^3

gives $c = \frac{0.100\ mol}{2.00\ dm^3} = \boxed{0.0500\ mol\ dm^{-3}}$

(b) $c = \frac{n}{V} = \frac{0.0100\ mol}{1.00\ dm^3} = \boxed{0.100\ mol\ dm^{-3}}$

(c) $c = \frac{n}{V} = \frac{0.100\ mol}{0.500\ dm^3} = \boxed{0.200\ mol\ dm^{-3}}$

(d) $c = \frac{n}{V} = \frac{0.100\ mol}{0.250\ dm^3} = \boxed{0.400\ mol\ dm^{-3}}$

(e) $c = \frac{n}{V} = \frac{0.100\ mol}{0.100\ dm^3} = \boxed{1.00\ mol\ dm^{-3}}$

Exercise 26

(a) Substituting into the expression

$$n = \frac{m}{M}$$

where m = 8.50 g and M = 169.9 g mol^{-1}

gives $n = \dfrac{8.50 \text{ g}}{169.9 \text{ g mol}^{-1}} = 0.0500$ mol

Substituting into the expression

$$c = \frac{n}{V}$$

where n = 0.0500 mol and V = 1.00 dm^3

gives $c = \dfrac{0.0500 \text{ mol}}{1.00 \text{ dm}^3} = 0.0500$ mol dm^{-3}

Or, substituting $n = \frac{m}{M}$ into $c = \frac{n}{V}$

$$c = \frac{m}{MV} = \frac{8.50 \text{ g}}{169.9 \text{ g mol}^{-1} \times 1.00 \text{ dm}^3} = \boxed{0.0500 \text{ mol dm}^{-3}}$$

(b) $c = \dfrac{m}{MV} = \dfrac{10.7 \text{ g}}{214.0 \text{ g mol}^{-1} \times 0.250 \text{ dm}^3} = \boxed{0.200 \text{ mol dm}^{-3}}$

(c) $c = \dfrac{m}{MV} = \dfrac{11.2 \text{ g}}{331.2 \text{ g mol}^{-1} \times 0.050 \text{ dm}^3} = \boxed{0.676 \text{ mol dm}^{-3}}$

(d) $c = \dfrac{m}{MV} = \dfrac{14.3 \text{ g}}{294.2 \text{ g mol}^{-1} \times 0.250 \text{ dm}^3} = \boxed{0.194 \text{ mol dm}^{-3}}$

(e) $c = \dfrac{m}{MV} = \dfrac{11.9 \text{ g}}{249.7 \text{ g mol}^{-1} \times 0.500 \text{ dm}^3} = \boxed{0.0953 \text{ mol dm}^{-3}}$

Exercise 27

(a) Substituting into the expression

$c = \frac{n}{V}$ in the form $n = cV$

where c = 5.00 mol dm^{-3} and V = 4.00 dm^3

gives n = 5.00 mol dm^{-3} × 4.00 dm^3 = $\boxed{20.0 \text{ mol}}$

(b) $n = cV$ = 2.50 mol dm^{-3} × 1.00 dm^3 = $\boxed{2.50 \text{ mol}}$

(c) $n = cV$ = 0.439 mol dm^{-3} × 0.020 dm^3 = $\boxed{8.78 \times 10^{-3} \text{ mol}}$

Exercise 28

(a) Substituting into the expression

$c = \frac{n}{V}$ in the form $n = cV$

where c = 0.100 mol dm^{-3} and V = 1.00 dm^3

gives n = 0.100 mol dm^{-3} × 1.00 dm^3 = 0.100 mol

Substituting into the expression

$n = \frac{m}{M}$ in the form $m = nM$

where n = 0.100 mol and M = 58.4 g mol^{-1}

gives m = 0.100 mol × 58.4 g mol^{-1} = 5.84 g

Or, combining $c = \frac{n}{V}$ with $n = \frac{m}{M}$ and substituting

$m = nM = cVM$ = 0.100 mol dm^{-3} × 1.00 dm^3 × 5.84 g mol^{-1} = $\boxed{5.84 \text{ g}}$

(b) $m = cVM$ = 1.00 mol dm^{-3} × 0.500 dm^3 × 110.9 g mol^{-1} = $\boxed{55.5 \text{ g}}$

(c) $m = cVM$ = 0.200 mol dm^{-3} × 0 250 dm^3 × 158.0 g mol^{-1} = $\boxed{7.90 \text{ g}}$

(d) $m = cVM$ = 0.117 mol dm^{-3} × 0.200 dm^3 × 40.0 g mol^{-1} = $\boxed{0.936 \text{ g}}$

Exercise 29 (a) 160 g (b) 1.07 g (c) 865 g (d) 37.0 g

Exercise 30
(a) 0.250 mol (b) 0.0166 mol (c) 1.25 mol (d) 15.0 mol

Exercise 31 132 g mol^{-1}

Exercise 32 430 g

Exercise 33 (a) 0.0500 mol dm^{-3} (b) 0.100 mol dm^{-3}

Exercise 34 (a) 0.0500 mol dm^{-3} (b) 0.150 mol dm^{-3}

Exercise 35 x = 18

Exercise 36 $C_5H_{11}NO$

Exercise 37 (a) $C_{11}H_{14}O_2$ (b) $C_{11}H_{14}O_2$

Exercise 38 (a) C_3H_8O $M_r = 60.1$ (b) C_3H_8O

Exercise 39 C_6H_6

Exercise 40

(a) $Ba(OH)_2(aq) + 2HCl(aq) \rightarrow BaCl_2(aq) + 2H_2O(l)$

Let A refer to HCl and B to $Ba(OH)_2$

Substituting into the expression

$$\frac{c_A V_A}{c_B V_B} = \frac{a}{b}$$

where $c_A = 0.0600 \text{ mol dm}^{-3}$ $c_B = ?$

$V_A = 25.0 \text{ cm}^3$ $V_B = 20.0 \text{ cm}^3$

$a = 2$ $b = 1$

gives $\dfrac{0.0600 \text{ mol dm}^{-3} \times 25.0 \text{ cm}^3}{c_B \times 20.0 \text{ cm}^3} = \dfrac{2}{1}$

Solving for c_B gives

$$c_B = \frac{0.0600 \text{ mol dm}^{-3} \times 25.0 \text{ cm}^3}{2 \times 20.0 \text{ cm}^3} = \boxed{0.0375 \text{ mol dm}^{-3}}$$

(b) It is not necessary to convert from cm^3 to dm^3 because the units of volume cancel in the final expression.

Exercise 41

(a) $NaOH(aq) + HNO_3(aq) \rightarrow NaNO_3(aq) + H_2O(l)$

Let A refer to NaOH and B to HNO_3

Substituting into the expression

$$\frac{c_A V_A}{c_B V_B} = \frac{a}{b}$$

where $c_A = 0.500 \text{ mol dm}^{-3}$ $c_B = 0.100 \text{ mol dm}^{-3}$

$V_A = ?$ $V_B = 50.0 \text{ cm}^3$

gives $\dfrac{0.500 \text{ mol dm}^{-3} \times V_A}{0.100 \text{ mol dm}^{-3} \times 50.0 \text{ cm}^3} = \dfrac{1}{1}$

$$\therefore V_A = \frac{0.100 \text{ mol dm}^{-3} \times 50.0 \text{ cm}^3}{0.500 \text{ mol dm}^{-3}} = \boxed{10.0 \text{ cm}^3}$$

(b) $\dfrac{c_A V_A}{c_B V_B} = \dfrac{a}{b}$ (A refers to NaOH, B to H_2SO_4)

$$\therefore \frac{0.500 \text{ mol dm}^{-3} \times V_A}{0.262 \text{ mol dm}^{-3} \times 22.5 \text{ cm}^3} = \frac{2}{1}$$

and $V_A = \dfrac{2 \times 0.262 \times 22.5 \text{ cm}^3}{0.500} = \boxed{23.6 \text{ cm}^3}$

Exercise 42

Substituting into the expression:

$$\frac{c_A V_A}{c_B V_B} = \frac{a}{b}$$

where $c_A = 0.50 \text{ mol dm}^{-3}$ $c_B = 0.20 \text{ mol dm}^{-3}$

$V_A = 25.0 \text{ cm}^3$ $V_B = 31.3 \text{ cm}^3$

$$\frac{a}{b} = \frac{0.50 \text{ mol dm}^{-3} \times 25.0 \text{ cm}^3}{0.20 \text{ mol dm}^{-3} \times 31.3 \text{ cm}^3} = \frac{2}{1}$$

$$\therefore \boxed{a = 2,\ b = 1}$$

Exercise 43

(a) +1 (b) -2 (c) +3 (d) +1 (e) -1

Exercise 44

(a)	KCl is ionic $K^+ + Cl^-$	$Ox(K) = +1$	$Ox(Cl) = -1$
(b)	Na_2O is ionic $2Na^+ + O^{2-}$	$Ox(Na) = +1$	$Ox(O) = -2$
(c)	BaF_2 is ionic $Ba^{2+} + 2F^-$	$Ox(Ba) = +2$	$Ox(F) = -1$
(d)	Na_3P is ionic $3Na^+ + P^{3-}$	$Ox(Na) = +1$	$Ox(P) = -3$
(e)	Mg_3N_2 is ionic $3Mg^{2+} + 2N^{3-}$	$Ox(Mg) = +2$	$Ox(N) = -3$
(f)	CsH is ionic $Cs^+ + H^-$	$Ox(Cs) = +1$	$Ox(H) = -1$

Exercise 45

(a) N is more electronegative than H

$\overset{\delta+}{H}-\overset{\delta-}{N}(-\overset{\delta+}{H})-\overset{\delta+}{H}$ imagined as $H^+\ N^{3-}\ H^+\ H^+$

$\therefore\ Ox(N) = -3,\quad Ox(H) = +1$

(b) O is more electronegative than Cl

$\overset{\delta+}{Cl}-\overset{\delta-}{O}-\overset{\delta+}{Cl}$ imagined as $Cl^+\ O^{2-}\ Cl^+$

$\therefore\ Ox(O) = -2,\quad Ox(Cl) = +1$

Exercise 46

(a) The O—O bond is non-polar, i.e. neither oxygen atom can be regarded as having gained or lost an electron with respect to the other. However, the O—H bonds are polar, and each oxygen atom is regarded as having gained an electron from a hydrogen atom.

$\overset{\delta+}{H}-\overset{\delta-}{O}-\overset{\delta-}{O}-\overset{\delta+}{H}$ imagined as $H^+\ O^-\ O^-\ H^+$

$\therefore\ Ox(O) = -1,\quad Ox(H) = +1$

(b) Fluorine is more electronegative than oxygen.

$\overset{\delta-}{F}-\overset{\delta+}{O}-\overset{\delta-}{F}$ imagined as $F^-\ O^{2+}\ F^-$

$\therefore\ Ox(O) = +2,\quad Ox(F) = -1$

Exercise 47

(a) $Ox(As) + Ox(\text{all other atoms}) = 0$

$\therefore\ Ox(As) = -Ox(\text{all other atoms}) = -\big(3Ox(H) + 4Ox(O)\big)$
$= -\big(3(+1) + 4(-2)\big) = -3 + 8 = \boxed{+5}$

(b) $2Ox(Cr) + Ox(\text{all other atoms}) = -1$

$\therefore\ 2Ox(Cr) = -1 - Ox(\text{all other atoms}) = -1 - \big(Ox(H) + 7Ox(O)\big)$
$= -1 - \big(+1 + 7(-2)\big) = -1 - (-13) = +12$

$\therefore\ Ox(Cr) = \boxed{+6}$

Exercise 48

(a) $2Ox(\text{Cr}) + Ox(\text{all other atoms}) = 0$

$\therefore\ Ox(\text{Cr}) = -\frac{1}{2}Ox(\text{all other atoms})$

$= -\frac{1}{2}\big(2Ox(\text{K}) + 7Ox(\text{O})\big) = -\frac{1}{2}\big(2(+1) + 7(-2)\big) = \boxed{+6}$

(b) $PbSO_4$ contains the SO_4^{2-} ion

$\therefore\ Ox(\text{S}) + 4Ox(\text{O}) = -2$

$\therefore\ Ox(\text{S}) = -2 - 4Ox(\text{O}) = -2 - 4(-2) = -2 + 8 = \boxed{+6}$

(c) $Ox(\text{N}) = -\big(Ox(\text{H}) + 3Ox(\text{O})\big) = -\big(+1 + 3(-2)\big) = -(1 - 6) = \boxed{+5}$

(d) $Ox(\text{S}) = -\frac{1}{2}\big(2Ox(\text{Na}) + 3Ox(\text{O})\big) = -\frac{1}{2}\big(2(+1) + 3(-2)\big) = -\frac{1}{2}(2 - 6) = \boxed{+2}$

(e) $Ox(\text{C}) = -\big(Ox(\text{Ca}) + 3Ox(\text{O})\big) = -\big(+2 + 3(-2)\big) = -(2 - 6) = \boxed{+4}$

(f) $Ox(\text{Ta}) = -\frac{1}{6}\big(8Ox(\text{Na}) + 19Ox(\text{O})\big) = -\frac{1}{6}\big(8(+1) + 19(-2)\big) = -\frac{1}{6}(8 - 38) = \boxed{+5}$

(g) There are two methods, depending on whether we consider the formula as one entity or as two ions.

(i) Considering the whole formula:

$Ox(\text{N}) = -\frac{1}{2}\big(4Ox(\text{H}) + 3Ox(\text{O})\big) = -\frac{1}{2}\big(4(+1) + 3(-2)\big) = -\frac{1}{2}(4 - 6) = \boxed{+1}$

(ii) Considering the ion NH_4^+:

$Ox(\text{N}) + 4Ox(\text{H}) = +1$

$\therefore\ Ox(\text{N}) = +1 - 4Ox(\text{H}) = +1 - 4(+1) = \boxed{-3}$

Considering the ion NO_3^-:

$Ox(\text{N}) + 3Ox(\text{O}) = -1$

$\therefore\ Ox(\text{N}) = -1 - 3Ox(\text{O}) = -1 - 3(-2) = \boxed{+5}$

Since the oxidation number concept is merely a 'book-keeping' device, both methods and, therefore, both answers may be considered correct. However, since the substance is known to be ionic, the second method is preferred. Note that the first answer is the mean of the second pair of answers. $\big(+1 = \frac{1}{2}(-3 + 5)\big)$

(h) $Ox(\text{S}) = -\frac{1}{4}\big(2Ox(\text{Na}) + 6Ox(\text{O})\big) = -\frac{1}{4}\big(2(+1) + 6(-2)\big) = -\frac{1}{4}(2 - 12) = \boxed{2\frac{1}{2}}$

Note that the fractional oxidation numbers sometimes appear in these calculations. This is usually because some atoms of the element concerned have one integral oxidation number and other atoms have another integral oxidation number.

Only if the structure is known can these integral oxidation numbers be allotted. In this case, the structure of the $S_4O_6^{2-}$ ion, with individual oxidation numbers added, is

(i) $Ox(\text{Fe}) = -\frac{1}{3}\big(4Ox(\text{O})\big) = -\frac{1}{3}\big(4(-2)\big) = \boxed{\frac{8}{3}}$

This fraction arises because Fe_3O_4 contains both Fe^{2+} and Fe^{3+} ions - Fe^{2+}, $2Fe^{3+}$, $4O^{2-}$.

The sum of the oxidation numbers for the three iron atoms is $+2 +3 +3 = 8$. The mean $Ox(\text{Fe})$ is therefore $\frac{8}{3}$.

Exercise 49

(a) $Fe(s) + CuSO_4(aq) \rightarrow FeSO_4(aq) + Cu(s)$
0 +2 +2 0

Iron is oxidized from Fe(0) to Fe(II)
Copper is reduced from Cu(II) to Cu(0)

$\therefore$ Reaction is redox

(b) $2KBr(aq) + Cl_2(aq) \rightarrow 2KCl(aq) + Br_2(aq)$
+1 -1 0 +1 -1 0

Bromine is oxidized from Br(-I) to Br(0)
Chlorine is reduced from Cl(0) to Cl(-I)

$\therefore$ Reaction is redox

(c) $CuSO_4(aq) + Pb(NO_3)_2(aq) \rightarrow Cu(NO_3)_2(aq) + PbSO_4(s)$
+2 +2 +2 +2

No oxidation or reduction, i.e. reaction is not redox.

(d) $C_2H_4(g) + H_2(g) \rightarrow C_2H_6(g)$
-2 +1 0 -3 +1

Some hydrogen is oxidized from H(0) to H(+I)
Carbon is reduced from C(-II) to C(-III)

$\therefore$ Reaction is redox.

Note that the oxidation number concept is not generally useful in organic chemistry except in the simplest compounds.

(e) $MnO_4^-(aq) + 5Fe^{2+}(aq) + 8H^+(aq) \rightarrow Mn^{2+}(aq) + 5Fe^{3+}(aq) + 4H_2O(l)$
+7 +2 +2 +3

Iron is oxidized from Fe(II) to Fe(III)
Manganese is reduced from Mn(VII) to Mn(II)

$\therefore$ Reaction is redox.

Exercise 50

(a) $\overset{+5}{IO_3}{}^-(aq) + \overset{-1}{I}{}^-(aq) + H^+(aq) \rightarrow \overset{0}{I_2}(aq) + H_2O(l)$

$\therefore$ oxidized species is I^- and reduced species is IO_3^-.

The balanced half-equations are:

$2IO_3^-(aq) + 12H^+(aq) + 10e^- \rightarrow I_2(aq) + 6H_2O(l)$

$2I^-(aq) \rightarrow I_2 + 2e^-$ (Multiply this equation by 5.)

Combining the two half-equations we get

$2IO_3^-(aq) + 12H^+(aq) + 10I^-(aq) \rightarrow 5I_2(aq) + I_2(aq) + 6H_2O(l)$

or $IO_3^-(aq) + 6H^+(aq) + 5I^-(aq) \rightarrow 3I_2(aq) + 3H_2O(l)$

(b) $\overset{+2}{S_2O_3}{}^{2-}(aq) + \overset{0}{I_2}(aq) \rightarrow \overset{-1}{I}{}^-(aq) + \overset{+2.5}{S_4O_6}{}^{2-}(aq)$

$\therefore$ oxidized species is $S_2O_3^{2-}$ and reduced species is I_2.

The balanced half-equations are:

$2S_2O_3^{2-}(aq) \rightarrow S_4O_6^{2-}(aq) + 2e^-$

$I_2(aq) + 2e^- \rightarrow 2I^-(aq)$

Combining the two half-equations we get

$2S_2O_3^{2-}(aq) + I_2(aq) \rightarrow S_4O_6^{2-}(aq) + 2I^-(aq)$

(c) $\overset{+7}{MnO_4}{}^-(aq) + H^+(aq) + \overset{+2}{Fe}{}^{2+}(aq) \rightarrow \overset{+2}{Mn}{}^{2+}(aq) + \overset{+3}{Fe}{}^{3+}(aq) + H_2O(l)$

$MnO_4^-(aq) + 8H^+(aq) + 5e^- \rightarrow Mn^{2+}(aq) + 4H_2O(l)$

$Fe^{2+}(aq) \rightarrow Fe^{3+}(aq) + e^-$ (Multiply by 5 and add)

$MnO_4^-(aq) + 8H^+(aq) + 5Fe^{2+}(aq) \rightarrow Mn^{2+}(aq) + 4H_2O(l) + 5Fe^{3+}(aq)$

(d) $\overset{+7}{MnO_4}{}^-(aq) + H^+(aq) + \overset{+3}{C_2O_4}{}^{2-}(aq) \rightarrow \overset{+2}{Mn}{}^{2+}(aq) + 2\overset{+4}{CO_2}(g) + H_2O(l)$

$MnO_4^-(aq) + 8H^+(aq) + 5e^- \rightarrow Mn^{2+}(aq) + 4H_2O(l)$ (x 2)

$C_2O_4^{2-}(aq) \rightarrow 2CO_2(g) + 2e^-$ (x 5)

$2MnO_4^-(aq) + 16H^+(aq) + 5C_2O_4^{2-}(aq) \rightarrow 2Mn^{2+}(aq) + 8H_2O(l) + 10CO_2(g)$

(e) $\overset{+7}{MnO_4}{}^-(aq) + H^+(aq) + \overset{-1}{H_2O_2}(aq) \rightarrow \overset{+2}{Mn}{}^{2+}(aq) + \overset{0}{O_2}(g) + \overset{-2}{H_2O}(l)$

$MnO_4^-(aq) + 8H^+(aq) + 5e^- \rightarrow Mn^{2+}(aq) + 4H_2O(l)$ (x 2)

$H_2O_2(aq) \rightarrow 2H^+(aq) + O_2(g) + 2e^-$ (x 5)

$2MnO_4^-(aq) + 6H^+(aq) + 5H_2O_2(aq) \rightarrow 2Mn^{2+}(aq) + 8H_2O(l) + 5O_2(g)$

(Note the simplification by deducting $10H^+$ from each side.)

Exercise 51

$M_r[FeSO_4 \cdot (NH_4)_2SO_4 \cdot 6H_2O] = 55.8 + 2(32.1) + 14(16.0) + 2(14.0) + 20(1.01)$
$= 392.2$

$\therefore c = \frac{n}{V} = \frac{9.80\ g/392.2\ g\ mol^{-1}}{0.250\ dm^3} = 0.100\ mol\ dm^{-3}$

Equation: $5Fe^{2+}(aq) + MnO_4^-(aq) + 8H^+(aq) \rightarrow 5Fe^{3+}(aq) + Mn^{2+}(aq) + 4H_2O(l)$

Substituting into the expression:

$\frac{c_A V_A}{c_B V_B} = \frac{a}{b}$ where A refers to Fe^{2+} and B to MnO_4^-

$\frac{0.100\ mol\ dm^{-3} \times 25.0\ cm^3}{c_B \times 24.6\ cm^3} = \frac{5}{1}$

$\therefore c_B = \frac{0.100\ mol\ dm^{-3} \times 25.0\ cm^3}{5 \times 24.6\ cm^3} = \boxed{0.0203\ mol\ dm^{-3}}$

Exercise 52

Equation: $H_2C_2O_4(aq) + 2NaOH(aq) \rightarrow Na_2C_2O_4(aq) + 2H_2O(l)$

$\frac{c_A V_A}{c_B V_B} = \frac{a}{b}$ where A refers to NaOH and B to $H_2C_2O_4$

i.e. $\frac{0.100\ mol\ dm^{-3} \times 14.8\ cm^3}{c_B \times 25.0\ cm^3} = \frac{2}{1}$

$\therefore c_B = \frac{0.100\ mol\ dm^{-3} \times 14.8\ cm^3}{2 \times 25.0\ cm^3} = 0.0296\ mol\ dm^{-3}$

$M_r(H_2C_2O_4) = 2(1.01) + 2(12.0) + 4(16.0) = 90.0$

$\therefore$ mass of $H_2C_2O_4$ per $dm^3 = 0.0296\ mol\ dm^{-3} \times 90.0\ g\ mol^{-1}$

$= \boxed{2.66\ g\ dm^{-3}}$

Equation: $2\tfrac{1}{2}C_2O_4^{2-}(aq) + MnO_4^-(aq) + 8H^+(aq) \rightarrow 5CO_2(g) + Mn^{2+}(aq) + 4H_2O(l)$

$\frac{c_A V_A}{c_B V_B} = \frac{a}{b}$ where A refers to MnO_4^- and B to $C_2O_4^{2-}$

i.e. $\frac{0.0200\ mol\ dm^{-3} \times 29.5\ cm^3}{c_B \times 25.0\ cm^3} = \frac{1}{2.50}$

$\therefore c_B = \frac{0.0200\ mol\ dm^{-3} \times 29.5\ cm^3 \times 2.50}{25.0\ cm^3} = 0.0590\ mol\ dm^{-3}$

$[C_2O_4^{2-}(aq)]$ from $Na_2C_2O_4$ = total $[C_2O_4^{2-}(aq)] - [C_2O_4^{-}(aq)]$ from $H_2C_2O_4$
$= 0.0590\ mol\ dm^{-3} - 0.0296\ mol\ dm^{-3}$
$= 0.0294\ mol\ dm^{-3}$

$M_r(Na_2C_2O_4) = 2(23.0) + 2(12.0) + 4(16.0) = 134$

$\therefore$ mass of $Na_2C_2O_4$ per $dm^{-3} = 0.0294\ mol\ dm^{-3} \times 134\ g\ mol^{-1}$

$= \boxed{3.94\ g\ dm^{-3}}$

Exercise 53

(a) Relative molecular mass of KIO_3 = 39 + 127 + 48 = 214

i.e. 1 dm³ of $\frac{M}{60}$ KIO_3 contains $\frac{214}{60}$ g = 3.57 g

$\therefore$ 250 cm³ require $\frac{3.57}{4}$ g = $\boxed{0.89\ \text{g}}$

(b) From the equations given,

1 mol of IO_3^- produces 3 mol of I_2 which react with 6 mol of $S_2O_3^{2-}$.

$\therefore$ 1 mol of IO_3^- is equivalent to 6 mol of $S_2O_3^{2-}$

Let A refer to IO_3^- and B refer to $S_2O_3^{2-}$

Substituting into the expression,

$$\frac{c_A V_A}{c_B V_B} = \frac{a}{b}$$

where $c_A = \frac{1}{60}$ mol dm⁻³ $\quad c_B = ?\quad a = 1$

V_A = 25.0 cm³ $\quad V_B$ = 20.0 cm³ $\quad b = 6$

gives $\dfrac{\frac{1}{60}\ \text{mol dm}^{-3} \times 25.0\ \text{cm}^3}{c_B \times 20.0\ \text{cm}^3} = \dfrac{1}{6}$

$\therefore c_B = \dfrac{\frac{1}{60}\ \text{mol dm}^{-3} \times 25.0\ \text{cm}^3 \times 6}{20.0\ \text{cm}^3} = \dfrac{2.5\ \text{mol dm}^{-3}}{20} = \boxed{0.125\ \text{mol dm}^{-3}}$

(c) To a 25 cm³ aliquot of HCl(aq) add excess KI(aq) and KIO_3(aq). Titrate the liberated I_2 with sodium thiosulphate solution of known concentration to the usual 'decolorization of starch indicator' end point.

From the chemical equations, 1 mol of H^+ is equivalent to 1 mol of $S_2O_3^{2-}$. The concentration of H^+ can be calculated by substituting in the equation:

$$\frac{c_A V_A}{c_B V_B} = \frac{1}{1}$$ where A refers to H^+ and B refers to $S_2O_3^{2-}$

(d) (i) $S_2O_3^{2-}$: Ox(S) = +2

(ii) From the equations for the reactions,

2 mol of $S_2O_3^{2-}$ reacts with 1 mol of I_2 released by 1 mol of Br_2

$\therefore$ 20 cm³ of 0.1 M $Na_2S_2O_3$ reacts with 10 cm³ of 0.1 M I_2 released by 10 cm³ of 0.1 M Br_2

If only 10 cm³ of 0.1 M Br_2 remains, then 50-10 = 40 cm³ must have reacted, i.e.

40 cm³ of 0.1 M Br_2 reacts with 10 cm³ of 0.1 M $S_2O_3^{2-}$

$\boxed{\therefore\ \text{4 mol of } Br_2 \text{ reacts with 1 mol } S_2O_3^{2-}}$

(iii) In reactants, Ox(Br) = 0 and Ox(S) = +2

In products, Ox(Br) = -1 (assumed to be Br^-)

8 Br atoms change oxidation number by -1 each. Total change = -8

$\therefore$ total change for S atoms must be +8, i.e. +4 each for 2 atoms

$\therefore$ Ox(S) in products = +2 +(+4) = $\boxed{+6}$

Exercise 53 (continued)

(iv) The most likely product with Ox(S) = 6 is SO_4^{2-}

$\therefore$ half-equations can be written:

$4Br_2(aq) + 8e^- \rightarrow 8Br^-(aq)$

$S_2O_3^{2-}(aq) + 5H_2O(l) \rightarrow 2SO_4^{2-}(aq) + 10H^+(aq) + 8e^-$

Combining these half-equations gives:

$$\boxed{S_2O_3^{2-}(aq) + 4Br_2(aq) + 5H_2O(l) \rightarrow 2SO_4^{2-}(aq) + 10H^+(aq) + 8Br^-(aq)}$$

Exercise 54

(a) Amount of Fe = $\dfrac{0.8\ \text{g}}{56\ \text{g mol}^{-1}} = \dfrac{1}{70}$ mol

Amount of HNO_3 = $\dfrac{0.9\ \text{g}}{63\ \text{g mol}^{-1}} = \dfrac{1}{70}$ mol

$\therefore$ molar ratio of Fe : HNO_3 is 1:1

(b) $$\underbrace{1Fe(s) + 1\ \overbrace{HNO_3(aq) \rightarrow 1Fe^{3+}(aq)}^{+3e^-} + }_{-3e^-}\left[\begin{array}{l} N_2O \\ \text{or } NO \\ \text{or } N_2O_4 \end{array}\right]$$

(c) Nitrogen in HNO_3 is in the +5 oxidation state. If the nitric acid gains three electrons, the final oxidation state of the nitrogen must be +2.

$\therefore$ the oxide produced must be $\boxed{NO}$

(d) $1Fe(s) - 3e^- \rightarrow 1Fe^{3+}(aq)$

$3H^+(aq) + 1HNO_3(aq) + 3e^- \rightarrow 1NO(g) + 2H_2O(l)$

Addition of these two equations gives the overall equation:

$$\boxed{Fe(s) + 3H^+(aq) + HNO_3(aq) \rightarrow Fe^{3+}(aq) + NO(g) + 2H_2O(l)}$$

Exercise 55

(a) 1. $Ag^+(aq) + Cl^-(aq) \rightarrow AgCl(s)$

Amount of Cl^- in 25 cm^3 of solution = amount of Ag^+ used = cV

$= 0.100 \text{ mol dm}^{-3} \times \frac{21.0}{1000} \text{ dm}^3 = \boxed{2.10 \times 10^{-3} \text{ mol}}$

2. Amount of Cl^- in whole sample $= 2.10 \times 10^{-3} \text{ mol} \times \frac{250 \text{ cm}^3}{25.0 \text{ cm}^3}$

$= 2.0 \times 10^{-2}$ mol

3. Mass of chloride ion in whole sample = $n \times M$

$= 2.10 \times 10^{-2} \text{ mol} \times 35.5 \text{ g mol}^{-1}$

$= 0.746$ g

4. Mass of S in the whole sample

= 1.420 g - 0.746 g

= 0.674 g

5. Amount of S in the whole sample $= \frac{m}{M}$

$= \frac{0.674 \text{ g}}{32.1 \text{ g mol}^{-1}}$

$= 2.10 \times 10^{-2}$ mol

There are equal amounts of chloride ion and sulphur, therefore one mole of chloride is combined with each mole of sulphur.

(b) The molar mass of the chloride is 135.2.

Since S and Cl atoms occur in equal numbers the molecular formula is $(SCl)_n$.

For n = 1 molar mass = 67.6

for n = 2 molar mass = 135.2

The molecular formula is $\boxed{S_2Cl_2}$

Exercise 56

(a) $Ag^+(aq) + Cl^-(aq) \rightarrow AgCl(s)$

Amount of Cl^- in 25 cm^3 of solution = amount of Ag^+ used = cV

$= 0.100 \text{ mol dm}^{-3} \times \frac{18.4}{1000} \text{ dm}^3 = \boxed{1.84 \times 10^{-3} \text{ mol}}$

(b) Amount of Cl^- in whole sample $= 1.84 \times 10^{-3} \text{ mol} \times \frac{250 \text{ cm}^3}{25.0 \text{ cm}^3}$

$= \boxed{1.84 \times 10^{-2} \text{ mol}}$

Mass of chloride ion in whole sample = $n \times M$

$= 1.84 \times 10^{-2} \text{ mol} \times 35.5 \text{ g mol}^{-1}$

$= \boxed{0.653 \text{ g}}$

(c) Mass of P in the whole sample = (mass of sample) - (mass of Cl^- in sample)

= 0.767 g - 0.653 g

= 0.114 g

∴ amount of P in the whole sample $= \frac{m}{M}$

$= \frac{0.114 \text{ g}}{31.0 \text{ g mol}^{-1}} = \boxed{3.68 \times 10^{-3} \text{ mol}}$

(d) Amount of chlorine combined with one mole of phosphorus:

3.68×10^{-3} mol of P combines with 1.84×10^{-2} mol of Cl^-

∴ 1 mol of P combines with $\frac{1.84 \times 10^{-2} \text{ mol}}{3.68 \times 10^{-3}}$

$= \boxed{5.00 \text{ mol of Cl atoms}}$

(e) The simplest formula of the chloride is $\boxed{PCl_5}$

Exercise 57 9.19 mol dm^{-3}

Exercise 58 $0.628 \text{ mol dm}^{-3}$

Exercise 59 (a) 0.0150 mol (b) 7.5 cm^3

Exercise 60 m = 2, n = 6

Exercise 61

(a) (i) 0.010 mol (ii) 0.025 mol (iii) 2.5 mol

(b) (i) +2 to 0 (ii) +5 (iii) 3

(c) $P_4(s) + 10CuSO_4(aq) + 12H_2O(l) \rightarrow 10\,Cu(s) + 4HPO_3(aq) + 10H_2SO_4(aq)$

Exercise 62

(a) (i) 0.010 (ii) 0.18 g

(b) (i) 0.010 (ii) 0.96 g

(c) (i) 0.0050 (ii) 0.32 g

(d) (i) 0.54 g (ii) 0.030

(e) $Cu(NH_4)_2(SO_4)_2 \cdot 6H_2O$ or $CuSO_4 \cdot (NH_4)_2SO_4 \cdot 6H_2O$

Exercise 63 73.9% Fe^{2+} 26.1% Fe^{3+}

Exercise 64

(a) Height of lithium-6 peak = 0.50 cm

Height of lithium-7 peak = 6.30 cm

Substituting into the expression:

$$\% \text{ abundance} = \frac{\text{amount of isotope}}{\text{total amount of all isotopes}} \times 100$$

$$\% \text{ abundance of } ^6\text{Li} = \frac{0.50}{6.30 + 0.50} \times 100 = \boxed{7.4\%}$$

$$\% \text{ abundance of } ^7\text{Li} = \frac{6.30}{6.30 + 0.50} \times 100 = \boxed{92.6\%}$$

(b) Total mass of a hundred atoms = (6 x 7.4) + (7 x 92.6) amu

= 44.4 + 648.2 amu = 692.6 amu

= 693 amu (3 significant figures)

substituting into the expression:

$$\text{Average mass} = \frac{\text{total mass}}{\text{number of atoms}}$$

$$\text{Average mass} = \frac{693}{100} \text{ amu} = 6.93 \text{ amu}$$

$\therefore$ relative atomic mass = $\boxed{6.93}$

Exercise 65

The mass/charge ratios refer to the singly-charged ions from the isotopes present in the sample: $^{20}_{10}Ne$; $^{21}_{10}Ne$; $^{22}_{10}Ne$

The relative intensities show the relative abundance of each isotope in the sample.

Substituting into the expression:

$$\% \text{ abundance} = \frac{\text{amount of isotope}}{\text{total amount of all isotopes}} \times 100$$

$$\% \text{ abundance of } ^{20}\text{Ne} = \frac{0.910}{0.910 + 0.0026 + 0.088} \times 100 = 91\%$$

$$\% \text{ abundance of } ^{21}\text{Ne} = \frac{0.0026}{0.910 + 0.0026 + 0.088} \times 100 = 0.26\%$$

$$\% \text{ abundance of } ^{22}\text{Ne} = \frac{0.088}{0.910 + 0.0026 + 0.088} \times 100 = 8.8\%$$

(Did you notice that the sum of the relative intensities is 1?)

Total mass of 100 atoms = (20 x 91) + (21 x 0.26) + (22 x 8.8)

= 1820 + 5.46 + 193.6 = 2.02×10^3 amu

$$\text{Average mass} = \frac{\text{total mass}}{\text{number of atoms}} = \frac{2.02 \times 10^3 \text{ amu}}{100} = 20.2 \text{ amu}$$

$\therefore$ relative atomic mass = $\boxed{20.2}$

Exercise 66

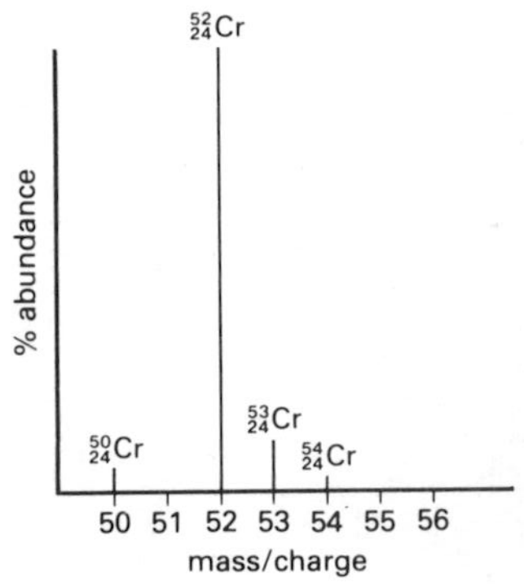

Total mass of 100 atoms

= (50 x 4.31) + (52 x 83.76)

+ (53 x 9.55) + (54 x 2.38)

= 215.5 + 4355.2 + 506.15 + 128.52

= 5.21×10^3 amu

$$\therefore \text{average mass} = \frac{5.21 \times 10^3 \text{ amu}}{100}$$

= 52.1 amu

and relative atomic mass = $\boxed{52.1}$

Exercise 67

Mass/charge ratio	Species
35	$^{35}_{17}Cl^{+}$ and $(^{35}_{17}Cl\ ^{35}_{17}Cl)^{2+}$
37	$^{37}_{17}Cl^{+}$ and $(^{37}_{17}Cl\ ^{37}_{17}Cl)^{2+}$
70	$(^{35}_{17}Cl—^{35}_{17}Cl)^{+}$
72	$(^{35}_{17}Cl—^{37}_{17}Cl)^{+}$
74	$(^{37}_{17}Cl—^{37}_{17}Cl)^{+}$

Exercise 68

(a) $^{212}_{84}Po \rightarrow {}^{4}_{2}He + {}^{208}_{82}Pb$

(b) $^{220}_{86}Rn \rightarrow {}^{4}_{2}He + {}^{216}_{84}Po$

Exercise 69

(a) $^{212}_{84}Po \rightarrow {}^{0}_{-1}e + {}^{212}_{85}At$

(b) $^{24}_{11}Na \rightarrow {}^{0}_{-1}e + {}^{24}_{12}Mg$

(c) $^{108}_{47}Ag \rightarrow {}^{0}_{-1}e + {}^{108}_{48}Cd$

Exercise 70

(a) For α-emission, the atomic number decreases by two and the mass number decreases by four.

(b) For β-emission, the atomic number increases by one and the mass number remains the same.

(i) $^{226}_{88}Ra \rightarrow {}^{4}_{2}He + {}^{222}_{86}Rn$ (ii) $^{43}_{19}K \rightarrow {}^{0}_{-1}e + {}^{43}_{20}Ca$

(iii) $^{43}_{20}Ca \rightarrow {}^{4}_{2}He + {}^{39}_{18}Ar$

Exercise 71

(a) $^{209}_{83}Bi$

Note that all nuclei with $Z > 83$ are unstable, regardless of the neutron to proton ratio.

(b) (i) $N/Z = \frac{209-83}{83} = \frac{126}{83} = \boxed{1.52}$

(ii) $N/Z = \frac{108-48}{48} = \frac{60}{48} = \boxed{1.25}$

(iii) $N/Z = \frac{12-6}{6} = \frac{6}{6} = \boxed{1.00}$

(c) The neutron to proton ratio for stable nuclei increases steadily from about 1.0 for the lightest nuclei to about 1.5 for the heaviest nuclei.

(d) (i) $^{11}_{6}C$ $N/Z = \frac{5}{6} < 1$ This is too low for stability. The graph confirms that $^{11}_{6}C$ is unstable.

(ii) $^{60}_{30}Zn$ $N/Z = \frac{30}{30} = 1.00$ This is too low for stability. The graph confirms that $^{60}_{30}Zn$ is unstable.

(iii) $^{106}_{47}Ag$ $N/Z = \frac{59}{47} = 1.26$ This is about right for stability. However, the graph does not show $^{106}_{47}Ag$ because it is, in fact, unstable.

Note that a favourable N/Z ratio does not guarantee stability.

(iv) $^{143}_{56}Ba$ $N/Z = \frac{87}{56} = 1.55$ This is too high for stability. The graph confirms that $^{143}_{56}Ba$ is unstable.

(v) $^{206}_{82}Pb$ $N/Z = \frac{124}{82} = 1.51$ This is about right for stability. The graph confirms that $^{206}_{82}Pb$ is stable.

Exercise 72

(a) $^{14}_{6}C \rightarrow {}^{14}_{7}N + {}^{0}_{-1}e$

(b) Reactant: $N/Z = \frac{14-6}{6} = \boxed{1.33}$

Product: $N/Z = \frac{14-7}{7} = \boxed{1.00}$

(c) The N/Z ratio for the reactant is too high for a light stable nucleus. The emission of an electron reduces N, the number of neutrons, by one but increases Z, the number of protons, also by one. In this way, β-emission reduces the N/Z ratio to a suitable value for stability.

Exercise 73 35.45

Exercise 74 6.95

Exercise 75

(a) Mass no.: 2 4 81 83 162

Ion : $^{2}_{1}H^{+}$ $^{2}_{1}H-^{2}_{1}H$ $^{81}_{35}Br^{+}$ $^{2}_{1}H-^{81}_{35}Br^{+}$ $^{81}_{35}Br^{+}-^{81}_{35}Br^{+}$

(b) 81

Exercise 76 69.7

Exercise 77

(a) 63.6 (b) $^{220}_{86}Rn \rightarrow ^{216}_{84}Po + ^{4}_{2}He$ $^{214}_{82}Pb \rightarrow ^{214}_{83}Bi + ^{0}_{-1}e$

Exercise 78

(a) $C(s) + O_2(g) \rightarrow CO_2(g);\ \Delta H^{\ominus} = -394\ kJ\ mol^{-1}$

(b) (i) Part (a) shows the enthalpy change for 1.00 mol of carbon;for 10.0 mol the enthalpy change is

$$10.0\ mol \times (-394\ kJ\ mol^{-1}) = \boxed{-3940\ kJ}$$

(ii) For 0.25 mol of carbon

$$\Delta H = 0.25\ mol \times (-394\ kJ\ mol^{-1}) = \boxed{-98.5\ kJ}$$

(iii) Amount of carbon $= \frac{m}{M} = \frac{18.0\ g}{12.0\ g\ mol^{-1}} = 1.5\ mol$

$$\therefore \Delta H = 1.5\ mol \times (-394\ kJ\ mol^{-1}) = \boxed{-591\ kJ}$$

(c) (i) You may be able to see at a glance that 197 is half 394, so that the amount of carbon is ½ mol and the mass is 6 g. Otherwise, since the quantity of heat depends directly on the amount of carbon burned;

$$\frac{1.00\ mol\ of\ C}{-394\ kJ} = \frac{x\ mol\ of\ C}{-197\ kJ}$$

$$\therefore x = \frac{197}{394} = 0.500$$

$$\therefore \text{mass of carbon} = 0.500\ mol \times 12.0\ g\ mol^{-1} = \boxed{6.00\ g}$$

(ii) $\frac{1.00\ mol\ of\ C}{-394\ kJ} = \frac{y\ mol\ of\ C}{-1000\ kJ}$

$$\therefore y = \frac{1000}{394} = 2.54$$

$$\therefore \text{mass of carbon} = 2.54\ mol \times 12.0\ g\ mol^{-1} = \boxed{30.5\ g}$$

Exercise 79

(a)(b)

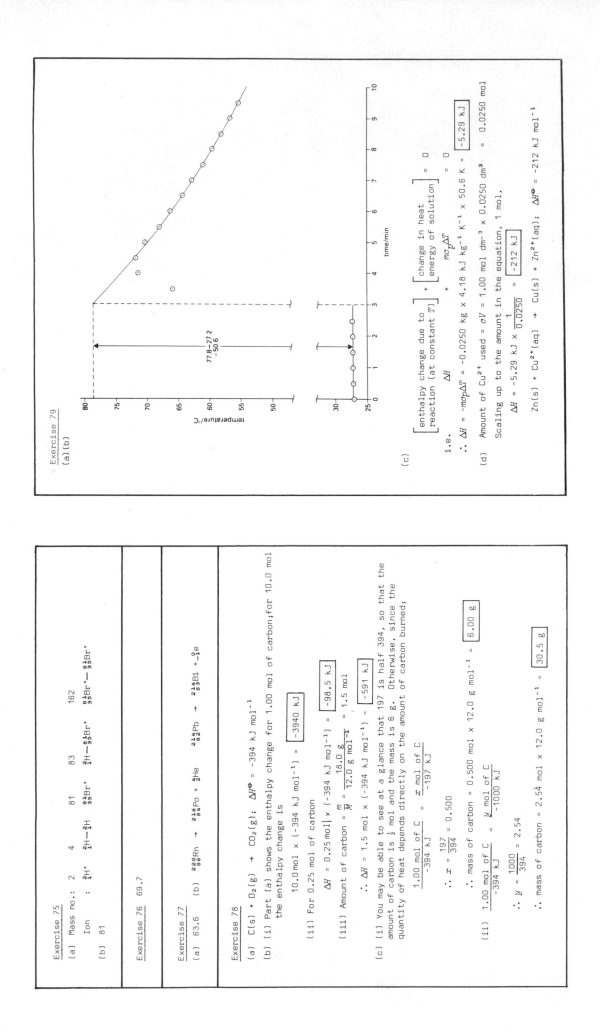

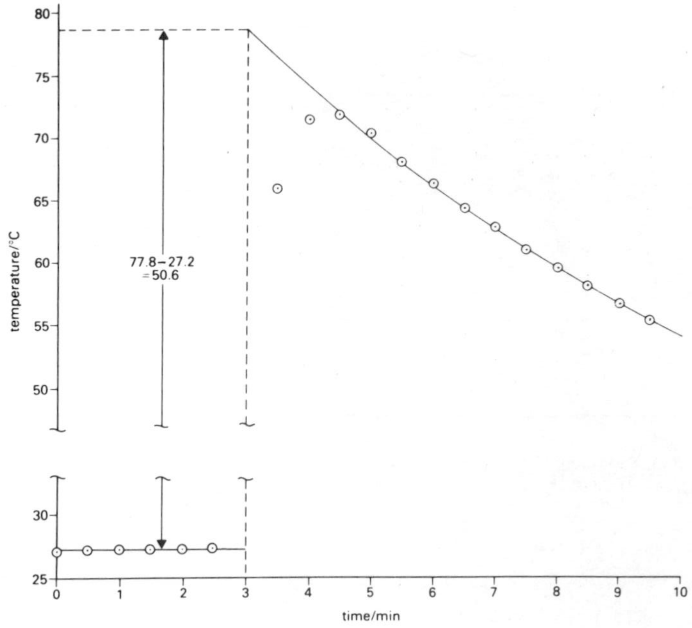

(c)

$$\begin{bmatrix}\text{enthalpy change due to}\\ \text{reaction (at constant } T)\end{bmatrix} + \begin{bmatrix}\text{change in heat}\\ \text{energy of solution}\end{bmatrix} = 0$$

i.e. $\Delta H + mc_p\Delta T = 0$

$$\therefore \Delta H = -mc_p\Delta T = -0.0250\ kg \times 4.18\ kJ\ kg^{-1}\ K^{-1} \times 50.6\ K = \boxed{-5.29\ kJ}$$

(d) Amount of Cu^{2+} used $= cV = 1.00\ mol\ dm^{-3} \times 0.0250\ dm^{3} = 0.0250\ mol$

Scaling up to the amount in the equation, 1 mol,

$$\Delta H = -5.29\ kJ \times \frac{1}{0.0250} = \boxed{-212\ kJ}$$

$$Zn(s) + Cu^{2+}(aq) \rightarrow Cu(s) + Zn^{2+}(aq);\ \Delta H^{\ominus} = -212\ kJ\ mol^{-1}$$

Exercise 80

ΔT = final temperature - initial temperature = 22.7 °C - 24.5 °C = -1.8 K

$$\begin{bmatrix}\text{enthalpy change on}\\ \text{dissolving } NH_4Cl\end{bmatrix} + \begin{bmatrix}\text{change in heat}\\ \text{energy of solution}\end{bmatrix} = 0$$

$$\Delta H \quad + \quad mc_p\Delta T \quad = \quad 0$$

$\therefore \Delta H = -mc_p\Delta T = -(0.090 \text{ kg} \times 4.18 \text{ kJ kg}^{-1} \text{ K}^{-1} \times (-1.8 \text{ K})) = +0.68 \text{ kJ}$

Scaling up from 0.050 mol to 1 mol,

$\Delta H = +0.68 \text{ kJ} \times \frac{1}{0.050} = +14 \text{ kJ}$

$NH_4Cl(s) + 100H_2O(l) \rightarrow NH_4Cl(aq,100H_2O)$; $\Delta H^\ominus$ = **+14 kJ mol⁻¹**

Note that only two significant figures are justified here because ΔT is small.

Exercise 81

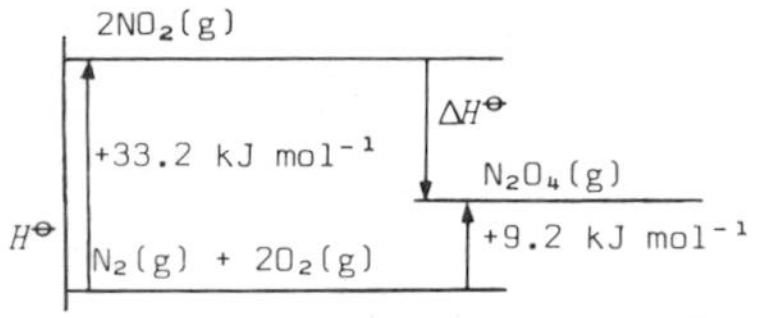

$+33.2 \text{ kJ mol}^{-1} = +9.2 \text{ kJ mol}^{-1} + (-\Delta H^\ominus)$

$\therefore \Delta H^\ominus = (9.2 - 33.2) \text{ kJ mol}^{-1}$ = **-24.0 kJ mol⁻¹**

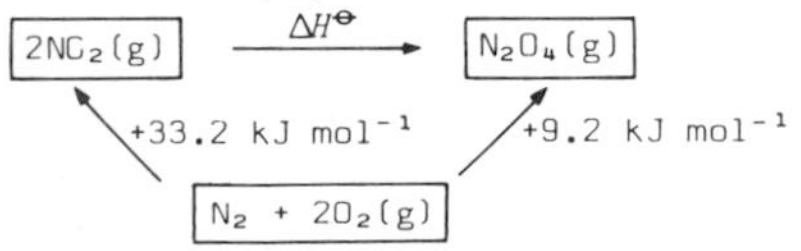

$+33.2 \text{ kJ mol}^{-1} + \Delta H^\ominus = +9.2 \text{ kJ mol}^{-1}$

$\therefore \Delta H^\ominus = (9.2 - 33.2) \text{ kJ mol}^{-1}$ = **-24.0 kJ mol⁻¹**

Exercise 82

(a) $NH_3(g) + HCl(g) + \cancel{NH_4Cl(s)} + 200H_2O(l)$

$\rightarrow \cancel{NH_4Cl(s)} + NH_4Cl(aq,200H_2O)$; $\Delta H^\ominus = (-175.3 + 16.3) \text{ kJ mol}^{-1}$

$NH_3(g) + HCl(g) + 200H_2O(l) \rightarrow NH_4Cl(aq,200H_2O)$; $\Delta H^\ominus = -159.0 \text{ kJ mol}^{-1}$

(b) $NH_3(g) + 100H_2O(l) + HCl(g) + 100H_2O(l) + NH_3\cancel{(aq,100H_2O)} + HCl\cancel{(aq,100H_2O)}$;

$\rightarrow NH_3\cancel{(aq,100H_2O)} + HCl\cancel{(aq,100H_2O)} + NH_4Cl(aq,200H_2O)$;

$\Delta H^\ominus = (-35.6 - 73.2 - 50.2) \text{ kJ mol}^{-1}$

$NH_3(g) + HCl(g) + 200H_2O(l) \rightarrow NH_4Cl(aq,200H_2O)$; $\Delta H^\ominus = -159.0 \text{ kJ mol}^{-1}$

(c) The enthalpy change for the overall process is the same by either Method 1 or Method 2.

(d)

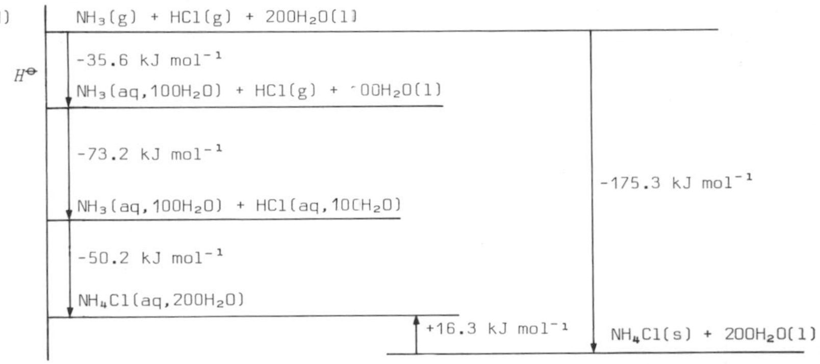

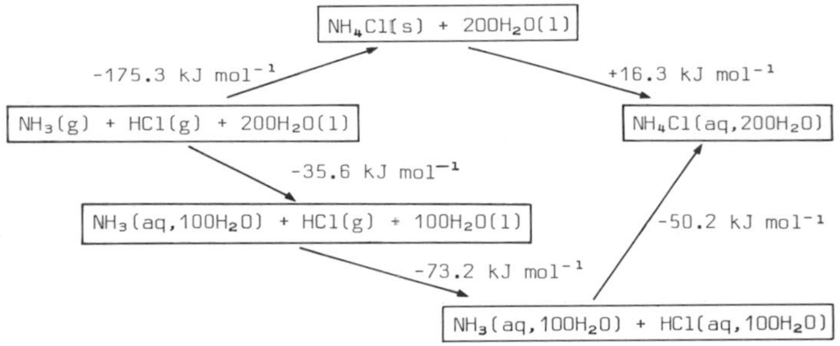

Exercise 83

(a) $\Delta H = -mc_p\Delta T = -(0.0450\ \text{kg} \times 4.18\ \text{kJ kg}^{-1}\ \text{K}^{-1} \times 11.3\ \text{K}) = -2.13\ \text{kJ}$

Scaling up, $\Delta H = -2.13\ \text{kJ} \times \dfrac{1\ \text{mol}}{0.0250\ \text{mol}} = -85.2\ \text{kJ}$

Thus $MgSO_4(s) + 100H_2O(l) \rightarrow MgSO_4(aq,100H_2O)$; $\Delta H^{\ominus} = \boxed{-85.2\ \text{kJ mol}^{-1}}$

(b) $\Delta H = -mc_p\Delta T = -(0.04185\ \text{kg} \times 4.18\ \text{kJ kg}^{-1}\ \text{K}^{-1} \times (-1.4\ \text{K}) = +0.24\ \text{kJ}$

Scaling up, $\Delta H = +0.24\ \text{kJ} \times \dfrac{1\ \text{mol}}{0.0250\ \text{mol}} = +9.6\ \text{kJ}$

Thus $MgSO_4\cdot 7H_2O(s) + 93H_2O(l) \rightarrow MgSO_4(aq,100H_2O)$; $\Delta H^{\ominus} = \boxed{+9.6\ \text{kJ mol}^{-1}}$

(c) $\boxed{93H_2O(l)} + \boxed{MgSO_4(s) + 7H_2O(l)} \xrightarrow{\Delta H^{\ominus}} \boxed{MgSO_4\cdot 7H_2O(s)} + \boxed{93H_2O(l)}$

ΔH_1: $MgSO_4(s) + 7H_2O(l) \rightarrow \boxed{MgSO_4(aq,100H_2O)}$; ΔH_2: $MgSO_4\cdot 7H_2O(s) \rightarrow \boxed{MgSO_4(aq,100H_2O)}$

$\Delta H^{\ominus} = \Delta H_1 - \Delta H_2 = (-85.2 - 9.6)\ \text{kJ mol}^{-1} = \boxed{-94.8\ \text{kJ mol}^{-1}}$

(d)

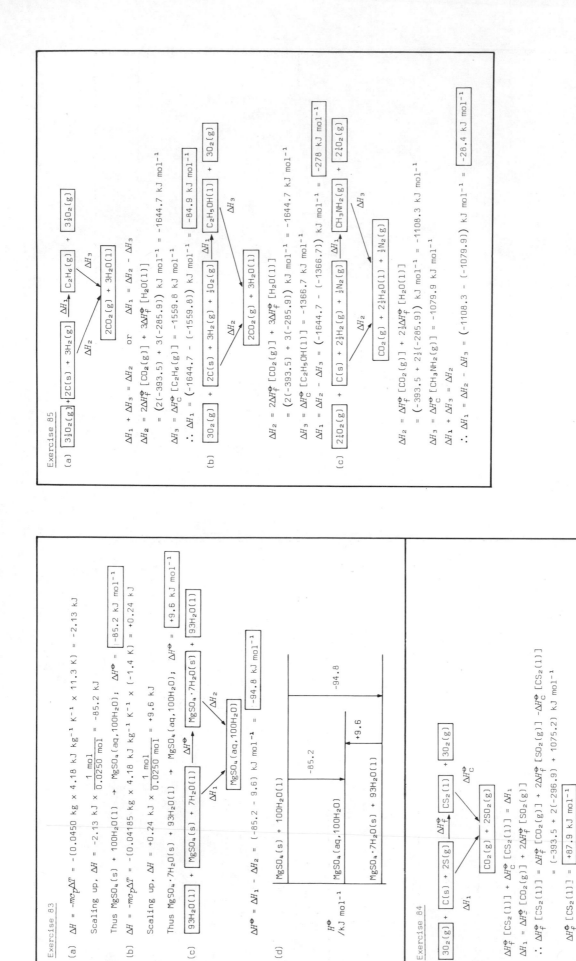

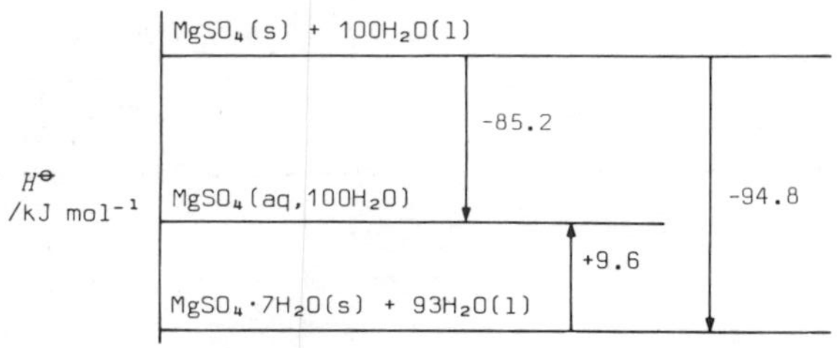

Exercise 84

$\boxed{3O_2(g)} + \boxed{C(s) + 2S(g)} \xrightarrow{\Delta H_f^{\ominus}} \boxed{CS_2(l)} + \boxed{3O_2(g)}$

ΔH_1: $C(s) + 2S(g) \rightarrow \boxed{CO_2(g) + 2SO_2(g)}$; $\Delta H_c^{\ominus}$: $CS_2(l) \rightarrow \boxed{CO_2(g) + 2SO_2(g)}$

$\Delta H_f^{\ominus}[CS_2(l)] + \Delta H_c^{\ominus}[CS_2(l)] = \Delta H_1$

$\Delta H_1 = \Delta H_f^{\ominus}[CO_2(g)] + 2\Delta H_f^{\ominus}[SO_2(g)]$

$\therefore \Delta H_f^{\ominus}[CS_2(l)] = \Delta H_f^{\ominus}[CO_2(g)] + 2\Delta H_f^{\ominus}[SO_2(g)] - \Delta H_c^{\ominus}[CS_2(l)]$

$= (-393.5 + 2(-296.9) + 1075.2)\ \text{kJ mol}^{-1}$

$\Delta H_f^{\ominus}[CS_2(l)] = \boxed{+87.9\ \text{kJ mol}^{-1}}$

Exercise 85

(a) $\boxed{3\tfrac{1}{2}O_2(g)} + \boxed{2C(s) + 3H_2(g)} \xrightarrow{\Delta H_1} \boxed{C_2H_6(g)} + \boxed{3\tfrac{1}{2}O_2(g)}$

ΔH_2: $2C(s) + 3H_2(g) \rightarrow \boxed{2CO_2(g) + 3H_2O(l)}$; ΔH_3: $C_2H_6(g) \rightarrow \boxed{2CO_2(g) + 3H_2O(l)}$

$\Delta H_1 + \Delta H_3 = \Delta H_2$ or $\Delta H_1 = \Delta H_2 - \Delta H_3$

$\Delta H_2 = 2\Delta H_f^{\ominus}[CO_2(g)] + 3\Delta H_f^{\ominus}[H_2O(l)]$

$= (2(-393.5) + 3(-285.9))\ \text{kJ mol}^{-1} = -1644.7\ \text{kJ mol}^{-1}$

$\Delta H_3 = \Delta H_c^{\ominus}[C_2H_6(g)] = -1559.8\ \text{kJ mol}^{-1}$

$\therefore \Delta H_1 = (-1644.7 - (-1559.8))\ \text{kJ mol}^{-1} = \boxed{-84.9\ \text{kJ mol}^{-1}}$

(b) $\boxed{3O_2(g)} + \boxed{2C(s) + 3H_2(g) + \tfrac{1}{2}O_2(g)} \xrightarrow{\Delta H_1} \boxed{C_2H_5OH(l)} + \boxed{3O_2(g)}$

ΔH_2: $2C(s) + 3H_2(g) + \tfrac{1}{2}O_2(g) \rightarrow \boxed{2CO_2(g) + 3H_2O(l)}$; ΔH_3: $C_2H_5OH(l) \rightarrow \boxed{2CO_2(g) + 3H_2O(l)}$

$\Delta H_2 = 2\Delta H_f^{\ominus}[CO_2(g)] + 3\Delta H_f^{\ominus}[H_2O(l)]$

$= (2(-393.5) + 3(-285.9))\ \text{kJ mol}^{-1} = -1644.7\ \text{kJ mol}^{-1}$

$\Delta H_3 = \Delta H_c^{\ominus}[C_2H_5OH(l)] = -1366.7\ \text{kJ mol}^{-1}$

$\Delta H_1 = \Delta H_2 - \Delta H_3 = (-1644.7 - (-1366.7))\ \text{kJ mol}^{-1} = \boxed{-278\ \text{kJ mol}^{-1}}$

(c) $\boxed{2\tfrac{1}{4}O_2(g)} + \boxed{C(s) + 2\tfrac{1}{2}H_2(g) + \tfrac{1}{2}N_2(g)} \xrightarrow{\Delta H_1} \boxed{CH_3NH_2(g)} + \boxed{2\tfrac{1}{4}O_2(g)}$

ΔH_2: $C(s) + 2\tfrac{1}{2}H_2(g) + \tfrac{1}{2}N_2(g) \rightarrow \boxed{CO_2(g) + 2\tfrac{1}{2}H_2O(l) + \tfrac{1}{2}N_2(g)}$; ΔH_3: $CH_3NH_2(g) \rightarrow \boxed{CO_2(g) + 2\tfrac{1}{2}H_2O(l) + \tfrac{1}{2}N_2(g)}$

$\Delta H_2 = \Delta H_f^{\ominus}[CO_2(g)] + 2\tfrac{1}{2}\Delta H_f^{\ominus}[H_2O(l)]$

$= (-393.5 + 2\tfrac{1}{2}(-285.9))\ \text{kJ mol}^{-1} = -1108.3\ \text{kJ mol}^{-1}$

$\Delta H_3 = \Delta H_c^{\ominus}[CH_3NH_2(g)] = -1079.9\ \text{kJ mol}^{-1}$

$\Delta H_1 + \Delta H_3 = \Delta H_2$

$\therefore \Delta H_1 = \Delta H_2 - \Delta H_3 = (-1108.3 - (-1079.9))\ \text{kJ mol}^{-1} = \boxed{-28.4\ \text{kJ mol}^{-1}}$

Exercise 86

(a) $CH_3OH(l) + 1\frac{1}{2}O_2(g) \rightarrow CO_2(g) + 2H_2O(l)$

$\Delta H^\ominus_f$/kJ mol^{-1} -238.9   0   -393.5   -2(285.9)

$\Delta H^\ominus = (-571.8 - 393.5 - (-238.9))$ kJ mol^{-1} = **-726.4 kJ mol^{-1}**

(b) $2CO(g) + O_2(g) \rightarrow 2CO_2(g)$

$\Delta H^\ominus_f$/kJ mol^{-1} -2(110.5)   0   -2(393.5)

$\Delta H^\ominus = (-787.0 - (-221.0))$ kJ mol^{-1} = **-566.0 kJ mol^{-1}**

(c) $ZnCO_3(s) \rightarrow ZnO(s) + CO_2(g)$

$\Delta H^\ominus_f$/kJ mol^{-1} -812.5   -348.0   -393.5

$\Delta H^\ominus = (-393.5 - 348.0 - (-812.5))$ kJ mol^{-1} = **+71.0 kJ mol^{-1}**

(d) $2Al(s) + Fe_2O_3(s) \rightarrow 2Fe(s) + Al_2O_3(s)$

$\Delta H^\ominus_f$/kJ mol^{-1} 0   -822.2   0   -1675.7

$\Delta H^\ominus = (-1675.7 - (-822.2))$ kJ mol^{-1} = **-853.5 kJ mol^{-1}**

Exercise 87

$CH_4(g) + H_2O(g) \rightarrow CO(g) + 3H_2(g)$

$\Delta H^\ominus_f$/kJ mol^{-1} -75   -242   -110   0

$\Delta H^\ominus = (-110 - (-242) - (-75))$ kJ mol^{-1} = **+207 kJ mol^{-1}**

Exercise 88

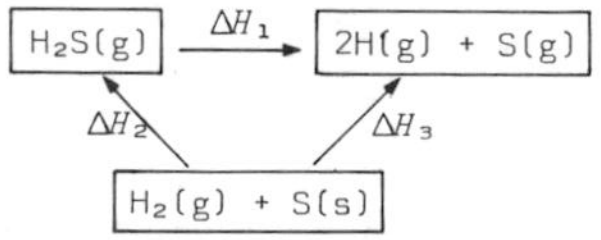

$\Delta H_3 = 2\Delta H^\ominus_{at}[H(g)] + \Delta H^\ominus_{at}[S(s)] = (2(218.0) + 238.1)$ kJ mol^{-1} = 674.1 kJ mol^{-1}

$\Delta H_2 = \Delta H^\ominus_f[H_2S(g)] = -20.6$ kJ mol^{-1}

$\Delta H_1 = -\Delta H_2 + \Delta H_3 = (20.6 + 674.1)$ kJ mol^{-1} = 694.7 kJ mol^{-1}

$\therefore \bar{E}(H\text{-}S) = \frac{694.7}{2}$ kJ mol^{-1} = **347.4 kJ mol^{-1}**

Exercise 89

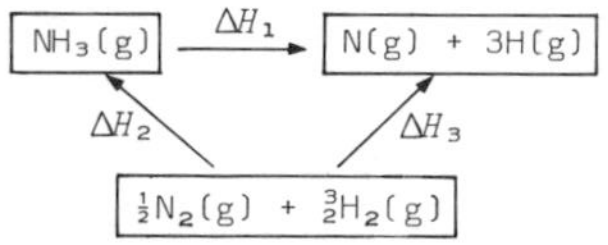

$\Delta H_3 = \Delta H^\ominus_{at}[N(g)] + 3\Delta H^\ominus_{at}[H(g)] = (473 + 3(218))$ kJ mol^{-1} = 1127 kJ mol^{-1}

$\Delta H_2 = -46$ kJ mol^{-1}

$\therefore \Delta H_1 = -\Delta H_2 + \Delta H_3 = (46 + 1127)$ kJ mol^{-1} = 1173 kJ mol^{-1}

$\bar{E}(N\text{—}H) = \frac{1173}{3}$ kJ mol^{-1} = **391 kJ mol^{-1}**

Exercise 90

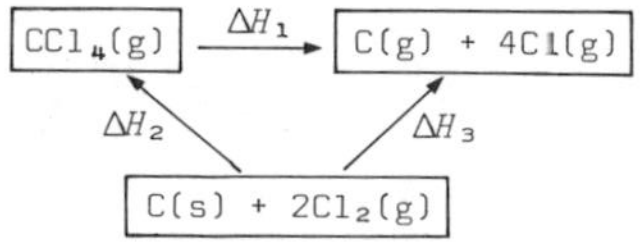

$\Delta H_2 = \Delta H^\ominus_f[CCl_4(l)] + \Delta H^\ominus_{vap}[CCl_4(l)] = (-135.5 + 30.5)$ kJ mol^{-1}

$= -105.0$ kJ mol^{-1}

$\Delta H_3 = \Delta H^\ominus_{at}[C(g)] + 4\Delta H^\ominus_{at}[Cl(g)] = (715.0 + 4(121.1))$ kJ mol^{-1} = 1199.4 kJ mol^{-1}

$\Delta H_1 = -\Delta H_2 + \Delta H_3 = (105.0 + 1199.4)$ kJ mol^{-1} = 1304.4 kJ mol^{-1}

$\therefore \bar{E}(C\text{—}Cl)$ $\frac{1304.4}{4}$ kJ mol^{-1} = **326 kJ mol^{-1}**

Exercise 91

Let x kJ mol^{-1} = $\bar{E}(C\text{—}C)$ and y kJ mol^{-1} = $\bar{E}(C\text{—}H)$

$C_6H_{14}(g) \rightarrow 6C(g) + 14H(g)$; $\Delta H^\ominus = +7512$ kJ mol^{-1}

$C_7H_{16}(g) \rightarrow 7C(g) + 16H(g)$; $\Delta H^\ominus = +8684$ kJ mol^{-1}

$\therefore 5x + 14y = 7512$ - - - - - - - - - (1)

and $6x + 16y = 8684$ - - - - - - - - - (2)

Multiplying equation (1) by 6 and equation (2) by 5 and subtracting:

$$\begin{aligned} 30x + 84y &= 45072 \\ 30x + 80y &= 43420 \\ 4y &= 1652 \end{aligned} \qquad \therefore y = 413$$

Substituting in equation (1)

$5x + 5782 = 7512$ $\therefore x = \frac{7512-5782}{5} = 346$

i.e. $\bar{E}(C\text{—}C)$ = **346 kJ mol^{-1}** and $\bar{E}(C\text{—}H)$ = **413 kJ mol^{-1}**

Exercise 92

(a) $H_2(g) + Cl_2(g) \rightarrow 2H{-}Cl(g)$

Bonds broken: $\bar{E}(H{-}H) + \bar{E}(Cl{-}Cl) = (436 + 242)\ kJ\ mol^{-1} = 678\ kJ\ mol^{-1}$

Bonds made: $-2\bar{E}(H{-}Cl) = -2(431)\ kJ\ mol^{-1} = -862\ kJ\ mol^{-1}$

$\therefore \Delta H = (678 - 862)\ kJ\ mol^{-1} = \boxed{-184\ kJ\ mol^{-1}}$

(b) $N_2(g) + 3H_2(g) \rightarrow 2NH_3(g)$

Bonds broken: $\bar{E}(N{\equiv}N) + 3\bar{E}(H{-}H) = (945 + 3(436))\ kJ\ mol^{-1} = 2253\ kJ\ mol^{-1}$

Bonds formed: $-6\bar{E}(N{-}H) = -6(389)\ kJ\ mol^{-1} = -2334\ kJ\ mol^{-1}$

$\Delta H = (2253 - 2334)\ kJ\ mol^{-1} = \boxed{-81\ kJ\ mol^{-1}}$

Exercise 93

(a) Bond energy terms is the average value of the enthalpy changes for the dissociation of a particular type of bond.

(b) (i) $6H(g) + 3C(g) \rightarrow$ Bonds formed: $6(C{-}H)$, $-6\bar{E}(C{-}H) = -6(415)\ kJ\ mol^{-1}$

$1(C{=}C)$, $-\bar{E}(C{=}C) = -598\ kJ\ mol^{-1}$

$1(C{-}C)$, $-\bar{E}(C{-}C) = -356\ kJ\ mol^{-1}$

$\therefore \Delta H = -(6(415) + 598 + 356)\ kJ\ mol^{-1} = \boxed{-3444\ kJ\ mol^{-1}}$

(ii) $3C(g) + 6H(g) + 2Br(g) \rightarrow$ (g)

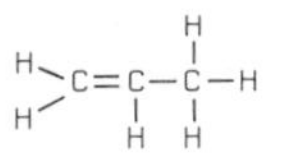

Bonds formed: $6C{-}H$; $-6\bar{E}(C{-}H) = -6(415\ kJ\ mol^{-1})$

$2C{-}Br$; $-2\bar{E}(C{-}Br) = -2(284\ kJ\ mol^{-1})$

$2C{-}C$; $-2\bar{E}(C{-}C) = -2(356\ kJ\ mol^{-1})$

$\therefore \Delta H = -(6(415) + 2(284) + 2(356))\ kJ\ mol^{-1} = \boxed{-3770\ kJ\ mol^{-1}}$

(c) $\boxed{CH_2{=}CH{-}CH_3(g) + Br_2(g)} \xrightarrow{\Delta H_1} \boxed{CH_2BrCHBrCH_3(g)}$

ΔH_2: $\boxed{3C(g) + 6H(g) + 2Br(g)} \rightarrow \boxed{CH_2{=}CH{-}CH_3(g) + Br_2(g)}$; ΔH_3: $\boxed{3C(g) + 6H(g) + 2Br(g)} \rightarrow \boxed{CH_2BrCHBrCH_3(g)}$

$\Delta H_2 = -3444\ kJ\ mol^{-1} - \bar{E}(Br{-}Br) = (-3444 - 193)\ kJ\ mol^{-1} = -3637\ kJ\ mol^{-1}$

$\Delta H_1 = -\Delta H_2 + \Delta H_3 = (3637 + (-3770))\ kJ\ mol^{-1} = \boxed{-133\ kJ\ mol^{-1}}$

Exercise 94

(a) $\boxed{C_2H_6(g) + Cl_2(g)} \xrightarrow{\Delta H_1} \boxed{C_2H_5Cl(g) + HCl(g)}$

ΔH_2: $\boxed{C_2H_6(g) + Cl_2(g)} \rightarrow \boxed{2C(g) + 6H(g) + 2Cl(g)}$; ΔH_3: $\boxed{C_2H_5Cl(g) + HCl(g)} \rightarrow \boxed{2C(g) + 6H(g) + 2Cl(g)}$

$\Delta H_2 = \bar{E}(C{-}C) + 6\bar{E}(C{-}H) + \bar{E}(Cl{-}Cl)$

$= (346 + 6(413) + 242))\ kJ\ mol^{-1} = 3066\ kJ\ mol^{-1}$

$\Delta H_3 = \bar{E}(C{-}C) + 5\bar{E}(C{-}H) + \bar{E}(C{-}Cl) + \bar{E}(H{-}Cl)$

$= (346 + 5(413) + 339 + 431)\ kJ\ mol^{-1} = 3181\ kJ\ mol^{-1}$

$\Delta H_1 = \Delta H_2 - \Delta H_3 = (3066 - 3181)\ kJ\ mol^{-1} = \boxed{-115\ kJ\ mol^{-1}}$

Or, more simply,

$\boxed{C_2H_6(g) + Cl_2(g)} \xrightarrow{\Delta H_1} \boxed{C_2H_5Cl(g) + HCl(g)}$

ΔH_2: $\boxed{C_2H_6(g) + Cl_2(g)} \rightarrow \boxed{C_2H_5(g) + H(g) + 2Cl(g)}$; ΔH_3: $\boxed{C_2H_5Cl(g) + HCl(g)} \rightarrow \boxed{C_2H_5(g) + H(g) + 2Cl(g)}$

$\Delta H_2 = \bar{E}(C{-}H) + \bar{E}(Cl{-}Cl) = (413 + 242)\ kJ\ mol^{-1} = 655\ kJ\ mol^{-1}$

$\Delta H_3 = \bar{E}(C{-}Cl) + \bar{E}(H{-}Cl) = (339 + 431)\ kJ\ mol^{-1} = 770\ kJ\ mol^{-1}$

$\Delta H_1 = \Delta H_2 - \Delta H_3 = (655 - 770)\ kJ\ mol^{-1} = \boxed{-115\ kJ\ mol^{-1}}$

(b)

	$C_2H_6(g)$	$+\ Cl_2(g)$	$\rightarrow\ C_2H_5Cl(g)$	$+\ HCl(g)$
$\Delta H_f^{\ominus}/kJ\ mol^{-1}$	-84.6	0	-136.5	-92.3

$\Delta H^{\ominus} = (-136.5 - 92.3 - (-84.6))\ kJ\ mol^{-1} = \boxed{-144.2\ kJ\ mol^{-1}}$

(c) The bond energy terms $\bar{E}(C{-}H)$ and $\bar{E}(C{-}Cl)$ used in the first calculation are average values and are not quite the same as the bond dissociation energies for the particular compounds C_2H_6 and C_2H_5Cl. If the bond dissociation energies were known and used in the calculation they would give a more accurate result, much closer to $-144.2\ kJ\ mol^{-1}$. Even then, however, the two answers might not quite agree because there is some uncertainty in data book values for $\Delta H_f^{\ominus}$ and bond dissociation energies.

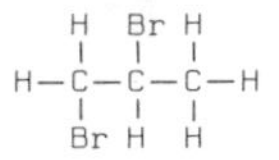

Exercise 95

(a)

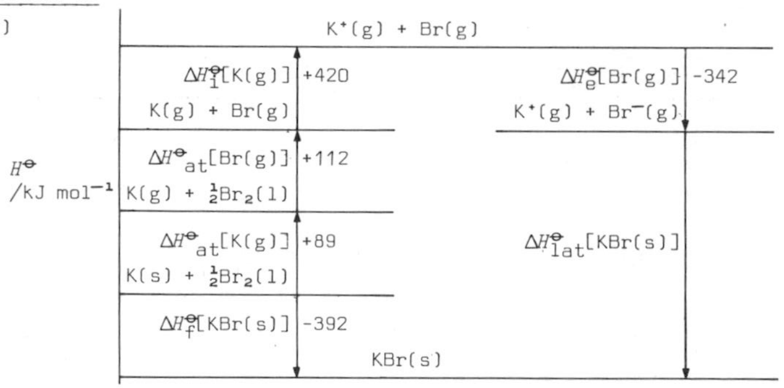

$\Delta H^{\ominus}_{lat}[KBr(s)] = -[392 + 89 + 112 + 420 - 342]\text{kJ mol}^{-1} = \boxed{-671 \text{ kJ mol}^{-1}}$

(b)

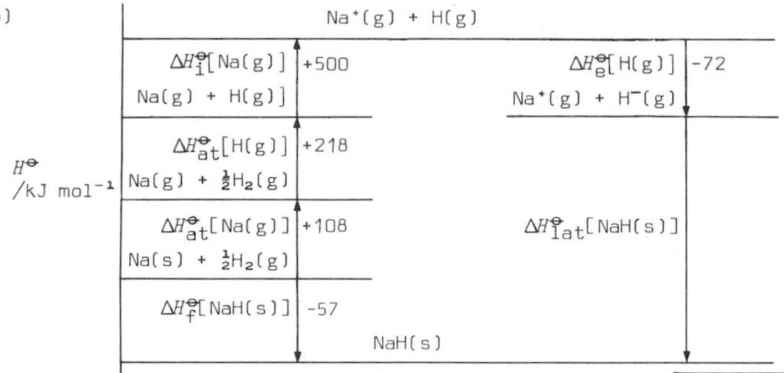

$\Delta H^{\ominus}_{lat}[NaH(s)] = -[57 + 108 + 218 + 500 - 72]\text{kJ mol}^{-1} = \boxed{-811 \text{ kJ mol}^{-1}}$

Exercise 96

(a)

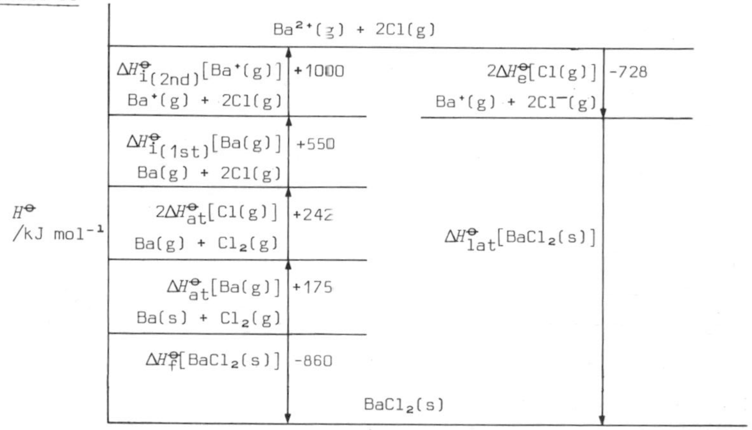

$\Delta H^{\ominus}_{lat}[BaCl_2(s)] = -(860 + 175 + 242 + 500 + 1000 - 728)\text{kJ mol}^{-1}$

$= \boxed{-2049 \text{ kJ mol}^{-1}}$

(b)

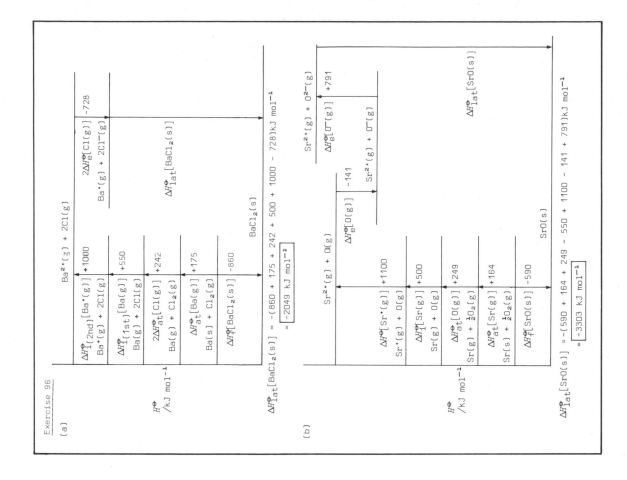

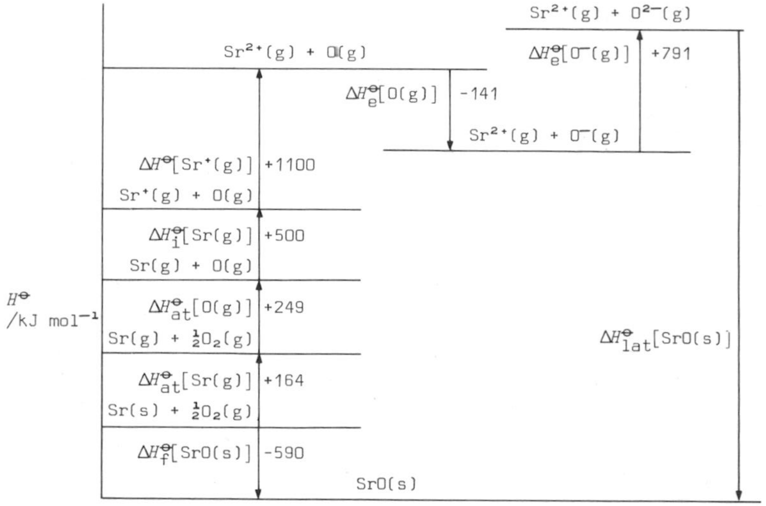

$\Delta H^{\ominus}_{lat}[SrO(s)] = -(590 + 164 + 249 - 550 + 1100 - 141 + 791)\text{kJ mol}^{-1}$

$= \boxed{-3303 \text{ kJ mol}^{-1}}$

Exercise 97

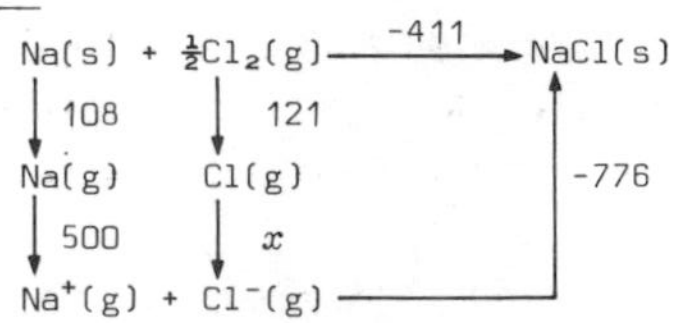

All values in kJ mol^{-1}

$-411 = 108 + 500 + 121 + x - 776$

$\therefore\ x = -411 - 108 - 500 - 121 + 776 = -364$

and $\Delta H^{\ominus}$ = $\boxed{-364\ \text{kJ mol}^{-1}}$

Exercise 98

(a) (i) B is the sum of the first and second ionization energies of calcium.

(ii) C is twice the enthalpy change of atomization of chlorine.

(iii) E is the lattice energy of $CaCl_2(s)$.

(iv) F is the enthalpy change of formation of $CaCl_2(s)$.

(b) F = A + B + C + D + E

$\therefore$ D = F - A - B - C - E = (-795 - 177 - 1690 - 242 + 2197)kJ mol^{-1}

= $\boxed{-707\ \text{kJ mol}^{-1}}$

(This is twice the electron affinity of chlorine.)

Exercise 99

(a) MgCl

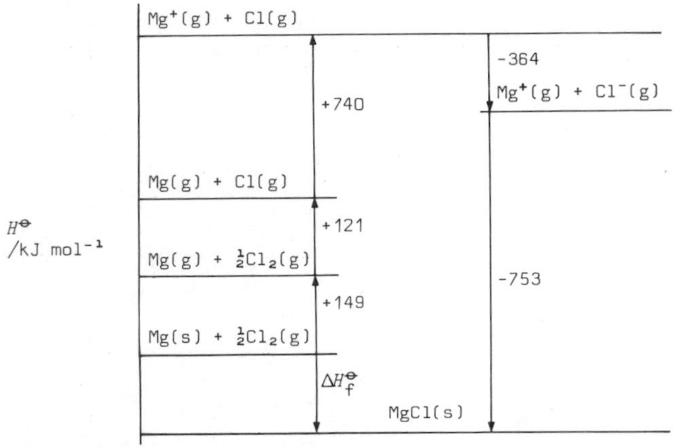

Exercise (Continued)

MgCl$_2$

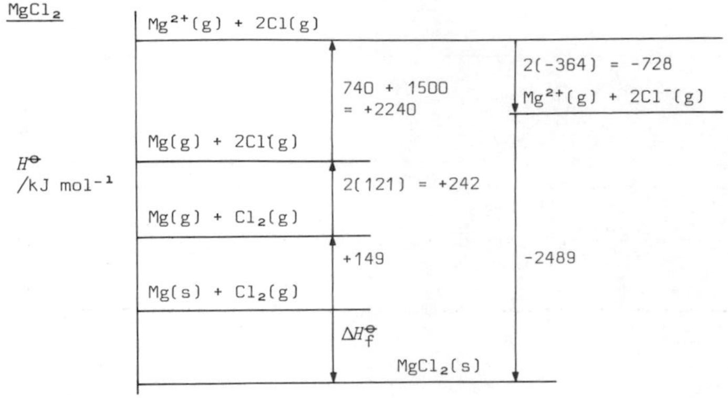

MgCl$_3$

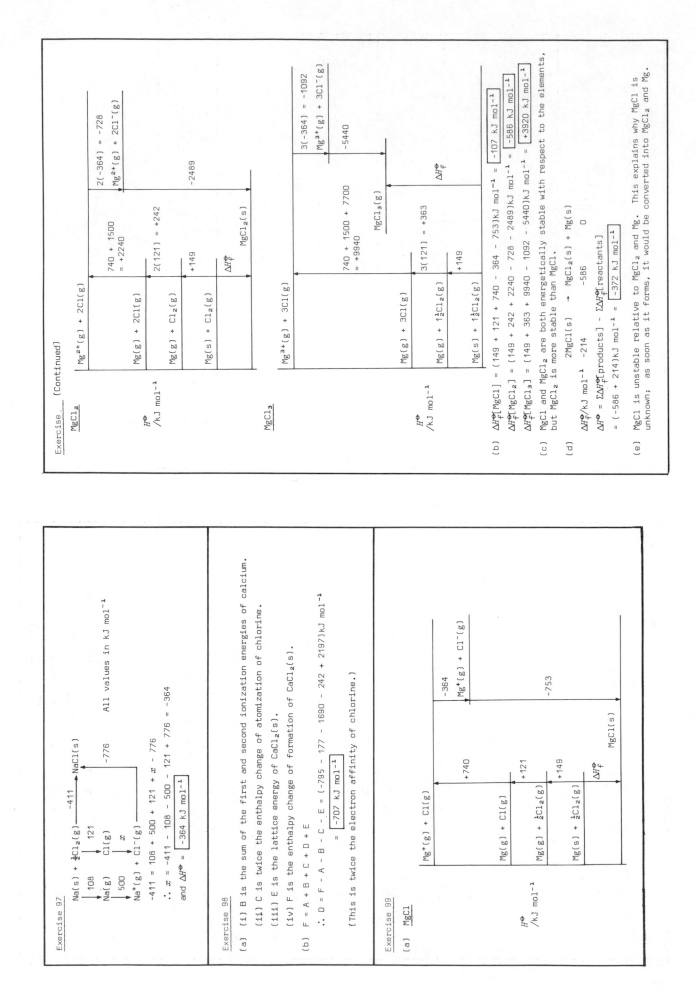

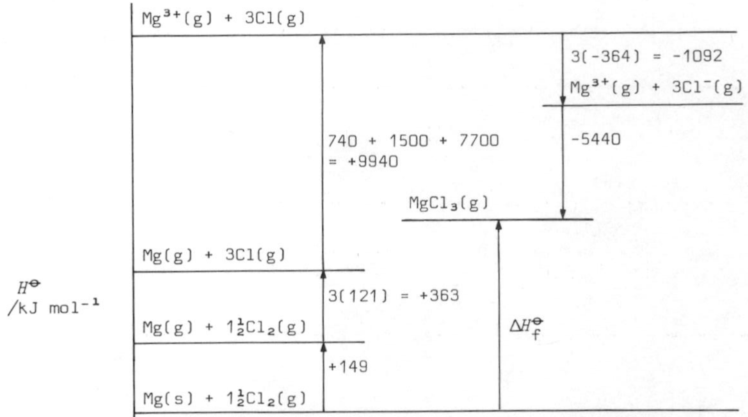

(b) $\Delta H^{\ominus}_f[MgCl]$ = (149 + 121 + 740 - 364 - 753)kJ mol^{-1} = $\boxed{-107\ \text{kJ mol}^{-1}}$

$\Delta H^{\ominus}_f[MgCl_2]$ = (149 + 242 + 2240 - 728 - 2489)kJ mol^{-1} = $\boxed{-586\ \text{kJ mol}^{-1}}$

$\Delta H^{\ominus}_f[MgCl_3]$ = (149 + 363 + 9940 - 1092 - 5440)kJ mol^{-1} = $\boxed{+3920\ \text{kJ mol}^{-1}}$

(c) MgCl and $MgCl_2$ are both energetically stable with respect to the elements, but $MgCl_2$ is more stable than MgCl.

(d)

	$2MgCl(s)$	$\rightarrow$	$MgCl_2(s)$ +	$Mg(s)$
$\Delta H^{\ominus}_f$/kJ mol^{-1}	-214		-586	0

$\Delta H^{\ominus} = \Sigma\Delta H^{\ominus}_f[\text{products}] - \Sigma\Delta H^{\ominus}_f[\text{reactants}]$

= (-586 + 214)kJ mol^{-1} = $\boxed{-372\ \text{kJ mol}^{-1}}$

(e) MgCl is unstable relative to $MgCl_2$ and Mg. This explains why MgCl is unknown; as soon as it forms, it would be converted into $MgCl_2$ and Mg.

Exercise 100

(a)(b)

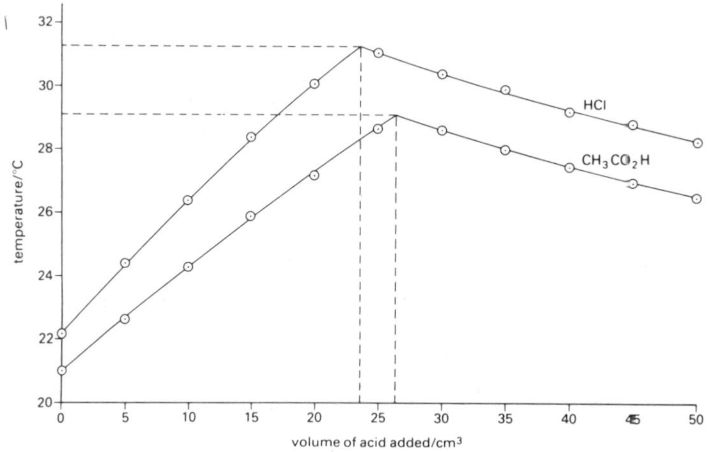

(c) $HCl(aq) + NaOH(aq) \rightarrow NaCl(aq) + H_2O(l)$

Amount of HCl = amount of NaOH

$c \times \frac{23.5}{1000}\,dm^3 = 1.00\,mol\,dm^{-3} \times \frac{50.0}{1000}\,dm^3$

$\therefore c = 1.00\,mol\,dm^{-3} \times \frac{50.0}{23.5} = \boxed{2.13\,mol\,dm^{-3}}$

$CH_3CO_2H(aq) + NaOH(aq) \rightarrow CH_3CO_2Na(aq) + H_2O(l)$

Amount of CH_3CO_2H = amount of NaOH

$\therefore c \times \frac{26.5}{1000}\,dm^3 = 1.00\,mol\,dm^{-3} \times \frac{50.0}{1000}\,dm^3$

$\therefore c = 1.00\,mol\,dm^{-3} \times \frac{50.0}{26.5} = \boxed{1.89\,mol\,dm^{-3}}$

(d) Volume of mixture when reaction is complete = $(50.0 + 23.5)cm^3 = 73.5\,cm^3$

$\Delta T = (31.3 - 22.2)K = 9.1\,K$

$\Delta H = -mc_p\Delta T = -0.0735\,kg \times 4.18\,kJ\,kg^{-1}\,K^{-1} \times 9.1\,K = -2.80\,kJ$

Amount of NaOH used = cV = $1.00\,mol\,dm^{-3} \times 0.0500\,dm^3 = 0.0500\,mol$

Scaling up to 1 mol, $\Delta H = -2.80\,kJ \times \frac{1}{0.0500} = -56.0\,kJ$

$\therefore HCl(aq) + NaOH(aq) \rightarrow NaCl(aq) + H_2O(l);\quad \Delta H^{\ominus} = \boxed{-56.0\,kJ\,mol^{-1}}$

(e) Volume of mixture when reaction is complete = $(50.0 + 26.5)cm^3$

$= 76.5\,cm^3$

$\Delta T = (29.1 - 21.0)K = 8.1\,K$

$\Delta H = -mc_p\Delta T = -0.0765\,kg \times 4.18\,kJ\,kg^{-1}\,K^{-1} \times 8.1\,K = -2.59\,kJ$

Scaling up to 1 mol, $\Delta H = -2.59\,kJ \times \frac{1}{0.0500} = -51.8\,kJ$

$\therefore CH_3CO_2H(aq) + NaOH(aq) \rightarrow CH_3CO_2Na(aq) + H_2O(l);$

$\Delta H^{\ominus} = \boxed{-51.8\,kJ\,mol^{-1}}$

Exercise 101 $\boxed{-75\,kJ\,mol^{-1}}$

Exercise 102 $\boxed{+800\,kJ\,mol^{-1}}$

Exercise 103 $\boxed{-1031\,kJ\,mol^{-1}}$

Exercise 104 $\boxed{+464\,kJ\,mol^{-1}}$

Exercise 105 $\boxed{-367\,kJ\,mol^{-1}}$

Exercise 106 $\boxed{-487\,kJ\,mol^{-1}}$

Exercise 107 $\boxed{+284\,kJ\,mol^{-1}}$

Exercise 108

(b) $\boxed{-3325(\pm 10)\,kJ\,mol^{-1}}$ (c) $\boxed{-279\,kJ\,mol^{-1}}$ (d) $\boxed{-1830\,kJ\,mol^{-1}}$

Exercise 109 (a) $\boxed{+51\,kJ\,mol^{-1}}$ (b) $\boxed{+587\,kJ\,mol^{-1}}$

Exercise 110

(a) A = $Mg^{2+}(g)$; $O^{2-}(g)$ B = $Mg^{2+}(g)$; $\frac{1}{2}O_2(g)$ [or $Mg(g)$; $O(g)$ with C & D reversed]

C = $+250\,kJ\,mol^{-1}$ D = $+2178\,kJ\,mol^{-1}$

E = $+153\,kJ\,mol^{-1}$ F = $-610\,kJ\,mol^{-1}$ G = lattice energy

(b) $\boxed{-3939\,kJ\,mol^{-1}}$

Exercise 111 $\boxed{-57\,kJ\,mol^{-1}}$ (i.e. per mole of water produced)

Exercise 112

(a) $p_1V_1 = p_2V_2$ (Boyle's law)

$\therefore V_2 = V_1 \times \frac{p_1}{p_2} = 25.0\ \text{dm}^3 \times \frac{150\ \text{atm}}{1.00\ \text{atm}} = \boxed{3.75 \times 10^3\ \text{dm}^3}$

(b) $\frac{V_1}{T_1} = \frac{V_2}{T_2}$ (Charles' law)

$\therefore T_2 = T_1 \times \frac{V_2}{V_1} = 288\ \text{K} \times \frac{4.00 \times 10^3\ \text{dm}^3}{3.75 \times 10^3\ \text{dm}^3} = \boxed{307\ \text{K (or 34 °C)}}$

Exercise 113

$p_2 = p_1 \times \frac{V_1}{V_2} = 1.00\ \text{atm} \times \frac{100\ \text{cm}^3}{37.0\ \text{cm}^3} = \boxed{2.70\ \text{atm}}$

Exercise 114

$V_2 = V_1 \times \frac{p_1}{p_2} = 73.0\ \text{cm}^3 \times \frac{1456\ \text{mmHg}}{760\ \text{mmHg}} = \boxed{140\ \text{cm}^3}$

Exercise 115

Combining Boyle's law and Charles's law gives:

$V_2 = V_1 \times \frac{p_1}{p_2} \times \frac{T_2}{T_1}$

$= 38.2\ \text{cm}^3 \times \frac{765\ \text{mmHg}}{760\ \text{mmHg}} \times \frac{273\ \text{K}}{(273 + 18)\text{K}} = \boxed{36.1\ \text{cm}^3}$

Exercise 116

$\frac{p_1V_1}{T_1} = \frac{p_2V_2}{T_2}$

or $V_2 = V_1 \times \frac{p_1}{p_2} \times \frac{T_2}{T_1} = 79.0\ \text{cm}^3 \times \frac{756\ \text{mm}}{760\ \text{mm}} \times \frac{273\ \text{K}}{294\ \text{K}} = \boxed{73.0\ \text{cm}^3}$

Exercise 117

In each case, $V_2 = V_1 \times \frac{p_1}{p_2} \times \frac{T_2}{T_1}$

(a) $V_2 = 29.2\ \text{cm}^3 \times \frac{762}{760} \times \frac{273}{298} = \boxed{26.8\ \text{cm}^3}$

(b) $V_2 = 5.13\ \text{dm}^3 \times \frac{1.02}{1.00} \times \frac{273}{335} = \boxed{4.26\ \text{dm}^3}$

(c) $V_2 = 132\ \text{cm}^3 \times \frac{100}{99.2} \times \frac{298}{273} = \boxed{145\ \text{cm}^3}$

(d) $V_2 = 42.1\ \text{cm}^3 \times \frac{98.7}{27.5} \times \frac{293}{304} = \boxed{146\ \text{cm}^3}$

Exercise 118

$T_2 = T_1 \times \frac{p_2}{p_1} \times \frac{V_2}{V_1} = 300\ \text{K} \times \frac{2}{1} \times \frac{2}{1} = \boxed{1.20 \times 10^3\ \text{K}}$

Exercise 119

$p_2 = p_1 \times \frac{V_1}{V_2} \times \frac{T_2}{T_1} = (28.5 + 14.7)\text{lb in}^{-2} \times \frac{21.0}{21.6} \times \frac{315}{288} = 45.9\ \text{in}^{-2}$

$\therefore$ gauge reads $(45.9 - 14.7)\text{lb in}^{-2} = \boxed{31.2\ \text{lb in}^{-2}}$

Exercise 120

(a)

$2N_2O(g)$	$\rightarrow\ 2N_2(g)$	$+\ O_2(g)$	
2 mol	2 mol	1 mol	
2 volumes	2 volumes	1 volume	(since equal amounts occupy equal volumes)

$\therefore$ volume of N_2O = volume of $N_2 = \boxed{15\ \text{cm}^3}$

and volume of $O_2 = \frac{1}{2}$ volume of $N_2 = \boxed{7.5\ \text{cm}^3}$

(b)

$4NH_3(g)$	$+\ 3O_2(g)$	$\rightarrow\ 2N_2(g)$	$+\ 6H_2O(g)$
4 mol	3 mol	2 mol	6 mol

$\therefore$ 4 volumes + 3 volumes → 8 volumes (since equal amounts occupy equal volumes)

or $\frac{4}{8}$ volume + $\frac{3}{8}$ volume → 1 volume

$\therefore$ volume of $NH_3 = \frac{4}{8} \times$ volume of products

$= \frac{4}{8} \times 128\ \text{cm}^3 = \boxed{64.0\ \text{cm}^3}$

and volume of $O_2 = \frac{3}{8} \times 128\ \text{cm}^3 = \boxed{48.0\ \text{cm}^3}$

Exercise 121

$H_2(g) + Cl_2(g) \rightarrow 2HCl(g)$

$\therefore$ volume of HCl = 2 × volume of H_2 = $2 \times 75\ \text{cm}^3 = \boxed{150\ \text{cm}^3}$

Exercise 122

Volume of $NH_3 = 15\ \text{cm}^3 \times \frac{4}{5} = \boxed{12\ \text{cm}^3}$

Volume of products $= 15\ \text{cm}^3 \times \frac{(4+6)}{5} = \boxed{30\ \text{cm}^3}$

Percentage change $= \frac{\text{volume increase}}{\text{volume at start}} \times 100 = \frac{30\ \text{cm}^3 - 27\ \text{cm}^3}{27\ \text{cm}^3} \times 100 = \boxed{11\%}$

Exercise 123

$C_2H_5OH(g) + 3O_2(g) \rightarrow 2CO_2(g) + 3H_2O(g)$

Volume of O_2 = 3 × volume of C_2H_5OH = $3 \times 78.5\ \text{cm}^3 = \boxed{236\ \text{cm}^3}$

Volume of products $= (3 + 2) \times 78.5\ \text{cm}^3 = \boxed{393\ \text{cm}^3}$

Exercise 124

Volume of oxygen used = starting volume - remainder

= 60 cm^3 - 15 cm^3 = 45 cm^3

Volume of carbon dioxide formed = volume absorbed by NaOH

= 45 cm^3 - 15 cm^3 = 30 cm^3

∴ hydrocarbon(g)	+	oxygen(g)	→	carbon dioxide(g)	+	water(l)
15 cm^3		45 cm^3		30 cm^3		-
1 volume		3 volumes		2 volumes		-
1 mol		3 mol		2 mol		? mol

$\therefore C_xH_y(g) + 3O_2(g) \rightarrow 2CO_2(g) + zH_2O(l)$

	Left-hand side of equation	Right-hand side of equation	Calculation
C atoms	x	2	$x = 2$
O atoms	3 x 2 = 6	(2 x 2) + z = 4 + z	6 = 4 + z ∴ $z = 2$
H atoms	y	$2z$	$y = 2z$ ∴ $y = 4$

The hydrocarbon C_xH_y is therefore C_2H_4.

Exercise 125

Since the hydrocarbon is C_3H_x,

1 mol of C_3H_x gives 3 mol of CO_2 and $\frac{x}{2}$ mol of H_2O

10 cm^3 of C_3H_x gives 30 cm^3 of CO_2 and $5x$ cm^3 of H_2O

∴ the equation can be written:

$C_3H_x(g)$	+	$yO_2(g)$	→	$3CO_2(g)$	+	$\frac{x}{2}H_2O(g)$
10 cm^3		$10y$ cm^3		30 cm^3		$5x$ cm^3

But the products occupy 5 cm^3 more than the reactants (this is true however much excess oxygen there is)

$\therefore 10 + 10y + 5 = 30 + 5x$

$10y = 15 + 5x \quad \text{or } y = \frac{3 + x}{2}$

The equation now becomes:

$C_3H_x(g) + (\frac{3 + x}{2})\ O_2(g) \rightarrow 3CO_2(g) + \frac{x}{2}H_2O(g)$

	Left-hand side of equation	Right-hand side of equation	Calculation
C atoms	3	3	None!
O atoms	$\frac{3 + x}{2} \times 2 = 3 + x$	$(3 \times 2) + \frac{x}{2} = 6 + \frac{x}{2}$	$3 + x = 6 + \frac{x}{2}$ ∴ $\frac{x}{2} = 3$ or $x = 6$
H atoms	x	x	None!

Thus the formula of the hydrocarbon is C_3H_6 i.e. x = 6

The volume of oxygen = 10 y cm^3 = 10 x $(\frac{3 + x}{2})$ cm^3 = 45 cm^3

Exercise 126

Volume of CO_2 = 150 cm^3 - 70 cm^3 = 80 cm^3 = 4 x volume of C_xH_y

Volume of O_2 = 200 cm^3 - 70 cm^3 = 130 cm^3 = $6\frac{1}{2}$ x volume of C_xH_y

$C_xH_y(g) + 6\frac{1}{2}O_2(g) \rightarrow 4CO_2(g) + zH_2O(l)$

C atoms: $x = 4$ O atoms: $13 = 8 + z$ ∴ $z = 5$

H atoms: $y = 2z = 10$ ∴ formula is C_4H_{10}

Exercise 127

Volume of CO_2 = 30 cm^3 = 2 x volume of C_xH_y

Volume of H_2O = 30 cm^3 = 2 x volume of C_xH_y

$C_xH_y(g) + zO_2(g) \rightarrow 2CO_2(g) + 2H_2O(l)$

C atoms: $x = 2$ O atoms: $2z = 4 + 2$ ∴ $z = 3$

H atoms: $y = 4$ ∴ formula is C_2H_4

Exercise 128

Volume of H_2O = 90 cm³ = 3 x volume of C_3H_x

$C_3H_x(g) + yO_2(g) \rightarrow zCO_2(g) + 3H_2O(l)$

C atoms: $3 = z$ O atoms: $2y = 2z + 3$ $\therefore\ y = 4\frac{1}{2}$

Volume of O_2 = $4\frac{1}{2}$ x volume of C_3H_x = $4\frac{1}{2}$ x 30 cm³ = 135 cm³

H atoms: $x = 3 \times 2 = 6$ $\boxed{x = 6}$

Exercise 129

(a) The volume of 0.122 g of H_2 at s.t.p. is given by:

$$V_2 = V_1 \times \frac{p_1}{p_2} \times \frac{T_2}{T_1}$$

$$= 0.211\ \text{dm}^3 \times \frac{7.00\ \text{atm}}{1.00\ \text{atm}} \times \frac{273\ \text{K}}{(273 + 20)\ \text{K}} = 1.38\ \text{dm}^3$$

The amount of H_2 is given by:

$$n = \frac{m}{M} = \frac{0.122\ \text{g}}{2.02\ \text{g mol}^{-1}} = 0.0604\ \text{mol}$$

Also $n = \frac{V}{V_m}$

$$\therefore\ V_m = \frac{V}{n} = \frac{1.38\ \text{dm}^3}{0.0604\ \text{mol}} = \boxed{22.8\ \text{dm}^3\ \text{mol}^{-1}}$$

(b) The volume of 1.10 g of C_4H_{10} at s.t.p. is given by:

$$V_2 = V_1 \times \frac{p_1}{p_2} \times \frac{T_2}{T_1}$$

$$= 34.4\ \text{dm}^3 \times \frac{30.0\ \text{mmHg}}{760\ \text{mmHg}} \times \frac{273\ \text{K}}{(273 + 600)\ \text{K}} = 0.425\ \text{dm}^3$$

The amount of C_4H_{10} is given by:

$$n = \frac{m}{M} = \frac{1.10\ \text{g}}{58.1\ \text{g mol}^{-1}} = 0.0189\ \text{mol}$$

Also, $n = \frac{V}{V_m}$

$$\therefore\ V_m = \frac{V}{n} = \frac{0.425\ \text{dm}^3}{0.0189\ \text{mol}} = \boxed{22.5\ \text{dm}^3}$$

(c) The volume of 2.00 g of O_2 at s.t.p. is given by:

$$V_2 = V_1 \times \frac{p_1}{p_2} \times \frac{T_2}{T_1}$$

$$= 51.9\ \text{dm}^3 \times \frac{5.00\ \text{kPa}}{100\ \text{kPa}} \times \frac{273\ \text{K}}{500\ \text{K}} = 1.42\ \text{dm}^3$$

The amount of O_2 is given by:

$$n = \frac{m}{M} = \frac{2.00\ \text{g}}{32.0\ \text{g mol}^{-1}} = 0.0625\ \text{mol}$$

Also, $n = \frac{V}{V_m}$

$$\therefore\ V_m = \frac{V}{n} = \frac{1.42\ \text{dm}^3}{0.0625\ \text{mol}} = \boxed{22.7\ \text{dm}^3\ \text{mol}^{-1}}$$

Exercise 130

(a) Let the volume of the gas at s.t.p. be V_2. Then

$$V_2 = V_1 \times \frac{p_1}{p_2} \times \frac{T_2}{T_1} = 10.0\ \text{dm}^3 \times \frac{350\ \text{mmHg}}{760\ \text{mmHg}} \times \frac{273\ \text{K}}{(273 + 27)\ \text{K}} = 4.19\ \text{dm}^3$$

Then $n = \frac{V_2}{V_m} = \frac{4.19\ \text{dm}^3}{22.4\ \text{dm}^3\ \text{mol}^{-1}} = \boxed{0.187\ \text{mol}}$

(b) Let the molar volume at 23 °C and 0.200 atm be V_2. Then

$$V_2 = V_1 \times \frac{p_1}{p_2} \times \frac{T_2}{T_1}$$

$$= 22.4\ \text{dm}^3\ \text{mol}^{-1} \times \frac{1.00\ \text{atm}}{0.200\ \text{atm}} \times \frac{(273 + 23)\ \text{K}}{273\ \text{K}} = 121\ \text{dm}^3\ \text{mol}^{-1}$$

The amount, n, of oxygen = $\frac{m}{M}$, but also, $n = \frac{V}{V_m}$

$$\therefore\ \frac{m}{M} = \frac{V}{V_m},\ \text{and}\ V = \frac{m}{M} V_m = \frac{8.00\ \text{g}}{32.0\ \text{g mol}^{-1}} \times 121\ \text{dm}^3\ \text{mol}^{-1} = \boxed{30.3\ \text{dm}^3}$$

Exercise 131

We have two molar volumes at different temperatures but constant pressure. Applying Charles's law,

$$\frac{V_1}{T_1} = \frac{V_2}{T_2}$$

$$\therefore\ T_2 = T_1 \times \frac{V_2}{V_1} = 273\ \text{K} \times \frac{24.0\ \text{dm}^3\ \text{mol}^{-1}}{22.4\ \text{dm}^3\ \text{mol}^{-1}} = \boxed{293\ \text{K}\ \text{or}\ 20\ ^\circ\text{C}}$$

Exercise 132

(a) $V = 86.1\ \text{cm}^3 \times \frac{1.25}{1.00} \times \frac{273}{290} = 101\ \text{cm}^3 = 0.101\ \text{dm}^3$

$$V_m = \frac{V}{n} = \frac{0.101\ \text{dm}^3}{0.0780\ \text{g} / 17.0\ \text{g mol}^{-1}} = \boxed{22.0\ \text{dm}^3\ \text{mol}^{-1}}$$

(b) $V = 90.2\ \text{cm}^3 \times \frac{298}{273} = 98.5\ \text{cm}^3 = 0.0985\ \text{dm}^3$

$$V_m = \frac{V}{n} = \frac{0.0985\ \text{dm}^3}{0.0591\ \text{g} / 16.0\ \text{g mol}^{-1}} = \boxed{26.7\ \text{dm}^3\ \text{mol}^{-1}}$$

Exercise 133

(a) V_m (288 K, 1.01 atm) = $22.4\ \text{dm}^3\ \text{mol}^{-1} \times \frac{1.00}{1.01} \times \frac{288}{273} = 23.4\ \text{dm}^3\ \text{mol}^{-1}$

$$n = \frac{V}{V_m} = \frac{0.500\ \text{dm}^3}{23.4\ \text{dm}^3\ \text{mol}^{-1}} = \boxed{0.0214\ \text{mol}}$$

(b) $n = \frac{V}{V_m} = \frac{0.152\ \text{dm}^3}{22.4\ \text{dm}^3\ \text{mol}^{-1}} = 0.00679\ \text{mol}$

$m = n \times M = 0.00679\ \text{mol} \times 32.0\ \text{g mol}^{-1} = \boxed{0.217\ \text{g}}$

(c) V_m (292 K, 92.8 kPa) = $22.4\ \text{dm}^3\ \text{mol}^{-1} \times \frac{100}{92.8} \times \frac{292}{273} = 25.8\ \text{dm}^3\ \text{mol}^{-1}$

$$V = n \times V_m = \frac{5.81\ \text{g}}{2.02\ \text{g mol}^{-1}} \times 25.8\ \text{dm}^3\ \text{mol}^{-1} = \boxed{74.2\ \text{dm}^3}$$

Exercise 134

(a) Radium and its compounds are radioactive. Radium nuclei emit α-particles, which are helium nuclei, and these pick up stray electrons to become helium atoms.

$$He^{2+} + 2e^- \rightarrow He$$

The amount of He produced in one hour is given by:

$$n = \frac{V}{V_m} = \frac{4.13 \times 10^{-4}\ cm^3}{22.4 \times 10^3\ cm^3\ mol^{-1}} = 1.84 \times 10^{-8}\ mol$$

∴ the number of atoms = 1.84×10^{-8} mol × L (L is the Avogadro constant)

The number of α-particles produced in one hour is:

$$3.07 \times 10^{12} \times 60 \times 60 = 1.11 \times 10^{16}$$

Each α-particle becomes a He atom

∴ $1.11 \times 10^{16} = 1.84 \times 10^{-8}\ mol \times L$

$$\therefore L = \frac{1.11 \times 10^{16}}{1.84 \times 10^{-8}\ mol} = \boxed{6.03 \times 10^{23}\ mol^{-1}}$$

Exercise 135

$pV = nRT$

$$\therefore R = \frac{pV}{nT} = \frac{1.00\ atm \times 22.4\ dm^3}{1.00\ mol \times 273\ K} = \boxed{0.0821\ atm\ dm^3\ K^{-1}\ mol^{-1}}$$

Exercise 136

(a) $R = 0.0821\ atm\ dm^{-3}\ K^{-1}\ mol^{-1}$

$= 0.0821\ (760\ mmHg)\ dm^{-3}\ K^{-1}\ mol^{-1} = \boxed{62.4\ dm^3\ mmHg\ K^{-1}\ mol^{-1}}$

(b) The value of p to use in the calculation is found by making successive substitutions as follows:

$p = 1.00\ atm = 100\ kN\ m^{-2} = 100\ (10^3\ N)\ m^{-2} = 1.00 \times 10^5\ N\ m^{-2}$

$= 1.00 \times 10^5\ (N\ m)\ m^{-3} = 1.00 \times 10^5\ J\ m^{-3}$

$= 1.00 \times 10^5\ J\ (10\ dm)^{-3} = 1.00 \times 10^2\ J\ dm^{-3}$

$$R = \frac{1.00 \times 10^2\ J\ dm^{-3} \times 22.4\ dm^3}{1.00\ mol \times 273\ K} = \boxed{8.21\ J\ mol^{-1}\ K^{-1}}$$

Exercise 137

(a) $pV = nRT$

$$\therefore V = \frac{nRT}{p} = \frac{0.500\ mol \times 0.0821\ atm\ dm^3\ K^{-1}\ mol^{-1} \times 273\ K}{1.00\ atm} = \boxed{11.2\ dm^3}$$

(b) $pV = nRT$ or $V = \frac{nRT}{p}$

where n, the amount of H_2, $= \frac{m}{M} = \frac{1.50\ g}{2.02\ g\ mol^{-1}} = 0.743\ mol$

and $p = 750\ mmHg = \frac{750}{760}\ atm = 0.987\ atm$

$$\therefore V = \frac{0.743\ mol \times 0.0821\ atm\ dm^3\ K^{-1}\ mol^{-1} \times 288\ K}{0.987\ atm} = \boxed{17.8\ dm^3}$$

(c) $pV = nRT$ or $T = \frac{pV}{nR}$

where $p = 760\ mmHg = 1.00\ atm$ and $n = \frac{4.71\ g}{28.0\ g\ mol^{-1}} = 0.168\ mol$

$$\therefore T = \frac{pV}{nR} = \frac{1.00\ atm \times 12.0\ dm^3}{0.168\ mol \times 0.0821\ atm\ dm^3\ K^{-1}\ mol^{-1}} = \boxed{870\ K\ or\ 597\ ^oC}$$

Exercise 138

$$Zn(s) + 2H^+(aq) \rightarrow Zn^{2+}(aq) + H_2(g)$$

∴ the amount of Zn = the amount of H_2

The amount n, of H_2 is given by:

$pV = nRT$ in the form $n = \frac{pV}{RT}$

where $p = \frac{755}{760}\ atm = 0.993\ atm$

$$\therefore n = \frac{0.993\ atm \times 2.00\ dm^3}{0.0821\ atm\ dm^3\ K^{-1}\ mol^{-1} \times 288\ K} = 0.0840\ mol$$

mass of zinc = amount of zinc × molar mass

$= 0.0840\ mol \times 65.4\ g\ mol^{-1} = \boxed{5.49\ g}$

Exercise 139

(a) $C_7H_5N_3O_6(s) + 5\tfrac{1}{4}O_2(g) = 7CO_2(g) + 2\tfrac{1}{2}H_2O(g) + 1\tfrac{1}{2}N_2(g)$

or $4C_7H_5N_3O_6(s) + 21O_2(g) = 28CO_2(g) + 10H_2O(g) + 6N_2(g)$

(b) From the equation, one mole of TNT gives $7 + 2\tfrac{1}{2} + 1\tfrac{1}{2} = 11$ mol of gas

$pV = nRT$ or $V = \dfrac{nRT}{p}$

$\therefore V = \dfrac{11 \text{ mol} \times 0.0821 \text{ atm dm}^3 \text{ mol}^{-1} \text{ K}^{-1} \times 673 \text{ K}}{1.00 \text{ atm}} = \boxed{608 \text{ dm}^3}$

(c) Ratio of volumes $= \dfrac{608 \text{ dm}^3}{0.50 \text{ dm}^3} = \dfrac{1216}{1}$

As a percentage this is $1216 \times 100 = \boxed{1.2 \times 10^5\%}$

(d) $pV = nRT$ or $p = \dfrac{nRT}{V}$

$\therefore p = \dfrac{11.0 \text{ mol} \times 0.0821 \text{ atm dm}^3 \text{ mol}^{-1} \text{ K}^{-1} \times 873 \text{ K}}{2.00 \text{ dm}^3} = \boxed{394 \text{ atm}}$

Exercise 140

(a) $V = \dfrac{nRT}{p} = \dfrac{0.250 \text{ mol} \times 0.0821 \text{ atm dm}^3 \text{ K}^{-1} \text{ mol}^{-1} \times 290 \text{ K}}{(750/760) \text{ atm}} = \boxed{6.03 \text{ dm}^3}$

(b) $V = \dfrac{nRT}{p} = \dfrac{0.521 \text{ g}}{17.0 \text{ g mol}^{-1}} \times \dfrac{0.0821 \text{ atm dm}^3 \text{ K}^{-1} \text{ mol}^{-1} \times 323 \text{ K}}{(762/760) \text{ atm}} = \boxed{0.811 \text{ dm}^3}$

Exercise 141

(a) $n(CaCO_3 \text{ \& } CO_2) = \dfrac{7.31 \text{ g}}{100 \text{ g mol}^{-1}} = 0.0731$ mol

$V = \dfrac{nRT}{p} = \dfrac{0.0731 \text{ mol} \times 0.0821 \text{ atm dm}^3 \text{ K}^{-1} \text{ mol}^{-1} \times 298 \text{ K}}{(764/760) \text{ atm}} = \boxed{1.78 \text{ dm}^3}$

(b) $n(Zn \text{ \& } H_2) = \dfrac{1.52 \text{ g}}{65.4 \text{ g mol}^{-1}} = 0.0232$ mol

$T = \dfrac{pV}{nR} = \dfrac{1.01 \text{ atm} \times 0.557 \text{ dm}^3}{0.0232 \text{ mol} \times 0.0821 \text{ atm dm}^3 \text{ K}^{-1} \text{ mol}^{-1}} = \boxed{295 \text{ K or } 22\ ^\circ\text{C}}$

(c) $n(H_2) = \tfrac{1}{2}n(HCl) = \tfrac{1}{2} \times 0.050 \text{ dm}^3 \times 0.102 \text{ mol dm}^{-3} = 0.00255$ mol

$V = \dfrac{nRT}{p} = \dfrac{0.00255 \text{ mol} \times 0.0821 \text{ atm dm}^{-3} \text{ K}^{-1} \text{ mol}^{-1} \times 287 \text{ K}}{(751/760) \text{ atm}} = \boxed{0.0608 \text{ dm}^3}$

Exercise 142

$p = \dfrac{nRT}{V} = \dfrac{5.00 \text{ g}}{18.0 \text{ g mol}^{-1}} \times \dfrac{0.0821 \text{ atm dm}^3 \text{ K}^{-1} \text{ mol}^{-1} \times 573 \text{ K}}{0.150 \text{ dm}^3} = \boxed{87.1 \text{ atm}}$

Exercise 143

(a) $n = \dfrac{m}{M}$

(b) $pV = nRT = \dfrac{m}{M} \times RT \qquad \therefore M = \dfrac{mRT}{pV}$

(c) $M = \dfrac{mRT}{pV}$ (Don't try to remember this - when you need it derive it as above.)

where $p = \dfrac{740}{760}$ atm $= 0.974$ atm

$\therefore M = \dfrac{3.72 \text{ g} \times 0.0821 \text{ atm dm}^3 \text{ mol}^{-1} \text{ K}^{-1} \times 373 \text{ K}}{0.974 \text{ atm} \times 2.00 \text{ dm}^3} = \boxed{58.5 \text{ g mol}^{-1}}$

Exercise 144

$M = \dfrac{mRT}{pV} = \dfrac{0.198 \text{ g} \times 0.0821 \text{ atm dm}^3 \text{ K}^{-1} \text{ mol}^{-1} \times 294 \text{ K}}{(759/760) \text{ atm} \times 0.109 \text{ dm}^3} = \boxed{43.9 \text{ g mol}^{-1}}$

Exercise 145

(A) $M = \dfrac{mRT}{pV} = \dfrac{0.672 \text{ g} \times 0.0821 \text{ atm dm}^3 \text{ K}^{-1} \text{ mol}^{-1} \times 298 \text{ K}}{1.00 \text{ atm} \times 0.257 \text{ dm}^3} = \boxed{64.0 \text{ g mol}^{-1}}$

(B) $M = \dfrac{mRT}{pV} = \dfrac{0.128 \text{ g} \times 0.0821 \text{ atm dm}^3 \text{ K}^{-1} \text{ mol}^{-1} \times 290 \text{ K}}{(763/760) \text{ atm} \times 0.111 \text{ dm}^3} = \boxed{27.3 \text{ g mol}^{-1}}$

(C) $M = \dfrac{mRT}{pV} = \dfrac{1.60 \text{ g} \times 0.0821 \text{ atm dm}^3 \text{ K}^{-1} \text{ mol}^{-1} \times 294 \text{ K}}{(103/100) \text{ atm} \times 0.526 \text{ dm}^3} = \boxed{71.3 \text{ g mol}^{-1}}$

Exercise 146

Volume of CO_2 = volume of flask

$= \dfrac{\text{mass of water}}{\text{density of water}} = \dfrac{152.8 \text{ g} - 47.9 \text{ g}}{1.00 \text{ gm cm}^{-3}} = 104.9 \text{ cm}^3$

Mass of empty flask = mass of flask and air - mass of air

= 47.933 g - (volume of air x density of air)

= 47.933 g - (104.9 cm^3 x 0.00118 g cm^{-3})

= 47.933 g - 0.124 g = 47.809 g

Mass of CO_2 = mass of flask and CO_2 - mass of flask

= 47.998 g - 47.809 g = 0.189 g

$pV = nRT = \dfrac{mRT}{M} \qquad \therefore M = \dfrac{mRT}{pV}$

If we use $R = 0.0821 \text{ atm dm}^3 \text{ K}^{-1} \text{ mol}^{-1}$,

then p must be in atm, i.e. $\dfrac{757 \text{ atm}}{760 \text{ atm}} = 0.996$ atm

and V must be in dm^3, i.e. 0.105 dm^3

and T must be in K, i.e. (273 + 26) K = 299 K

$\therefore M = \dfrac{0.189 \text{ g} \times 0.0821 \text{ atm dm}^3 \text{ K}^{-1} \text{ mol}^{-1} \times 299 \text{ K}}{0.996 \text{ atm} \times 0.105 \text{ dm}^3} = \boxed{44.4 \text{ g mol}^{-1}}$

Exercise 147

$$M = \frac{mRT}{pV} = \frac{0.25\ cm^3 \times 0.91\ g\ cm^{-3} \times 0.0821\ atm\ dm^3\ K^{-1}\ mol^{-1} \times 433\ K}{(764/760)\ atm \times 0.081\ dm^3}$$

$$= \boxed{99\ g\ mol^{-1}}$$

Exercise 148

(D) $$M = \frac{mRT}{pV} = \frac{0.184\ g \times 0.0821\ atm\ dm^3\ K^{-1}\ mol^{-1} \times 373\ K}{0.984\ atm \times 0.0792\ dm^3} = \boxed{72.3\ g\ mol^{-1}}$$

(E) $$M = \frac{mRT}{pV} = \frac{0.295\ g \times 0.0821\ atm\ dm^3\ K^{-1}\ mol^{-1} \times 413\ K}{(747/760)\ atm \times 0.121\ dm^3} = \boxed{84.1\ g\ mol^{-1}}$$

(F) $$M = \frac{mRT}{pV} = \frac{0.163\ g \times 0.0821\ atm\ dm^3\ K^{-1}\ mol^{-1} \times 374\ K}{(105\ 100)\ atm \times 0.0650\ dm^3} = \boxed{73.3\ g\ mol^{-1}}$$

Exercise 149

(a) Density, ρ, $= \frac{\text{mass of sample}}{\text{volume of sample}} = \frac{M}{V_m}$

Since V_m is constant, $\frac{\rho_{SO_2}}{\rho_{H_2}} = \frac{M_{SO_2}}{M_{H_2}} = \frac{64.0}{2.02} = \boxed{\frac{31.7}{1} \text{ or } 31.7:1}$

(b)

$$\frac{\text{rate of effusion of } H_2}{\text{rate of effusion of } SO_2} = \sqrt{\frac{\rho_{SO_2}}{\rho_{H_2}}} = \sqrt{31.7} = \boxed{5.6}$$

i.e. hydrogen escapes 5.6 times as fast.

Exercise 150

$$\frac{\text{rate of effusion of } H_2}{\text{rate of effusion of CO}} = \sqrt{\frac{\text{density of CO}}{\text{density of } H_2}} = \sqrt{\frac{M_{CO}}{M_{H_2}}}$$

Now the rate of effusion $= \frac{\text{amount effused}}{\text{time taken}}$

$\therefore$ for a constant amount, rate $\propto \frac{1}{\text{time}}$

$$\therefore \frac{\text{time for effusion of CO}}{\text{time for effusion of } H_2} = \sqrt{\frac{M_{CO}}{M_{H_2}}}$$

or $$\left(\frac{t_{CO}}{t_{H_2}}\right)^2 = \frac{M_{CO}}{M_{H_2}}$$

$$\therefore M_{CO} = M_{H_2} \times \left(\frac{t_{CO}}{t_{H_2}}\right)^2 = 2.02\ g\ mol^{-1} \times \left(\frac{93s}{25s}\right)^2 = \boxed{28.0\ g\ mol^{-1}}$$

Exercise 151

$$\text{Ratio} = \sqrt{\frac{238 + 6(19.0)}{235 + 6(19.0)}} = \sqrt{\frac{352}{349}} = 1.004 \text{ or } \boxed{1.004:1}$$

Exercise 152

$$\frac{47\ s}{17\ s} = \sqrt{\frac{32\ g\ mol^{-1}}{M}} \quad \therefore M = 32\ g\ mol^{-1} \times \frac{17^2}{47^2} = \boxed{4.2\ g\ mol^{-1}}$$

$$\frac{47\ s}{t} = \sqrt{\frac{32\ g\ mol^{-1}}{17\ g\ mol^{-1}}} \quad \therefore t = 47\ s \times \sqrt{\frac{17}{32}} = \boxed{34\ s}$$

Exercise 153

$$\frac{28\ s}{24\ s} = \sqrt{\frac{44\ g\ mol^{-1}}{M}} \quad \therefore M = 44\ g\ mol^{-1} \times \frac{24^2}{28^2} = \boxed{32\ g\ mol^{-1}}$$

$$(X_{CO_2} \times 44) + (1 - X_{CO_2})(28) = 32 \quad \therefore X_{CO_2} = \frac{32 - 28}{44 - 28} = \boxed{0.25}$$

and $X_{CO} = \boxed{0.75}$

Exercise 154

After admission of H_2, p_{O_2} is still 0.30 atm and total pressure $= p_{O_2} + p_{H_2} = 0.80$ atm

$\therefore$ $p_{H_2} = 0.80$ atm $- 0.30$ atm $= 0.50$ atm

After admission of N_2, p_{O_2} is still 0.30 atm and p_{H_2} is still 0.50 atm and total pressure $= p_{O_2} + p_{H_2} + p_{N_2} = 0.90$ atm

$\therefore$ $p_{N_2} = 0.90$ atm $- 0.80$ atm $= 0.10$ atm

$$\boxed{p_{O_2} = 0.30\ atm,\ p_{H_2} = 0.50\ atm,\ p_{N_2} = 0.10\ atm}$$

Exercise 155

$$2NH_3(g) \rightarrow N_2(g) + 3H_2(g)$$
1 mol 3 mols

$$X_{N_2} = \frac{1\ mol}{1\ mol + 3\ mol} = \frac{1}{4} \text{ and } X_{H_2} = \frac{3\ mol}{1\ mol + 3\ mol} = \frac{3}{4}$$

$$p_{N_2} = pX_{N_2} = 760\ mmHg \times \frac{1}{4} = \boxed{190\ mmHg}$$

$$p_{H_2} = pX_{H_2} = 760\ mmHg \times \frac{3}{4} = \boxed{570\ mmHg}$$

Exercise 156

$$p_{O_2} = pX_{O_2} = p \times \frac{\text{amount of } O_2}{\text{total amount}}$$

$\therefore$ total amount = amount of $O_2 \times \frac{p}{p_{O_2}} = 0.25 \text{ mol} \times \frac{0.90 \text{ atm}}{0.30 \text{ atm}} = \boxed{0.75 \text{ mol}}$

$$p_{H_2} = pX_{H_2} = p \times \frac{\text{amount of } H_2}{\text{total amount}}$$

$\therefore$ amount of H_2 = total amount $\times \frac{p_{H_2}}{p} = 0.75 \text{ mol} \times \frac{0.50 \text{ atm}}{0.90 \text{ atm}} = \boxed{0.42 \text{ mol}}$

$$p_{N_2} = pX_{N_2} = p \times \frac{\text{amount of } N_2}{\text{total amount}}$$

$\therefore$ amount of N_2 = total amount $\times \frac{p_{N_2}}{p} = 0.75 \text{ mol} \times \frac{0.10 \text{ atm}}{0.90 \text{ atm}} = \boxed{0.08 \text{ mol}}$

(Check: 0.25 mol of O_2 + 0.42 mol of H_2 + 0.08 mol of N_2 = 0.75 mol)

Exercise 157

$X_{H_2O} = \frac{\text{amount of } H_2O}{\text{total amount}} = \frac{\text{volume of } H_2O}{\text{total volume}}$ (applying Avogadro's theory)

$$= \frac{85.0 \text{ cm}^3 - 82.0 \text{ cm}^3}{85.0 \text{ cm}^3} = 0.0353$$

$$p_{H_2O} = pX_{H_2O} = 1.00 \text{ atm} \times 0.0353 = \boxed{0.0353 \text{ atm}}$$

Exercise 158

$$p_{O_2} = pX_{O_2}$$

$$\therefore X_{O_2} = \frac{p_{O_2}}{p} = \frac{2.0 \times 10^4 \text{ N m}^{-2}}{5.0 \times 10^5 \text{ N m}^{-2}} = 4.0 \times 10^{-2}$$

By Avogadro's theory, mole fraction = volume fraction.

Also, volume percentage = volume fraction × 100 = $4.0 \times 10^{-2} \times 100 = \boxed{4.0\%}$

Exercise 159

$$p_{O_2} = 1.00 \text{ atm} \times \frac{200 \text{ cm}^3}{700 \text{ cm}^3} = \boxed{0.286 \text{ atm}}$$

$$p_{He} = 2.00 \text{ atm} \times \frac{500 \text{ cm}^3}{700 \text{ cm}^3} = \boxed{1.43 \text{ atm}}$$

$$p = (1.43 + 0.286) \text{ atm} = \boxed{1.72 \text{ atm}}$$

Exercise 160

$$p_{O_2} = \boxed{1.00 \text{ atm}}$$

$$p_{N_2} = p_{O_2} \times \frac{300}{500} = \boxed{0.600 \text{ atm}}$$

$$p_{CO_2} = 3.10 \text{ atm} - 0.60 \text{ atm} - 1.00 \text{ atm} = \boxed{1.50 \text{ atm}}$$

$$V = 500 \text{ cm}^3 \times \frac{1.50 \text{ atm}}{1.00 \text{ atm}} = \boxed{750 \text{ cm}^3}$$

Exercise 161

$$n = \frac{0.0232 \text{ g}}{28.0 \text{ g mol}^{-1}} + \frac{0.0417 \text{ g}}{28.0 \text{ g mol}^{-1}} = 0.00232 \text{ mol}$$

$$p = \frac{nRT}{V} = \frac{0.00232 \text{ mol} \times 0.0821 \text{ atm dm}^3 \text{ K}^{-1} \text{ mol}^{-1} \times 296 \text{ K}}{0.600 \text{ dm}^3} = \boxed{0.0940 \text{ atm}}$$

$$p_{N_2} = 0.0940 \text{ atm} \times \frac{0.000829 \text{ mol}}{0.00232 \text{ mol}} = \boxed{0.0336 \text{ atm}}$$

$$p_{CO} = 0.0940 \text{ atm} \times \frac{0.00149 \text{ mol}}{0.00232 \text{ mol}} = \boxed{0.0604 \text{ atm}}$$

Exercise 162

$$n = \frac{82.2 \text{ g}}{4.00 \text{ g mol}^{-1}} + \frac{27.4 \text{ g}}{32.0 \text{ g mol}^{-1}} = 20.55 \text{ mol} + 0.86 \text{ mol} = 21.4 \text{ mol}$$

$$V = \frac{nRT}{p} = \frac{21.4 \text{ mol} \times 0.0821 \text{ atm dm}^3 \text{ K}^{-1} \text{ mol}^{-1} \times 284 \text{ K}}{25.0 \text{ atm}} = \boxed{20.0 \text{ dm}^3}$$

$$p_{He} = 25.0 \text{ atm} \times \frac{20.6 \text{ mol}}{21.4 \text{ mol}} = \boxed{24.1 \text{ atm}}$$

$$p_{O_2} = 25.0 \text{ atm} - 24.1 \text{ atm} = \boxed{0.9 \text{ atm}}$$

Exercise 163

$$xCO(g) + yCH_4(g) + (x + 2y)\,O_2(g) \rightarrow (x + y)\,CO_2(g) + 2yH_2O(l)$$

Using relative volumes as stoichiometric coefficients,

volume of initial mixture = $(x + y)$ cm^3 = 15 cm^3 and

contraction in volume = volume of oxygen used up = $(x + 2y)$ cm^3 = 21 cm^3

Hence $y = 6$, $x = 9$

$\therefore X_{CO} = \frac{9}{15}$ and $p_{CO} = 1 \text{ atm} \times \frac{9}{15} = \boxed{0.60 \text{ atm}}$

$X_{CH_4} = \frac{6}{15}$ and $p_{CO} = 1 \text{ atm} \times \frac{6}{15} = \boxed{0.40 \text{ atm}}$

Exercise 164

(a)
$$pV = \frac{1}{3} mN\overline{c^2} \quad \text{and} \quad pV = nRT$$
$$\therefore \quad \frac{1}{3} mN\overline{c^2} = nRT$$
$$\overline{c^2} = \frac{nRT}{\frac{1}{3}mN} = \frac{3nRT}{mN}$$

but mN = mass of each molecule x number of molecules

= total mass of gas = amount of gas x molar mass = nM

Substituting $mN = nM$ in the expression for $\overline{c^2}$:
$$\overline{c^2} = \frac{3nRT}{mN} = \frac{3nRT}{nM} = \frac{3RT}{M}$$
$$\therefore \quad c_{rms} = \boxed{\sqrt{\frac{3RT}{M}}}$$

(b)
$$c_{rms} = \sqrt{\frac{3RT}{M}}$$
$$= \sqrt{\frac{3 \times 8.31 \text{ J K}^{-1} \text{ mol}^{-1} \times 298 \text{ K}}{2.02 \text{ g mol}^{-1}}} = \sqrt{3678 \text{ J g}^{-1}}$$

This would give a very strange unit for speed ($J^{\frac{1}{2}}$ $g^{\frac{1}{2}}$!) To obtain an answer in a more conventional unit, we make the substitution:

$J = kg\ m^2\ s^{-2} = 10^3\ g\ m^2\ s^{-2}$

$$\therefore \quad c_{rms} = \sqrt{3678 \text{ g}^{-1} \times 10^3 \text{ g m}^2 \text{ s}^{-2}} = \sqrt{3.678 \times 10^6 \text{ m}^2 \text{ s}^{-2}}$$
$$= \boxed{1.92 \times 10^3 \text{ m s}^{-1}}$$

(c)
$$c_{rms} = \sqrt{\frac{3RT}{M}} = \sqrt{\frac{3 \times 8.31 \text{ J K}^{-1} \text{ mol}^{-1} \times 1123 \text{ K}}{209 \text{ g mol}^{-1}}} = \sqrt{134.0 \text{ J g}^{-1}}$$
$$= \sqrt{134.0 \text{ g}^{-1} \times 10^3 \text{ g m}^2 \text{ s}^{-2}} = \sqrt{13.4 \times 10^4 \text{ m}^2 \text{ s}^{-2}} = \boxed{3.66 \times 10^2 \text{ m s}^{-1}}$$

Exercise 165

(a) In one revolution of the cylinder, T moves a distance equal to the circumference of the cylinder, i.e. πd or π x 12.0 cm.

In 1.00 s, the cylinder makes 125 revolutions.

∴ the speed of T = 125 x π x 12.0 cm s^{-1} = 4712 cm s^{-1} = $\boxed{47.1 \text{ m s}^{-1}}$

(b) Time taken $= \dfrac{\text{distance}}{\text{speed}} = \dfrac{1.50 \text{ cm}}{4712 \text{ cm s}^{-1}} = \boxed{3.18 \times 10^{-4} \text{ s}}$

(c) Speed $= \dfrac{\text{distance}}{\text{time}} = \dfrac{12.0 \text{ cm}}{3.18 \times 10^{-4} \text{ s}} = 3.77 \times 10^4 \text{ cm s}^{-1} = \boxed{377 \text{ m s}^{-1}}$

(d) The answers are comparable. However, the average speed, the root-mean-square speed and the most popular speed (mode speed) are not quite the same quantities. (You would not be expected to explain the difference unless you study statistics.)

Exercise 166 $\boxed{506 \text{ cm}^3}$

Exercise 167

(d) Gradient = $\boxed{4.50 \text{ atm dm}^3}$ Gradient = nRT

(e) $\boxed{27.9}$ (f) $\boxed{0.951 \text{ g dm}^{-3}}$

Exercise 168 (a) $\boxed{210 \text{ cm}^3}$ (b) $\boxed{270 \text{ cm}^3}$ (c) $\boxed{C_7H_8}$

Exercise 169 $\boxed{309 \text{ K } (36\ ^\circ\text{C})}$

Exercise 170 $\boxed{8.01 \times 10^{16}}$

Exercise 171 $\boxed{159}$

Exercise 172 (a) $\boxed{273 \text{ cm}^3}$ (b) $\boxed{44.8 \text{ g mol}^{-1}}$ (c) $\boxed{C_3H_6}$

Exercise 173 $p_{O_2} = \boxed{0.286 \text{ atm}}$ $p_{He} = \boxed{1.43 \text{ atm}}$ $p_T = \boxed{1.72 \text{ atm}}$

Exercise 174

(a) (i) $K_c = \dfrac{[H_2(g)][Br_2(g)]}{[HBr(g)]^2}$ no unit

(ii) $K_c = \dfrac{[SO_3(g)]^2}{[SO_2(g)]^2[O_2(g)]}$ $mol^{-1}\ dm^3$

(iii) $K_c = \dfrac{[Cu^{2+}(aq)][NH_3(aq)]^4}{[Cu(NH_3)_4{}^{2+}(aq)]}$ $mol^4\ dm^{-12}$

(iv) $K_c = \dfrac{[P_4(g)][F_2(g)]^{10}}{[PF_5(g)]^4}$ $mol^7\ dm^{-21}$

(v) $K_c = \dfrac{[CH_3CO_2C_2H_5(l)][H_2O(l)]}{[C_2H_5OH(l)][CH_3CO_2H(l)]}$ no unit

(b) In equations (i) and (v) the same number of moles appears on each side of the equation. This means that the same number of concentration terms appears at the top and bottom of the equilibrium constant expressions, so that the units of concentration cancel.

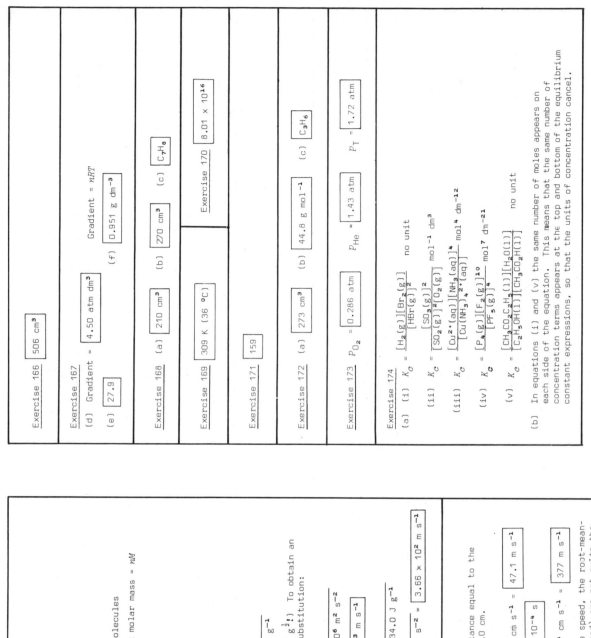

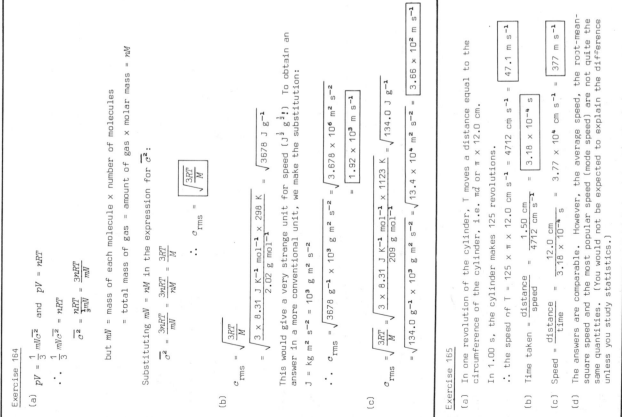

Exercise 175

(a) $K_c = \frac{[CO(g)][Cl_2(g)]}{[COCl_2(g)]}$ $\quad K_c' = \frac{[COCl_2(g)]}{[CO(g)][Cl_2(g)]}$

(b) $K_c = \frac{1}{K_c'}$ (or $K_c' = \frac{1}{K_c}$)

Exercise 176

(a) $K_c = \frac{[NO_2(g)]}{[N_2O_4(g)]^{\frac{1}{2}}}$ $\quad K_c' = \frac{[NO_2(g)]^2}{[N_2O_4(g)]}$

(b) Squaring K_c from (a) gives $K_c^2 = \frac{[NO_2(g)]^2}{[N_2O_4(g)]} = K_c'$

$\therefore$ at 100 $^{\circ}$C, $K_c = \sqrt{K_c'} = (0.490 \text{ mol dm}^{-3})^{\frac{1}{2}} = \boxed{0.700 \text{ mol}^{\frac{1}{2}} \text{ dm}^{-\frac{3}{2}}}$

and at 200 $^{\circ}$C, $K_c = \sqrt{K_c'} = (18.6 \text{ mol dm}^{-3})^{\frac{1}{2}} = \boxed{4.31 \text{ mol}^{\frac{1}{2}} \text{ dm}^{-\frac{3}{2}}}$

Exercise 177

$K_c = \frac{[NO_2]^2}{[N_2O_4]} = \frac{(0.010 \text{ mol dm}^{-3})^2}{0.021 \text{ mol dm}^{-3}} = \boxed{4.76 \times 10^{-3} \text{ mol dm}^{-3}}$

Exercise 178

(a) $[PCl_5(g)] = \frac{0.0042 \text{ mol}}{2.0 \text{ dm}^3} = 0.0021 \text{ mol dm}^{-3}$

$[Cl_2(g)] = [PCl_3(g)] = \frac{0.040 \text{ mol}}{2.0 \text{ dm}^3} = 0.020 \text{ mol dm}^{-3}$

$\therefore K_c = \frac{[PCl_3(g)][Cl_2(g)]}{[PCl_5(g)]} = \frac{(0.020 \text{ mol dm}^{-3})^2}{0.0021 \text{ mol dm}^{-3}} = \boxed{0.19 \text{ mol dm}^{-3}}$

(b) (i) When K_c is large, the concentrations of products are greater than the concentrations of reactants.

(ii) When K_c is small, the concentrations of products are smaller than the concentrations of reactants.

Exercise 179

$K_c = \frac{[PCl_3(g)][Cl_2(g)]}{[PCl_5(g)]} = \frac{(1.50 \times 10^{-2} \text{ mol dm}^{-3}) \times (1.50 \times 10^{-2} \text{ mol dm}^{-3})}{1.18 \times 10^{-3} \text{ mol dm}^{-3}}$

$K_c = \boxed{0.19 \text{ mol dm}^{-3}}$

Exercise 180

$K_c = \frac{[SO_3(g)]^2}{[SO_2(g)]^2 [O_2(g)]} = \frac{(0.92 \text{ mol dm}^{-3})^2}{(0.23 \text{ mol dm}^{-3})^2(1.37 \text{ mol dm}^{-3})} = \boxed{11.7 \text{ dm}^3 \text{ mol}^{-1}}$

Exercise 181

(a) $H_2(g) + I_2(g) \rightleftharpoons 2HI(g)$

(b) $K_c = \frac{[HI(g)]^2}{[H_2(g)][I_2(g)]}$

(c) Mixture 1. $K_c = \frac{(0.1715 \text{ mol}/1.00 \text{ dm}^3)^2}{(0.02265 \text{ mol}/1.00 \text{ dm}^3)\times(0.02840 \text{ mol}/1.00 \text{ dm}^3)}$

$= \frac{0.02941 \text{ mol}^2 \text{ dm}^{-6}}{6.433 \times 10^{-4} \text{ mol}^2 \text{ dm}^{-6}} = \boxed{45.72}$

Mixture 2. $K_c = \frac{(0.1779 \text{ mol}/1.00 \text{ dm}^3)^2}{(0.01699 \text{ mol}/1.00 \text{ dm}^3) \times (0.04057 \text{ mol}/1.00 \text{ dm}^3)}$

$= \frac{0.03165 \text{ mol}^2 \text{ dm}^{-6}}{6.893 \times 10^{-4} \text{ mol}^2 \text{ dm}^{-6}} = \boxed{45.92}$

(d) $K_c = \frac{(0.1715 \text{ mol}/2.00 \text{ dm}^3)^2}{(0.02265 \text{ mol}/2.00 \text{ dm}^3) \times (0.02840 \text{ mol}/2.00 \text{ dm}^3)}$

$= \frac{7.353 \times 10^{-3} \text{ mol}^2 \text{ dm}^{-6}}{1.608 \times 10^{-4} \text{ mol}^2 \text{ dm}^{-6}} = \boxed{45.72}$

(Did you notice the short cut? 2.00 dm^3 cancels, giving the same expression as in part (c).)

(e) $K_c = \frac{(0.1779 \text{ mol}/V \text{ dm}^3)^2}{(0.01699 \text{ mol}/V \text{ dm}^3) \times (0.04057 \text{ mol}/V \text{ dm}^3)}$

$= \frac{(0.1779 \text{ mol})^2}{0.01699 \text{ mol} \times 0.04057 \text{ mol}} \times \frac{V^2 \text{ dm}^6}{V^2 \text{ dm}^6} = \boxed{45.92}$

Exercise 182

Amount of HCl = amount of NaOH = cV = 0.974 mol dm^{-3} × 0.0105 dm^3

= 0.0102 mol

Mass of HCl (pure) = nM = 0.0102 mol × 36.5 g mol^{-1} = 0.372 g

Mass of H_2O in HCl(aq) = 5.17 g - 0.372 g = 4.80 g

Total mass of H_2O at start = 4.80 g + 0.99 g = 5.79 g

Amount of H_2O at start = m/M = 5.79 g/18.0 g mol^{-1} = 0.322 mol

Total eqm amount of acid = amount of NaOH = cV

= 0.974 mol dm^{-3} × 0.0392 dm^3 = 0.0382 mol

Eqm amount of CH_3CO_2H = total amount - amount of HCl

= (0.0382 - 0.0102) mol = 0.0280 mol

Eqm amount of C_2H_5OH = amount of CH_3CO_2H = 0.0280 mol

Initial amount of $CH_3CO_2C_2H_5$ = m/M = 3.64 g/88.1 g mol^{-1} = 0.0413 mol

Eqm amount of $CH_3CO_2C_2H_5$ = initial amount - amount reacted

= (0.0413 - 0.0280)mol = 0.0133 mol

Eqm amount of H_2O = initial amount - amount reacted

= (0.322 - 0.0280)mol = 0.294 mol

$K_c = \frac{[C_2H_5OH][CH_3CO_2H]}{[CH_3CO_2C_2H_5][H_2O]} = \frac{0.0280 \times 0.0280}{0.0133 \times 0.294}$ (units cancel)

$= \boxed{0.201}$

Exercise 183

	N_2O_4	$\rightleftharpoons$	$2NO_2$
Equilibrium concn/mol dm^{-3}	x		1.85×10^{-3}

$$K_c = \frac{[NO_2]^2}{[N_2O_4]}$$

$$\therefore \quad 1.06 \times 10^{-5}\ \text{mol dm}^{-3} = \frac{(1.85 \times 10^{-3}\ \text{mol dm}^{-3})^2}{x\ \text{mol dm}^{-3}}$$

$$\therefore \quad x = \frac{(1.85 \times 10^{-3})^2}{(1.06 \times 10^{-5})} = 0.323$$

and $[N_2O_4]$ = $\boxed{0.323\ \text{mol dm}^{-3}}$

Exercise 184

	$2H_2S(g)$	$\rightleftharpoons$	$2H_2(g)$	+	$S_2(g)$
Equilibrium concn/mol dm^{-3}	4.84×10^{-3}		x		2.33×10^{-3}

$$K_c = \frac{[H_2(g)]^2[S_2(g)]}{[H_2S(g)]^2}$$

$$2.25 \times 10^{-4}\ \text{mol dm}^{-3} = \frac{(x\ \text{mol dm}^{-3})^2 \times (2.33 \times 10^{-3}\ \text{mol dm}^{-3})}{(4.84 \times 10^{-3}\ \text{mol dm}^{-3})^2}$$

$$x^2 = \frac{(2.25 \times 10^{-4})(4.84 \times 10^{-3})^2}{(2.33 \times 10^{-3})} = 2.26 \times 10^{-6}$$

$$x = \sqrt{2.26 \times 10^{-6}} = 1.50 \times 10^{-3}$$

$$\therefore \quad [H_2(g)] = \boxed{1.50 \times 10^{-3}\ \text{mol dm}^{-3}}$$

Exercise 185

$$K_c = \frac{[HI(g)]^2}{[H_2(g)][I_2(g)]} \quad \therefore \quad 54.1 = \frac{(3.53 \times 10^{-3}\ \text{mol dm}^{-3})^2}{(0.48 \times 10^{-3}\ \text{mol dm}^{-3} \times x)}$$

$$x = [I_2(g)] = \frac{1.25 \times 10^{-5}\ \text{mol}^2\ \text{dm}^{-6}}{0.026\ \text{mol dm}^{-3}} = \boxed{4.8 \times 10^{-4}\ \text{mol dm}^{-3}}$$

Exercise 186

$$K_c = \frac{[CH_3CO_2C_2H_5(l)][H_2O(l)]}{[CH_3CO_2H(l)][C_2H_5OH(l)]}$$

Because the same number of moles appear on both sides of the equation, it is possible to substitute amounts rather than concentrations.

$$\text{i.e. } 4.0 = \frac{0.66\ \text{mol} \times 0.66\ \text{mol}}{0.33\ \text{mol} \times x}$$

$$x = \text{amount of } C_2H_5OH = \frac{0.436\ \text{mol}^2}{1.32\ \text{mol}} = \boxed{0.33\ \text{mol}}$$

Exercise 187

	$PCl_5(g)$	$\rightleftharpoons$	$PCl_3(g)$	+	$Cl_2(g)$
(a) Equilibrium amount/mol	x		0.15		0.090

$$K_c = \frac{[PCl_3(g)][Cl_2(g)]}{[PCl_5(g)]}$$

Substituting concentrations gives

$$0.19\ \text{mol dm}^{-3} = \frac{(0.15\ \text{mol}/2.0\ \text{cm}^3)(0.090\ \text{mol}/2.0\ \text{dm}^3)}{(x\ \text{mol}/2.0\ \text{dm}^3)}$$

$$\text{i.e. } 0.19 = \frac{0.075 \times 0.045}{x/\ 2.0} = \frac{0.075 \times 0.045 \times 2.0}{x}$$

$$\therefore \quad x = \frac{0.075 \times 0.045 \times 2.0}{0.19} = 0.0355$$

$$\therefore \quad \text{amount of } PCl_5 = \boxed{0.036\ \text{mol}}$$

(b) mass = amount x molar mass

$$= 0.0355\ \text{mol} \times 208.5\ \text{g mol}^{-1} = \boxed{7.40\ \text{g}}$$

Exercise 188

Amount of pentene at equilibrium = initial amount - amount reacted
= $(0.020 - 9.0 \times 10^{-3})$ mol = 0.011 mol

Amount of ethanoic acid at equilibrium = initial amount - amount reacted
= $(0.010 - 9.0 \times 10^{-3})$ mol = 1.0×10^{-3} mol

	CH_3CO_2H	+	C_5H_{10}	$\rightleftharpoons$	$CH_3CO_2C_5H_{11}$
Initial amount/mol	0.010		0.020		0
Equilibrium amount/mol	1.0×10^{-3}		0.011		9.0×10^{-3}

$$K_c = \frac{[CH_3CO_2C_5H_{11}]}{[CH_3CO_2H][C_5H_{10}]}$$

$$= \frac{(9.0 \times 10^{-3}\ \text{mol}/0.600\ \text{dm}^3)}{(1.0 \times 10^{-3}\ \text{mol}/0.600\ \text{dm}^3) \times (0.011\ \text{mol}/0.600\ \text{dm}^3)} = \boxed{491\ \text{dm}^3\ \text{mol}^{-1}}$$

Exercise 189

Amount of H_2 at equilibrium = initial amount - amount reacted.

The equation shows that 2 mol of HI requires 1 mol of H_2 to react.

$\therefore$ to produce 3.00 mol of HI, the amount of H_2 reacted = 1.50 mol

Amount of H_2 at equilibrium = 1.90 mol - 1.50 mol = 0.40 mol

The calculation for I_2 is the same. Since the equation has equal numbers of moles on each side, we can substitute amounts rather than concentrations in the equilibrium law expression. (Volume cancels.)

	$H_2(g)$	+	$I_2(g)$	$\rightleftharpoons$	$2HI(g)$
Initial amount/mol	1.90		1.90		0
Equilibrium amount/mol	0.40		0.40		3.00

$$K_c = \frac{[HI(g)]^2}{[H_2(g)][I_2(g)]} = \frac{(3.00\ \text{mol})^2}{(0.40\ \text{mol})(0.40\ \text{mol})} = \boxed{56}$$

Exercise 190

Initial amount of $CH_3CO_2H = \frac{6.0\ g}{60.1\ g\ mol^{-1}} = 0.10\ mol$

Initial amount of $C_2H_5OH = \frac{6.9\ g}{46.1\ g\ mol^{-1}} = 0.15\ mol$

Equilibrium amount of $CH_3CO_2C_2H_5 = \frac{7.0\ g}{88.1\ g} = 0.079\ mol$

The equation shows that the equilibrium amount of H_2O is also 0.079 mol and that this is formed from an equal amount of CH_3CO_2H or C_2H_5OH.

Equilibrium amount of CH_3CO_2H = initial amount - amount reacted.

$= 0.10\ mol - 0.079\ mol = 0.021\ mol$

Equilibrium amount of $C_2H_5OH = 0.15\ mol - 0.079\ mol = 0.071\ mol$

	CH_3CO_2H +	C_2H_5OH ⇌	$CH_3CO_2C_2H_5$ +	H_2O
Initial amount/mol	0.10	0.15	0	0
Equilibrium amount/mol	0.021	0.071	0.079	0.079

Since the equation has equal numbers of moles on each side, we can substitute amounts rather than concentrations in the equilibrium law expression.

$$K_c = \frac{[CH_3CO_2C_2H_5][H_2O]}{[CH_3CO_2H][C_2H_5OH]} = \frac{(0.079)^2}{0.021 \times 0.071} = \boxed{4.2}$$

Exercise 191

Amount of H_2 at equilibrium = initial amount - amount reacted

$= (20.57 - \frac{1}{2} \times 10.22)\ mol = 15.46\ mol$

Amount of I_2 at equilibrium = initial amount - amount reacted

$= (5.22 - \frac{1}{2} \times 10.22)\ mol = 0.11\ mol$

	$H_2(g)$ +	$I_2(g)$ ⇌	$2HI(g)$
Initial amount/mol	20.57	5.22	0
Equilibrium amount/mol	15.46	0.11	10.22

$$K_c = \frac{[HI(g)]^2}{[H_2(g)][I_2(g)]} = \frac{(10.22\ mol)^2}{(15.46\ mol)(0.11\ mol)} = \boxed{61} \qquad \text{(Volume cancels)}$$

Exercise 192

Equilibrium concentration of N_2O_4 = initial concn - concn reacted

$= (0.1307 - 0.0007)\ mol\ dm^{-3} = 0.1300\ mol\ dm^{-3}$

	$N_2O_4(l)$ ⇌	$2NO_2(l)$
Initial concn/mol dm^{-3}	0.1307	0
Equilibrium concn/mol dm^{-3}	0.1300	0.0014

$$K_c = \frac{[NO_2(l)]^2}{[N_2O_4(l)]} = \frac{(0.0014\ mol\ dm^{-3})^2}{(0.1300\ mol\ dm^{-3})} = \boxed{1.51 \times 10^{-5}\ mol\ dm^{-3}}$$

Exercise 193

	$CO(g)$ +	$H_2O(g)$ ⇌	$CO_2(g)$ +	$H_2(g)$
Initial amount/mol	3.0	3.0	0	0
Equilibrium amount/mol	$3.0 - x$	$3.0 - x$	x	x

$$K_c = \frac{[CO_2(g)][H_2(g)]}{[CO(g)][H_2O(g)]}$$

$$4.00 = \frac{(x\ mol)(x\ mol)}{(3.0 - x)\ mol\ (3.0 - x)\ mol} = \frac{x^2}{(3.0 - x)^2}$$

Taking the square root of both sides,

$$2.00 = \frac{x}{3.0 - x}$$

$\therefore\ 6.0 - 2.00\,x = x$ or $3x = 6.0$

$\therefore\ x = 2.0$ and amount of hydrogen = $\boxed{2.0\ mol}$

Exercise 194

Amount of $PCl_5 = \frac{2.085\ g}{208.5\ g\ mol^{-1}} = 0.0100\ mol$

	$PCl_5(g)$ ⇌	$PCl_3(g)$ +	$Cl_2(g)$
Initial amount/mol	0.0100	0	0
Equilibrium amount/mol	$0.0100 - x$	x	x

$$K_c = \frac{[PCl_3(g)][Cl_2(g)]}{[PCl_5(g)]}$$

$$0.19\ mol\ dm^{-3} = \frac{(\frac{x}{0.500}\ mol\ dm^{-3})(\frac{x}{0.500}\ mol\ dm^{-3})}{(\frac{0.0100 - x}{0.500}\ mol\ dm^{-3})}$$

$$0.19 = \frac{x^2}{0.500(0.0100 - x)} = \frac{x^2}{0.00500 - 0.500\,x}$$

$9.5 \times 10^{-4} - 0.095\,x = x^2$ or $x^2 + 0.095x - (9.5 \times 10^{-4}) = 0$

$$x = \frac{-b \pm \sqrt{b^2 - 4ac}}{2a} \quad \text{where } a = 1,\ b = 0.095 \text{ and } c = -(9.5 \times 10^{-4})$$

$$x = \frac{-0.095 \pm \sqrt{(0.095)^2 + (4 \times 9.5 \times 10^{-4})}}{2}$$

$$2x = -0.095 \pm \sqrt{9.025 \times 10^{-3} + 3.8 \times 10^{-3}}$$

$$= -0.095 \pm \sqrt{1.2825 \times 10^{-2}} = -0.095 \pm 0.1132$$

$\therefore\ x = 0.0091$ or -0.1041 (absurd root)

$$[PCl_5(g)] = \frac{(0.0100 - 0.0091)\ mol}{0.500\ dm^3} = \boxed{0.0018\ mol\ dm^{-3}}$$

$$[Cl_2(g)] = [PCl_3(g)] = \frac{0.0091\ mol}{0.500\ dm^3} = \boxed{0.0182\ mol\ dm^{-3}}$$

Exercise 195

	$H_2(g)$	+	$I_2(g)$	$\rightleftharpoons$	$2HI(g)$
Initial amount/mol	2.00		2.00		0
Equilibrium amount/mol	$2.00 - x$		$2.00 - x$		$2x$

$$K_c = \frac{[HI(g)]^2}{[H_2(g)][I_2(g)]}$$

$$49.0 = \frac{(2x)^2}{(2.00-x)(2.00-x)}$$

Taking square roots, $7.00 = \dfrac{2x}{2.00 - x}$

$14.0 - 7.00x = 2x$

$9.00x = 14.0$ and $x = 1.56$

$\therefore$ amount of HI = $2x$ mol = **3.12 mol**

amount of H_2 = amount of I_2 = $(2.00 - x)$ mol = **0.44 mol**

Exercise 196

Initial amount of ethanol = $\dfrac{50.0\ g}{46.1\ g\ mol^{-1}}$ = 1.08 mol

Initial amount of propanoic acid = $\dfrac{50.0\ g}{74.1\ g\ mol^{-1}}$ = 0.675 mol

	$C_2H_5OH(l)$	+	$C_2H_5CO_2H(l)$	$\rightleftharpoons$	$C_2H_5CO_2C_2H_5(l)$	+	$H_2O(l)$
Initial amount/mol	1.08		0.675		0		0
Equilibrium amount/mol	$1.08 - x$		$0.675 - x$		x		x

$$K_c = \frac{[C_2H_5CO_2C_2H_5(l)][H_2O(l)]}{[C_2H_5OH(l)][C_2H_5CO_2H(l)]}$$

$$7.5 = \frac{x \times x}{(1.08 - x)(0.675 - x)} = \frac{x^2}{0.729 - 1.755\,x + x^2}$$

$5.47 - 13.16\,x + 7.5\,x^2 = x^2$

$6.5\,x^2 - 13.16\,x + 5.47 = 0$

$$x = \frac{-b \pm \sqrt{b^2 - 4ac}}{2a}$$

$$= \frac{13.16 \pm \sqrt{(13.16)^2 - (4 \times 6.5 \times 5.47)}}{2 \times 6.5}$$

$13\,x = 13.16 \pm \sqrt{173.2 - 142.2}$

$= 13.16 \pm \sqrt{31.0} = 13.16 \pm 5.57$

$\therefore x = 0.584$ or $x = 1.44$ (absurd root)

$\therefore$ amount of ethyl propanoate = 0.584 mol

and mass of ethyl propanoate = 0.584 mol × 102.1 g mol^{-1} = **59.6 g**

Exercise 197

	$CH_3CO_2H(l)$	+	$C_2H_5OH(l)$	$\rightleftharpoons$	$CH_3CO_2C_2H_5(l)$	+	$H_2O(l)$
Initial amount/mol	8.0		6.0		0		0
Equilibrium amount/mol	$8.0 - x$		$6.0 - x$		x		x

$$K_c = \frac{[CH_3CO_2C_2H_5(l)][H_2O(l)]}{[CH_3CO_2H(l)][C_2H_5OH(l)]}$$

$$4.5 = \frac{x \times x}{(8.0 - x)(6.0 - x)} = \frac{x^2}{48.0 - 14.0x + x^2}$$

$3.5x^2 - 63.0\,x + 216.0 = 0$

$$x = \frac{-(-63.0) \pm \sqrt{(-63.0)^2 - (4 \times 3.5 \times 216.0)}}{2 \times 3.5}$$

$$= \frac{63.0 \pm \sqrt{945}}{7.0} = 4.61 \text{ or } 13.4 \text{ (absurd root)}$$

$\therefore$ equilibrium amount of water = **4.6 mol**

Exercise 198

	$PCl_5(g)$	$\rightleftharpoons$	$PCl_3(g)$	+	$Cl_2(g)$
Initial amount/mol	0.019		0		0
Equilibrium amount/mol	$0.019 - x$		x		x

$$K_c = \frac{[PCl_3(g)][Cl_2(g)]}{[PCl_5(g)]}$$

$$0.19\ mol\ dm^{-3} = \frac{(x\ mol/0.75\ dm^3) \times (x\ mol/0.75\ dm^3)}{(0.019 - x)\ mol/0.75\ dm^3}$$

$$0.19 = \frac{x^2}{(0.75)(0.019 - x)}$$

$x^2 + 0.1425x - 2.71 \times 10^{-3} = 0$

$$x = \frac{-0.1425 \pm \sqrt{(0.1425)^2 - (4)(1)(-2.71 \times 10^{-3})}}{2 \times 1} = \frac{-0.1425 \pm \sqrt{0.0311}}{2}$$

= 0.017 and −0.159 (absurd root)

Amount of PCl_5 at equilibrium = (0.019 − 0.017) mol = **2.0×10^{-3} mol**

Exercise 199

$$K_p = \frac{p^2_{SO_3}}{p^2_{SO_2} \times p_{O_2}} = \frac{(4.5\ atm)^2}{(0.090\ atm)^2 \times 0.083\ atm} = 3.01 \times 10^4\ atm^{-1}$$

Exercise 200

$$K_p = \frac{p^2_{NO_2}}{p_{N_2O_4}} = \frac{(0.67\ atm)^2}{0.33\ atm} = 1.36\ atm$$

Exercise 201

$$K_p = \frac{p^2_{HI}}{p_{H_2} \times p_{I_2}} = \frac{(0.40\ atm)^2}{(0.25\ atm) \times (0.16\ atm)} = \boxed{4.0}$$

Exercise 202

(a) Amount of $H_2 = \frac{13.0\ g}{2.02\ g\ mol^{-1}} = 6.40\ mol$

Amount of $NH_3 = \frac{25.1\ g}{17.0\ g\ mol^{-1}} = 1.48\ mol$

Amount of $N_2 = \frac{59.6\ g}{28.0\ g\ mol^{-1}} = 2.13\ mol$

Total amount = (6.40 + 1.48 + 2.13) mol = 10.0 mol

$$X_{H_2} = \frac{\text{amount of } H_2}{\text{total amount}} = \frac{6.40\ mol}{10.0\ mol} = \boxed{0.640}$$

$$X_{NH_3} = \frac{1.48\ mol}{10.0\ mol} = \boxed{0.148}$$

$$X_{N_2} = \frac{2.13\ mol}{10.0\ mol} = \boxed{0.213}$$

(b) Partial pressure = mole fraction x total pressure

$p_{H_2} = 0.640 \times 10.0\ atm = \boxed{6.40\ atm}$

$p_{NH_3} = 0.148 \times 10.0\ atm = \boxed{1.48\ atm}$

$p_{N_2} = 0.213 \times 10.0\ atm = \boxed{2.13\ atm}$

(c) $$K_p = \frac{p^2_{NH_3}}{p_{N_2} \times p^3_{H_2}} = \frac{(1.48\ atm)^2}{2.13\ atm \times (6.40\ atm)^3} = \boxed{3.92 \times 10^{-3}\ atm^{-2}}$$

Exercise 203

Total amount of gas = (0.33 + 0.67 + 0.67) mol = 1.67 mol

Partial pressure = mole fraction x total pressure

$$p_{PCl_5} = \frac{0.33\ mol}{1.67\ mol} \times 10.0\ atm = 1.98\ atm$$

$$p_{PCl_3} = p_{Cl_2} = \frac{0.67\ mol}{1.67\ mol} \times 10.0\ atm = 4.01\ atm$$

$$K_p = \frac{p_{PCl_3} \times p_{Cl_2}}{p_{PCl_5}} = \frac{(4.01\ atm)(4.01\ atm)}{(1.98\ atm)} = \boxed{8.12\ atm}$$

Exercise 204

Total amount of gas = (0.224 + 0.142) mol = 0.366 mol

Partial pressure = mole fraction x total pressure

$$p_{CO_2} = \frac{0.224}{0.366} \times 1.83\ atm = 1.12\ atm$$

$$p_{NH_3} = \frac{0.142}{0.366} \times 1.83\ atm = 0.710\ atm$$

$$K_p = p^2_{NH_3} \times p_{CO_2} = (0.710\ atm)^2 \times 1.12\ atm = \boxed{0.565\ atm^3}$$

Exercise 205

Total amount of gas = (0.560 + 0.060 + 1.27) mol = 1.89 mol

$$p_{H_2} = \frac{0.560}{1.89} \times 2.00\ atm = 0.593\ atm$$

$$p_{I_2} = \frac{0.060}{1.89} \times 2.00\ atm = 0.0635\ atm$$

$$p_{HI} = \frac{1.27}{1.89} \times 2.00\ atm = 1.34\ atm$$

$$K_p = \frac{p^2_{HI}}{p_{H_2} \times p_{I_2}} = \frac{(1.34\ atm)^2}{0.593\ atm \times 0.0635\ atm} = \boxed{47.7}$$

Exercise 206

Total amount of gas = (0.96 + 0.04 + 0.02) mol = 1.02 mol

$$p_{NO_2} = \frac{0.96}{1.02} \times p_T \qquad p_{NO} = \frac{0.04}{1.02} \times p_T \qquad p_{O_2} = \frac{0.02}{1.02} \times p_T$$

$$K_p \quad \frac{p^2_{NO} \times p_{O_2}}{p^2_{NO_2}}$$

$$6.8 \times 10^{-6}\ atm = \frac{(0.04/1.02)^2\ p^2_T\ (0.02/1.02)\ p_T}{(0.96/1.02)^2\ p^2_T}$$

$$p_T = \frac{6.8 \times 10^{-6}\ atm \times 0.886}{0.00154 \times 0.0196} = \boxed{0.2\ atm}$$

Exercise 207

Total amount of gas = (2.0 + 1.0 + 4.0) mol = 7.0 mol

$$K_p = \frac{p_{CO} \times p_{H_2}}{p_{H_2O}}$$

$$3.72\ atm = \frac{\frac{2.0}{7.0} \times p_T \times \frac{1.0}{7.0} \times p_T}{\frac{4.0}{7.0} \times p_T}$$

$$p_T = \frac{3.72\ atm \times (4.0/7.0)}{(2.0/7.0) \times (1.0/7.0)} = \boxed{52\ atm}$$

Exercise 208

Rearranging the expression $K_p = K_c(RT)^{\Delta n}$ and putting $\Delta n = 1$

$$K_c = \frac{K_p}{(RT)^{\Delta n}} = \frac{0.811 \text{ atm}}{0.0821 \text{ atm dm}^3 \text{ K}^{-1} \text{ mol}^{-1} \times 523 \text{ K}}$$

$$= \boxed{1.89 \times 10^{-2} \text{ mol dm}^{-3}}$$

Exercise 209

$$K_p = K_c(RT)^{\Delta n} = 2.0 \text{ mol}^{-2} \text{ dm}^6 \times (0.0821 \text{ atm dm}^3 \text{ K}^{-1} \text{ mol}^{-1} \times 620 \text{ K})^{-2}$$

$$= \boxed{7.72 \times 10^{-4} \text{ atm}^{-2}}$$

Exercise 210

Amount of SO_3 at equilibrium = $2.00 \text{ mol} \times \frac{20}{100} = 0.40 \text{ mol}$

Initial amount/mol	2.0		2.0		0
	$2SO_2(g)$	+	$O_2(g)$	$\rightleftharpoons$	$2SO_3(g)$
Equilibrium amount/mol	1.60		1.80		0.40

Total amount of gas = (1.60 + 1.80 + 0.40) mol = 3.80 mol

$$p_{SO_3} = X_{SO_3} p_T = \frac{0.40}{3.80} \times p_T$$

$$p_{SO_2} = \frac{1.60}{3.80} \times p_T \qquad p_{O_2} = \frac{1.80}{3.80} \times p_T$$

$$K_p = \frac{p^2_{SO_3}}{p^2_{SO_2} \times p_{O_2}}$$

$$0.13 \text{ atm}^{-1} = \frac{(0.40/3.80)^2 p_T^2}{(1.60/3.80)^2 p_T^2 (1.80/3.80) p_T} = \frac{0.0111}{0.177 \times 0.474\, p_T}$$

$$p_T = \frac{0.0111}{0.13 \text{ atm}^{-1} \times 0.177 \times 0.474} = \boxed{1.02 \text{ atm}}$$

Exercise 211

Initial amount/mol	2		1		0
	$2SO_2(g)$	+	$O_2(g)$	$\rightleftharpoons$	$2SO_3(g)$
Equilibrium amount/mol	4/3		2/3		2/3

Total amount of gas = 4/3 + 2/3 + 2/3 = 8/3

Partial pressure = mole fraction x total pressure

$$p_{SO_3} = \frac{(2/3)}{(8/3)} \times 9 \text{ atm} = 9/4 \text{ atm}$$

$$p_{SO_2} = \frac{(4/3)}{(8/3)} \times 9 \text{ atm} = 9/2 \text{ atm}$$

$$p_{O_2} = \frac{(2/3)}{(8/3)} \times 9 \text{ atm} = 9/4 \text{ atm}$$

$$K_p = \frac{p^2_{SO_3}}{p^2_{SO_2} \times p_{O_2}} = \frac{(9/4)^2 \text{ atm}^2}{(9/2)^2 \text{ atm}^2 \times 9/4 \text{ atm}} = \boxed{\frac{1}{9} \text{ atm}^{-1}}$$

Exercise 212

Amount of N_2O_4 at equilibrium = initial amount - amount reacted

= (1.00 - 0.66) mol = 0.34 mol

Amount of NO_2 formed = 2 x amount N_2O_4 reacted

= 2 x 0.66 mol = 1.32 mol

Initial amount/mol	1.00		0
	$N_2O_4(g)$	$\rightleftharpoons$	$2NO_2(g)$
Equilibrium amount/mol	0.34		1.32

Total amount of gas = (0.34 + 1.32) mol = 1.66 mol

Partial pressure = mole fraction x total pressure

$$p_{N_2O_4} = \frac{0.34 \text{ mol}}{1.66 \text{ mol}} \times 98.3 \text{ kPa} = 20.1 \text{ kPa}$$

$$p_{NO_2} = \frac{1.32 \text{ mol}}{1.66 \text{ mol}} \times 98.3 \text{ kPa} = 78.2 \text{ kPa}$$

$$K_p = \frac{p^2_{NO_2}}{p_{N_2O_4}} = \frac{(78.2 \text{ kPa})^2}{20.1 \text{ kPa}} = \boxed{304 \text{ kPa}}$$

Since 100 kPa = 760 mmHg = 1.00 atm

$$K_p = 304 \text{ kPa} \times \frac{760 \text{ mmHg}}{100 \text{ kPa}} = \boxed{2310 \text{ mmHg}}$$

$$\text{and } K_p = 304 \text{ kPa} \times \frac{1.00 \text{ atm}}{100 \text{ kPa}} = \boxed{3.04 \text{ atm}}$$

Exercise 213

(a)

	$2NO_2(g)$	$\rightleftharpoons$	$2NO(g)$	+	$O_2(g)$
Equilibrium amount/mol	0.96		0.040		0.020

Total amount of gas = (0.96 + 0.04 + 0.02) mol = 1.02 mol

$$p_{NO_2} = \frac{0.96}{1.02} \times 0.20 \text{ atm} = 0.19 \text{ atm}$$

$$p_{NO} = \frac{0.040}{1.02} \times 0.20 \text{ atm} = 0.0078 \text{ atm}$$

$$p_{O_2} = \frac{0.02}{1.02} \times 0.20 \text{ atm} = 0.0039 \text{ atm}$$

$$K_p = \frac{p^2_{NO} \times p_{O_2}}{p^2_{NO_2}} = \frac{0.0078^2 \text{ atm}^2 \times 0.0039 \text{ atm}}{0.19^2 \text{ atm}^2} = \boxed{6.6 \times 10^{-6} \text{ atm}}$$

(b) Average molar mass = $X_{NO_2} M_{NO_2} + X_{NO} M_{NO} + X_{O_2} M_{O_2}$

$$\left(\frac{0.96}{1.02} \times 46.0 \text{ g mol}^{-1}\right) + \left(\frac{0.040}{1.02} \times 30.0 \text{ g mol}^{-1}\right) + \left(\frac{0.020}{1.02} \times 32.0 \text{ g mol}^{-1}\right)$$

$$= (43.3 + 1.18 + 0.63) \text{g mol}^{-1} = \boxed{45.1 \text{ g mol}^{-1}}$$

Exercise 214

(a) $M_{avg} = X_{CO} M_{CO} + X_{CO_2} M_{CO_2}$

Since $X_{CO} + X_{CO_2} = 1$, $X_{CO_2} = 1 - X_{CO}$

$36 \text{ g mol}^{-1} = (X_{CO} \times 28 \text{ g mol}^{-1}) + (1 - X_{CO}) \times 44 \text{ g mol}^{-1}$

$36 = 28X_{CO} + 44 - 44X_{CO}$

$X_{CO} = \frac{44 - 36}{44 - 28} = \frac{8}{16} = \boxed{0.5}$

(b) $p_{CO} = X_{CO}\, p_T = 0.5 \times 12 \text{ atm} = 6 \text{ atm}$

$p_{CO_2} = X_{CO_2} p_T = 0.5 \times 12 \text{ atm} = 6 \text{ atm}$

$K_p = \frac{p^2_{CO}}{p_{CO_2}} = \frac{(6 \text{ atm})^2}{6 \text{ atm}} = \boxed{6 \text{ atm}}$

(c) Let $x = X_{CO}$ and $1 - x = X_{CO_2}$

$K_p = \frac{(X_{CO} \times p_T)^2}{X_{CO_2} \times p_T}$

$\therefore \quad 6 \text{ atm} = \frac{(x \times 2 \text{ atm})^2}{(1 - x) \times 2 \text{ atm}}$ or $6 = \frac{2x^2}{1-x}$

$\therefore \quad 2x^2 = 6 - 6x$

$x^2 + 3x - 3 = 0$

$x = \frac{-3 \pm \sqrt{9 + 12}}{2} = 0.8$ or -3.8 (absurd root)

$\therefore \quad X_{CO} = \boxed{0.8}$

Exercise 215

(a) Total amount of gas = (0.20 + 0.010 + 3.8) mol = 4.01 mol

$p_{PCl_5} = \frac{0.20}{4.01} \times 3.0 \text{ atm} = 0.15 \text{ atm}$

$p_{PCl_3} = \frac{0.010}{4.01} \times 3.0 \text{ atm} = 7.48 \times 10^{-3} \text{ atm}$

$p_{Cl_2} = \frac{3.8}{4.01} \times 3.0 \text{ atm} = 2.84 \text{ atm}$

$K_p = \frac{p_{PCl_3} \times p_{Cl_2}}{p_{PCl_5}} = \frac{7.48 \times 10^{-3} \text{ atm} \times 2.84 \text{ atm}}{0.15 \text{ atm}} = \boxed{0.14 \text{ atm}}$

(b) $M_{avg} = X_{PCl_5} M_{PCl_5} + X_{PCl_3} M_{PCl_3} + X_{Cl_2} M_{Cl_2}$

$= \frac{0.20}{4.01} \times 208.5 \text{ g mol}^{-1} + \frac{0.010}{4.01} \times 137.5 \text{ g mol}^{-1} + \frac{3.8}{4.01} \times 71.0 \text{ g mol}^{-1}$

$= (10.40 + 0.34 + 67.28) \text{ g mol}^{-1} = \boxed{78 \text{ g mol}^{-1}}$

Exercise 216

(a) $M_{avg} = X_{NO_2} M_{NO_2} + X_{N_2O_4} M_{N_2O_4}$

$72.4 \text{ g mol}^{-1} = X_{NO_2} \times 46.0 \text{ g mol}^{-1} + (1 - X_{NO_2}) \times 92.0 \text{ g mol}^{-1}$

$X_{NO_2} = \frac{(92.0 - 72.4) \text{g mol}^{-1}}{(92.0 - 46.0) \text{g mol}^{-1}} = \frac{19.6}{46.0} = \boxed{0.426}$

(b) $p_{NO_2} = 0.426 \times 1 \text{ atm} = 0.426 \text{ atm}$

$p_{N_2O_4} = (1 - 0.426) \times 1 \text{ atm} = 0.574 \text{ atm}$

$K_p = \frac{p^2_{NO_2}}{p_{N_2O_4}} = \frac{(0.426 \text{ atm})^2}{(0.574 \text{ atm})} = \boxed{0.316 \text{ atm}}$

(c)

	$N_2O_4(g)$	$\rightleftharpoons$	$2NO_2(g)$
Initial amount/mol	1.0		0
Equilibrium amount/mol	$1 - x$		$2x$

Total amount at equilibrium = $(1 - x + 2x)$ mol = $(1 + x)$ mol

$p_{NO_2} = \frac{\text{amount of } NO_2}{\text{total amount of gas}} \times p_T = \frac{2x}{1 + x} \times 6.00 \text{ atm}$

$p_{N_2O_4} = \frac{\text{amount of } N_2O_4}{\text{total amount of gas}} \times p_T = \frac{1 - x}{1 + x} \times 6.00 \text{ atm}$

$K_p = \frac{p^2_{NO_2}}{p_{N_2O_4}}$

$0.316 \text{ atm} = \frac{(\frac{2x}{1 + x})^2 \times (6.00 \text{ atm})^2}{(\frac{1 - x}{1 + x}) \times 6.00 \text{ atm}}$

$0.316 = \frac{6.00 \times 4x^2}{(1 + x)(1 - x)} = \frac{24.00\, x^2}{1 - x^2}$

$\therefore \quad 24.316x^2 = 0.316$ and $x = \sqrt{0.0130} = 0.114$

Mole fraction of NO_2 = $\frac{\text{amount of } NO_2}{\text{total amount}}$

$= \frac{2x}{1 + x} = \frac{2 \times 0.114}{1.114} = \boxed{0.205}$

Exercise 217

$\log \frac{K_2}{K_1} = \frac{\Delta H^\ominus}{2.30\, R} (\frac{1}{T_1} - \frac{1}{T_2})$

$\log K_2 - \log 0.0118 = \frac{177300 \text{ J mol}^{-1}}{2.30 \times 8.31 \text{ J K}^{-1} \text{ mol}^{-1}} (\frac{1}{1338 \text{ K}} - \frac{1}{1473 \text{ K}})$

$\log K_2 = -1.928 + 9276.4 \text{ K} (6.85 \times 10^{-5} \text{ K}^{-1}) = -1.293$

$\therefore \quad K_2 = \boxed{0.0510 \text{ atm}}$

Exercise 218

$$\log \frac{K_2}{K_1} = \frac{\Delta H^\ominus}{2.30\,R}\left(\frac{1}{T_1} - \frac{1}{T_2}\right)$$

$$\therefore \Delta H^\ominus = \frac{2.30\,R \times \log \frac{K_2}{K_1}}{\left(\frac{1}{T_1} - \frac{1}{T_2}\right)} = \frac{2.30 \times 8.31\ \text{J K}^{-1}\ \text{mol}^{-1} \times \log\left(\frac{3.74}{2.44}\right)}{\left(\frac{1}{1000\ \text{K}} - \frac{1}{1200\ \text{K}}\right)}$$

$$= \frac{3.545\ \text{J K}^{-1}\ \text{mol}^{-1}}{1.667 \times 10^{-4}\ \text{K}^{-1}} = 2.13 \times 10^{4}\ \text{J mol}^{-1} = \boxed{21.3\ \text{kJ mol}^{-1}}$$

Exercise 219

(a) $K_p = \dfrac{p^2_{NO}}{p_{N_2} \times p_{O_2}}$

(b) From the expression for K_p above, $p^2_{NO} = K_p \times p_{N_2} \times p_{O_2}$

The values of K_p and p_{N_2} remain constant, so that $p_{NO} \propto \sqrt{p_{O_2}}$

p_{O_2} has values of 0.05 atm in the fourth column and 0.2 atm in the third.
p_{NO} is proportional to the square roots of these values.

Since $0.05 = \frac{1}{4} \times 0.2$, $\sqrt{0.05} = \frac{1}{2} \times \sqrt{0.2}$

and the value of p_{NO} in the fourth column is half that in the third.

(c) The table shows that the value of K_p increases with temperature. This indicates a higher yield of NO with increased temperature.
Since there are the same number of molecules either side of the equation, changing the pressure will have no effect on the yield of NO.

(d) The reaction is endothermic, because K_p increases with temperature.
Le Chatelier's principle predicts that an increase in temperature will favour an endothermic reaction.

Exercise 220

$$\log \frac{K_2}{K_1} = \frac{\Delta H^\ominus}{2.30\,R}\left(\frac{1}{T_1} - \frac{1}{T_2}\right)$$

Taking the values of K_p at 1800 K and 2000 K

$$\therefore \Delta H^\ominus = \frac{2.30\,R \times \log \frac{K_2}{K_1}}{\frac{1}{T_1} - \frac{1}{T_2}} = \frac{2.30 \times 8.31\ \text{J K}^{-1}\ \text{mol}^{-1} \times \log \frac{4.08}{1.21}}{\frac{1}{1800} - \frac{1}{2000}}$$

$$= \frac{10.089\ \text{J K}^{-1}\ \text{mol}^{-1}}{5.56 \times 10^{-5}} = 1.81 \times 10^{5}\ \text{J mol}^{-1} = \boxed{181\ \text{kJ mol}^{-1}}$$

Exercise 221

(a)

	$CdCO_3(s)$	⇌	$Cd^{2+}(aq)$	+	$CO_3^{2-}(aq)$
Equilibrium concn/mol dm^{-3}			1.58×10^{-7}		1.58×10^{-7}

$K_s = [Cd^{2+}(aq)][CO_3^{2-}(aq)] = 1.58 \times 10^{-7}\ \text{mol dm}^{-3} \times 1.58 \times 10^{-7}\ \text{mol dm}^{-3}$
$= \boxed{2.50 \times 10^{-14}\ \text{mol}^2\ \text{dm}^{-6}}$

(b)

	$CaF_2(s)$	⇌	$Ca^{2+}(aq)$	+	$2F^-(aq)$
Equilibrium concn/mol dm^{-3}			2.15×10^{-4}		$2 \times 2.15 \times 10^{-4}$

$K_s = [Ca^{2+}(aq)][F^-(aq)]^2 = 2.15 \times 10^{-4}\ \text{mol dm}^{-3} \times (4.30 \times 10^{-4}\ \text{mol dm}^{-3})^2$
$= \boxed{3.98 \times 10^{-11}\ \text{mol}^3\ \text{dm}^{-9}}$

(c)

	$Cr(OH)_3(s)$	⇌	$Cr^{3+}(aq)$	+	$3OH^-(aq)$
Equilibrium concn/mol dm^{-3}			1.39×10^{-8}		$3 \times 1.39 \times 10^{-8}$

$K_s = [Cr^{3+}(aq)][OH^-(aq)]^3 = 1.39 \times 10^{-8}\ \text{mol dm}^{-3} \times (3 \times 1.39 \times 10^{-8}\ \text{mol dm}^{-3})^3$
$= \boxed{1.01 \times 10^{-30}\ \text{mol}^4\ \text{dm}^{-12}}$

Exercise 222

Let A refer to $H^+(aq)$ and B refer to $OH^-(aq)$

$$\frac{c_A V_A}{c_B V_B} = \frac{1}{1}$$

$$\frac{0.100\ \text{mol dm}^{-3} \times 11.45\ \text{cm}^3}{c_B \times 25.0\ \text{cm}^3} = 1$$

$$\therefore c_B = \frac{0.100\ \text{mol dm}^{-3} \times 11.45\ \text{cm}^3}{25.0\ \text{cm}^3} = \boxed{0.0458\ \text{mol dm}^{-3}}$$

$[Ca^{2+}(aq)] = \frac{1}{2}[OH^-(aq)] = \boxed{0.0229\ \text{mol dm}^{-3}}$

Solubility of $Ca(OH)_2$ = $[Ca^{2+}(aq)]$ = 0.0229 mol dm^{-3} (at 15 °C)

$\therefore K_s = [Ca^{2+}(aq)][OH^-(aq)]^2$
$= 0.0229\ \text{mol dm}^{-3} \times (0.0458\ \text{mol dm}^{-3})^2 = \boxed{4.80 \times 10^{-5}\ \text{mol}^3\ \text{dm}^{-9}}$

Exercise 223

(a)

	$CuS(s)$	$\rightleftharpoons$	$Cu^{2+}(aq)$	+	$S^{2-}(aq)$
Equilibrium concn/mol dm^{-3}			x		x

$K_s = [Cu^{2+}(aq)][S^{2-}(aq)]$

$6.3 \times 10^{-36}\ mol^2\ dm^{-6} = x\ mol\ dm^{-3} \times x\ mol\ dm^{-3}$

$x^2 = 6.3 \times 10^{-36}$ $\quad \therefore x = 2.5 \times 10^{-18}$

and solubility = $\boxed{2.5 \times 10^{-18}\ mol\ dm^{-3}}$

(b)

	$Fe(OH)_2(s)$	$\rightleftharpoons$	$Fe^{2+}(aq)$	+	$2OH^-(aq)$
Equilibrium concn/mol dm^{-3}			x		$2x$

$K_s = [Fe^{2+}(aq)][OH^-(aq)]^2$

$6.0 \times 10^{-15}\ mol^3\ dm^{-9} = x\ mol\ dm^{-3} \times (2x\ mol\ dm^{-3})^2$

$\therefore 4x^3 = 6.0 \times 10^{-15}$

$x = (1.5 \times 10^{-15})^{\frac{1}{3}} = 1.1 \times 10^{-5}$

and solubility = $\boxed{1.1 \times 10^{-5}\ mol\ dm^{-3}}$

(c)

	$Ag_3PO_4(s)$	$\rightleftharpoons$	$3Ag^+(aq)$	+	$PO_4^{3-}(aq)$
Equilibrium concn/mol dm^{-3}			$3x$		x

$K_s = [Ag^+(aq)]^3[PO_4^{3-}(aq)]$

$1.25 \times 10^{-20}\ mol^4\ dm^{-12} = (3x\ mol\ dm^{-3})^3 \times x\ mol\ dm^{-3}$

$\therefore 27x^4 = 1.25 \times 10^{-20}$ $\quad x = (4.63 \times 10^{-22})^{\frac{1}{4}} = 4.64 \times 10^{-6}$

and solubility = $\boxed{4.64 \times 10^{-6}\ mol\ dm^{-3}}$

Exercise 224

(a)

	$SrSO_4(s)$	$\rightleftharpoons$	$Sr^{2+}(aq)$	+	$SO_4^{2-}(aq)$
Initial concn/mol dm^{-3}			0		0.10
Equilibrium concn/mol dm^{-3}			x		$(0.10 + x)$

$K_s = [Sr^{2+}(aq)][SO_4^{2-}(aq)]$

$4.0 \times 10^{-7}\ mol^2\ dm^{-6} = (x\ mol\ dm^{-3})(0.10 + x)\ mol\ dm^{-3}$

Assume that $(0.10 + x) \approx 0.10$

$\therefore 4.0 \times 10^{-7} = x \times 0.10$

$\therefore x = 4.0 \times 10^{-6}$

and solubility = $\boxed{4.0 \times 10^{-6}\ mol\ dm^{-3}}$

(b)

	$MgF_2(s)$	$\rightleftharpoons$	$Mg^{2+}(aq)$	+	$2F^-(aq)$
Initial concn/mol dm^{-3}			0		0.20
Equilibrium concn/mol dm^{-3}			x		$(2x + 0.20)$

$K_s = [Mg^{2+}(aq)][F^-(aq)]^2$

$7.2 \times 10^{-9}\ mol^3\ dm^{-9} = (x\ mol\ dm^{-3})(2x + 0.20)^2\ mol^2\ dm^{-6}$

Assume that $(2x + 0.20) \approx 0.20$

$7.2 \times 10^{-9} = x \times (0.20)^2$

$\therefore x = 1.8 \times 10^{-7}$

and solubility = $\boxed{1.8 \times 10^{-7}\ mol\ dm^{-3}}$

Exercise 225

$K_s(SrCO_3) = 1.1 \times 10^{-10}\ mol^2\ dm^{-6}$

Since the volume is doubled

	$SrCO_3(s)$	$\rightleftharpoons$	$Sr^{2+}(aq)$	+	$CO_3^{2-}(aq)$
Initial concn/mol dm^{-3}					0.50
Equilibrium concn/mol dm^{-3}			x		$(x + 0.50)$

$K_s = [Sr^{2+}(aq)][CO_3^{2-}(aq)]$

$1.1 \times 10^{-10}\ mol^2\ dm^{-6} = x\ mol\ dm^{-3} \times (x + 0.50)\ mol\ dm^{-3}$

Assume that $x << 0.5$ i.e. $x + 0.50 \approx 0.50$

$\therefore x = \frac{1.1 \times 10^{-10}}{0.50} = 2.2 \times 10^{-10}$

and $[Sr^{2+}(aq)]$ = $\boxed{2.2 \times 10^{-10}\ mol\ dm^{-3}}$

Exercise 226

Before mixing, $[Ba^{2+}(aq)] = 0.10\ mol\ dm^{-3}$.

Mixing changes volume from 150 cm^3 to 200 cm^3.

Then $[Ba^{2+}(aq)] = 0.10\ mol\ dm^{-3} \times \frac{150}{200} = 0.075\ mol\ dm^{-3}$

Before mixing, $[F^-(aq)] = 0.050\ mol\ dm^{-3}$

Mixing changes volume from 50 cm^3 to 200 cm^3.

Then, $[F^-(aq)] = 0.050\ mol\ dm^{-3} \times \frac{50}{200} = 0.0125\ mol\ dm^{-3}$

Ion product = $[Ba^{2+}(aq)][F^-(aq)]^2$

$= 0.075\ mol\ dm^{-3} \times (0.0125\ mol\ dm^{-3})^2 = 1.2 \times 10^{-5}\ mol^3\ dm^{-9}$

Ion product is greater than K_s. Therefore **precipitation occurs**

Exercise 227

(a) Ag^+Cl^- : 2 concentration terms, ∴ unit = $(mol\ dm^{-3})^2 = mol^2\ dm^{-6}$

$Pb^{2+}2Br^-$: 3 concentration terms, ∴ unit = $(mol\ dm^{-3})^3 = mol^3\ dm^{-9}$

$Ag^+BrO_3^-$: 2 concentration terms, ∴ unit = $(mol\ dm^{-3})^2 = mol^2\ dm^{-6}$

$Mg^{2+}2OH^-$: 3 concentration terms, ∴ unit = $(mol\ dm^{-3})^3 = mol^3\ dm^{-9}$

(b) When equal volumes of the solution are mixed, their concentrations, before reaction, are halved: i.e. concentrations of all ions required for ion product calculations = $5.0 \times 10^{-4}\ mol\ dm^{-3}$

(i) Ion product = $[Ag^+(aq)][Cl^-(aq)] = (5.0 \times 10^{-4}\ mol\ dm^{-3})^2$

$= 2.5 \times 10^{-7}\ mol^2\ dm^{-6}$

$> K_s$ ∴ **precipitation occurs**

(ii) Ion product = $[Pb^{2+}(aq)][Br^-(aq)]^2 = (5.0 \times 10^{-4}\ mol\ dm^{-3})^3$

$= 1.25 \times 10^{-10}\ mol^3\ dm^{-9}$

$< K_s$ ∴ **no precipitation**

(iii) Ion product = $[Ag^+(aq)][BrO_3^-(aq)] = (5.0 \times 10^{-4}\ mol\ dm^{-3})^2$

$= 2.5 \times 10^{-7}\ mol^2\ dm^{-6}$

$< K_s$ ∴ **no precipitation**

(iv) Ion product = $[Mg^{2+}(aq)][OH^-(aq)]^2 = (5.0 \times 10^{-4}\ mol\ dm^{-3})^3$

$= 1.25 \times 10^{-10}\ mol^3\ dm^{-9}$

$> K_s$ ∴ **precipitation occurs**

Exercise 228

$K_d = \frac{[acid\ (aq)]}{[acid\ (ether)]}$

$K_d = \frac{0.0759\ mol\ dm^{-3}}{0.0114\ mol\ dm^{-3}} = 6.66$

$K_d = \frac{0.108\ mol\ dm^{-3}}{0.0162\ mol\ dm^{-3}} = 6.67$

$K_d = \frac{0.158\ mol\ dm^{-3}}{0.0237\ mol\ dm^{-3}} = 6.67$

$K_d = \frac{0.300\ mol\ dm^{-3}}{0.0451\ mol\ dm^{-3}} = 6.65$

Average K_d = **6.66**

Or, from the graph,

slope $= \frac{\Delta[acid\ (aq)]}{\Delta[acid\ (ether)]}$

$K_d = \frac{(0.268 - 0.050)}{(0.0400 - 0.0072)}$

$= \frac{0.218}{0.0328} =$ **6.65**

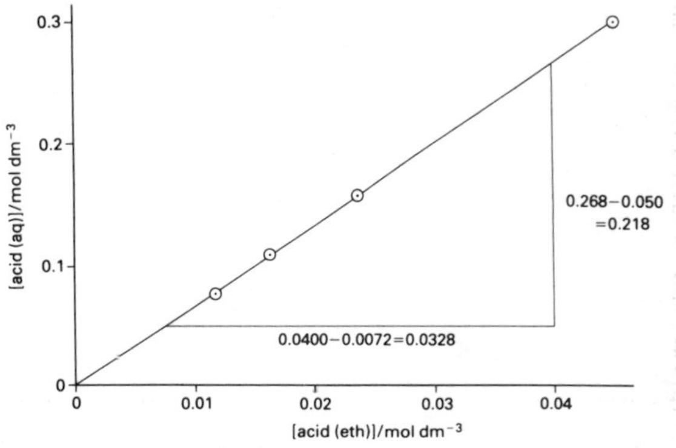

Exercise 229

(a)

	0.010	0
Initial amount/mol	0.010	0
	$NH_3(tce)$ ⇌	$NH_3(aq)$
Equilibrium amount/mol	x	$0.010 - x$

$$K_d = \frac{[NH_3(aq)]}{[NH_3(tce)]}$$

$$290 = \frac{(0.010 - x)\ \text{mol}/0.10\ \text{dm}^3}{x\ \text{mol}/0.10\ \text{dm}^3} = \frac{0.010 - x}{x}$$

$$290\,x = 0.010 - x$$

$$x = \frac{0.010}{291} = 3.4 \times 10^{-5}$$

∴ the amount of ammonia remaining = x mol = $\boxed{3.4 \times 10^{-5}\ \text{mol}}$

(b) After the first addition of water, the remaining amount, x mol of ammonia is given by the same method as in (a):

$$290 = \frac{(0.010 - x_1)\ \text{mol}/0.025\ \text{dm}^3}{x_1\ \text{mol}/0.10\ \text{dm}^3} = \frac{4(0.010 - x_1)}{x_1}$$

$$290\,x_1 = 0.040 - 4x_1$$

$$x_1 = \frac{0.040}{294} = 1.4 \times 10^{-4}$$

Note that this is $0.010 \times \frac{4}{294}$

or, remaining amount = original amount $\times \dfrac{V(tce)/V(aq)}{K_d + V(tce)/V(aq)}$

Thus, after the second addition,

$$x_2 = (0.010 \times \frac{4}{294}) \times \frac{4}{294} = 0.010 \times (\frac{4}{294})^2$$

and after the fourth addition,

$$x_4 = 0.010 \times (\frac{4}{294})^4 = 3.4 \times 10^{-10}$$

The amount finally remaining = x_4 mol = $\boxed{3.4 \times 10^{-10}\ \text{mol}}$

Note that the amount remaining is reduced by a factor of 10^{-5}: clearly, the extraction is made far more efficient by dividing the extracting liquid into several portions.

Exercise 230

If the molecular formula in trichloromethane is $(CH_3CO_2H)_n$, then

$$K_d = \frac{[CH_3CO_2H(aq)]^n}{[CH_3CO_2H(tcm)]} \qquad \text{(tcm = trichloromethane)}$$

Taking logarithms,

$$\log K_d = n \log [CH_3CO_2H(aq)] - \log [CH_3CO_2H(tcm)]$$

Rearranging this equation into the form $y = mx + c$

$$\log[CH_3CO_2H(tcm)] = n \log[CH_3CO_2H(aq)] - \log K_d$$

∴ a plot of $\log[CH_3CO_2H(tcm)]$ against $\log[CH_3CO_2H(aq)]$ should be a straight line with slope n. (Here we may use concentrations in g dm^{-3} because they are proportional to concentrations in mol dm^{-3}.)

$\log([CH_3CO_2H(tcm)]/\text{g dm}^{-3})$	1.24	1.64	1.93
$\log([CH_3CO_2H(aq)]/\text{g dm}^{-3})$	2.47	2.68	2.81

From the graph, the slope, $n = \dfrac{1.90 - 1.40}{2.80 - 2.55} = \dfrac{0.50}{0.25} = 2.0$

∴ the formula of ethanoic acid in trichloromethane is $\boxed{(CH_3CO_2H)_2}$

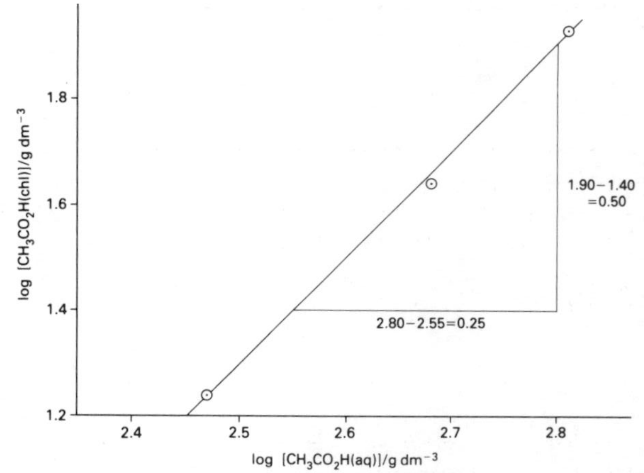

Exercise 231 $\frac{1}{3}$ or 0.333

Exercise 232 6 atm^{-1}

Exercise 233 6.5×10^{-31} $mol^5\ dm^{-15}$

Exercise 234 1.0×10^{-4} mol dm^{-3}

Exercise 235 7.79×10^{-3} mol dm^{-3}

Exercise 236 (a) 0.13 atm^{-2} (b) 22.2 g mol^{-1}

Exercise 237 3.89×10^{-5}

Exercise 238 1.6 atm

Exercise 239

$[HI(g)] = 1.9\ mol\ dm^{-3}$ $[H_2(g)] = 1.1\ mol\ dm^{-3}$ $[I_2(g)] = 0.07\ mol\ cm^{-3}$

Exercise 240

(a) $K_s = [Pb^{2+}(aq)][Cl^-(aq)]^2$

(b) (i) $1.6 \times 10^{-2}\ mol\ dm^{-3}$ (ii) $4.0 \times 10^{-3}\ (mol\ dm^{-3})$

Exercise 241 (a) 0.0500 atm (b) 27%

Exercise 242

(a) $K_c = \dfrac{[SO_4^{2-}(aq)]}{[I^-(aq)]^2}$

(b) The concentration of a solid is constant. Therefore, changing the amount of solid present has no effect on the equilibrium position.

(c) (i) 0.100 mol dm^{-3} (ii) 0.0620 mol dm^{-3}

(iii) 0.038 mol dm^{-3} (iv) 0.019 mol dm^{-3}

(v) 4.94 $dm^3\ mol^{-1}$

Exercise 243 25 (The reciprocal, 0.040, could be acceptable.)

Exercise 244

$K_w = [H^+(aq)][OH^-(aq)] = 1.0 \times 10^{-14}\ mol^2\ dm^{-6}$

$[H^+(aq)] = [OH^-(aq)] = \sqrt{1.0 \times 10^{-14}\ mol^2\ dm^{-6}} = 1.0 \times 10^{-7}\ mol\ dm^{-3}$

The fraction ionized is, therefore, $\dfrac{1.0 \times 10^{-7}\ mol\ dm^{-3}}{55.6\ mol\ dm^{-3}} = \boxed{1.80 \times 10^{-9}}$

(i.e. approx 18 molecules in 10 000 000 000)

Exercise 245

(a) In 0.010 M HCl, $[H^+(aq)] = 0.010\ mol\ dm^{-3}$

$K_w = [H^+(aq)]\,[OH^-(aq)]$

$\therefore [OH^-(aq)] = \dfrac{K_w}{[H^+(aq)]} = \dfrac{1.0 \times 10^{-14}\ mol^2\ dm^{-6}}{0.010\ mol\ dm^{-3}} = \boxed{1.0 \times 10^{-12}\ mol\ dm^{-3}}$

(b) In 0.10 M H_2SO_4, $[H^+(aq)] = 2 \times 0.10\ mol\ dm^{-3} = 0.20\ mol\ dm^{-3}$

$K_w = [H^+(aq)]\,[OH^-(aq)]$

$\therefore [OH^-(aq)] = \dfrac{K_w}{[H^+(aq)]} = \dfrac{1.0 \times 10^{-14}\ mol^2\ dm^{-6}}{0.20\ mol\ dm^{-3}} = \boxed{5.0 \times 10^{-14}\ mol\ dm^{-3}}$

Exercise 246

(a) In 0.010 M KOH, $[OH^-(aq)] = 0.010\ mol\ dm^{-3}$

$K_w = [H^+(aq)]\,[OH^-(aq)]$

$\therefore [H^+(aq)] = \dfrac{K_w}{[OH^-(aq)]} = \dfrac{1.0 \times 10^{-14}\ mol^2\ dm^{-6}}{0.010\ mol\ dm^{-3}} = \boxed{1.0 \times 10^{-12}\ mol\ dm^{-3}}$

(b) In 0.050 M $Ba(OH)_2$, $[OH^-(aq)] = 2 \times 0.050\ mol\ dm^{-3} = 0.10\ mol\ dm^{-3}$

$\therefore [H^+(aq)] = \dfrac{K_w}{[OH^-(aq)]} = \dfrac{1.0 \times 10^{-14}\ mol^2\ dm^{-6}}{0.10\ mol\ dm^{-3}} = \boxed{1.0 \times 10^{-13}\ mol\ dm^{-3}}$

Exercise 247

$[H^+(aq)] = 10\ mol\ dm^{-3} = 1.0 \times 10^1\ mol\ dm^{-3}$

$pH = -\log[H^+(aq)] = -\log(1 \times 10^1) = -(0 + 1) = \boxed{-1}$

Exercise 248

(a) $K_w = [H^+(aq)]\,[OH^-(aq)]$

$[H^+(aq)] = \dfrac{K_w}{[OH^-(aq)]} = \dfrac{1.0 \times 10^{-14}\ mol^2\ dm^{-6}}{1.0 \times 10^{-8}\ mol\ dm^{-3}} = 1.0 \times 10^{-6}\ mol\ dm^{-3}$

$pH = -\log[H^+(aq)] = -\log(1.0 \times 10^{-6}) = 0 - (-6.0) = \boxed{6.0}$

(b) $[H^+(aq)] = \dfrac{K_w}{[OH^-(aq)]} = \dfrac{1.0 \times 10^{-14}\ mol^2\ dm^{-6}}{0.10\ mol\ dm^{-3}} = 1.0 \times 10^{-13}\ mol\ dm^{-3}$

$pH = -\log[H^+(aq)] = -\log(1.0 \times 10^{-13}) = 0 - (-13.0) = \boxed{13.0}$

(c) $[H^+(aq)] = \dfrac{K_w}{[OH^-(aq)]} = \dfrac{1.0 \times 10^{-14}\ mol^2\ dm^{-6}}{1.0 \times 10^{-4}\ mol\ dm^{-3}} = 1.0 \times 10^{-10}\ mol\ dm^{-3}$

$pH = -\log[H^+(aq)] = -\log(1.0 \times 10^{-10}) = 0 - (-10.0) = \boxed{10.0}$

(d) $[H^+(aq)] = \dfrac{K_w}{[OH^-(aq)]} = \dfrac{1.0 \times 10^{-14}\ mol^2\ dm^{-6}}{10\ mol\ dm^{-3}} = 1.0 \times 10^{-15}\ mol\ dm^{3}$

$pH = -\log[H^+(aq)] = -\log(1.0 \times 10^{-15}) = 0 - (-15) = \boxed{15.0}$

Exercise 249

(a) $pH = -\log [H^+(aq)]$

$= -\log (7.0 \times 10^{-7}) = -(-7 + 0.8451) = -(-6.1549) = \boxed{6.2}$

(b) $pH = -\log [H^+(aq)] = -\log (3.2 \times 10^{-9}) = -(-9 + 0.5052)$

$= -(-8.4948) = \boxed{8.5}$

(c) $pH = -\log [H^+(aq)] = -\log (8.34 \times 10^{-3}) = -(-3 + 0.921) = \boxed{2.08}$

(d) $pH = -\log [H^+(aq)] = -\log (2.62 \times 10^{-8}) = -(-8 + 0.418) = \boxed{7.58}$

(e) $pH = -\log [H^+(aq)] = -\log 5.4 \times 10^{-1} = -(-1 + 0.732) = \boxed{0.27}$

(f) $pH = -\log [H^+(aq)] = -\log (2.23 \times 10^{-3}) = -(-3 + 0.348) = \boxed{2.65}$

Exercise 250

(a) $K_w = [H^+(aq)]\,[OH^-(aq)]$

In pure water, $[H^+(aq)] = [OH^-(aq)]$ so that $K_w = [H^+(aq)]^2$

$\therefore [H^+(aq)] = \sqrt{0.30 \times 10^{-14}\ \text{mol}^2\ \text{dm}^{-6}} = 5.48 \times 10^{-8}\ \text{mol dm}^{-3}$

$pH = -\log [H^+(aq)] = -\log (5.48 \times 10^{-8}) = -(-8 + 0.7388) = -(-7.2612)$

$= \boxed{7.3}$

(b) $K_w = [H^+(aq)]\,[OH^-(aq)]$

In pure water $[H^+(aq)] = [OH^-(aq)]$ so that $K_w = [H^+(aq)]^2$

$5.47 \times 10^{-14}\ \text{mol}^2\ \text{dm}^{-6} = [H^+(aq)]^2$

$[H^+(aq)] = \sqrt{5.47 \times 10^{-14}\ \text{mol}^2\ \text{dm}^{-6}} = 2.34 \times 10^{-7}\ \text{mol dm}^{-3}$

$pH = -\log [H^+(aq)] = -\log (2.34 \times 10^{-7}) = -(-7 + 0.3692) = -(-6.6308)$

$= \boxed{6.6}$

Exercise 251

(a) $pH = -\log [H^+(aq)] = 9.20$

$\log [H^+(aq)] = -9.20$

$\therefore [H^+(aq)] = \text{antilog}(-10 + 0.80) = \boxed{6.3 \times 10^{-10}\ \text{mol dm}^{-3}}$

(b) $pH = -\log [H^+(aq)] = 2.632$

$\log [H^+(aq)] = -2.632$

$\therefore [H^+(aq)] = \text{antilog}(-3 + 0.368) = \boxed{2.33 \times 10^{-3}\ \text{mol dm}^{-3}}$

(c) $pH = -\log [H^+(aq)]$

$\log [H^+(aq)] = -pH = -11.11$

$[H^+(aq)] = \text{antilog}(-12 + 0.89) = \boxed{7.76 \times 10^{-12}\ \text{mol dm}^{-3}}$

(d) $pH = -\log [H^+(aq)]$

$\log [H^+(aq)] = -pH = -1.11$

$[H^+(aq)] = \text{antilog}(-2 + 0.89) = \boxed{7.76 \times 10^{-2}\ \text{mol dm}^{-3}}$

(e) $pH = -\log [H^+(aq)]$

$\log [H^+(aq)] = -pH = -5.64$

$[H^+(aq)] = \text{antilog}(-6 + 0.36) = \boxed{2.29 \times 10^{-6}\ \text{mol dm}^{-3}}$

(f) $pH = -\log [H^+(aq)]$

$\log [H^+(aq)] = -pH = -7.71$

$[H^+(aq)] = \text{antilog}(-8 + 0.29) = \boxed{1.95 \times 10^{-8}\ \text{mol dm}^{-3}}$

Exercise 252

First, calculate $[H^+(aq)]$ from pH

$pH = -\log [H^+(aq)] = 4.17$

$\log [H^+(aq)] = -4.17$

$\therefore [H^+(aq)] = \text{antilog}\ -4.17 = \text{antilog}(-5 + 0.83) = 6.76 \times 10^{-5}\ \text{mol dm}^{-3}$

At equilibrium, the concentrations of hydrogen ion and hydrogencarbonate ion are the same ($6.76 \times 10^{-5}\ \text{mol dm}^{-3}$)

	$H_2CO_3(aq)$	$\rightleftharpoons$	$H^+(aq)$	+	$HCO_3^-(aq)$
Initial concn/mol dm^{-3}	0.010		0		0
Equilibrium concn/mol dm^{-3}	$(0.010 - 6.76 \times 10^{-5})$		6.76×10^{-5}		6.76×10^{-5}

$$K_a = \frac{[H^+(aq)]\,[HCO_3^-(aq)]}{[H_2CO_3(aq)]} = \frac{(6.76 \times 10^{-5}\ \text{mol dm}^{-3})^2}{(0.010 - 6.76 \times 10^{-5}\ \text{mol dm}^{-3})}$$

Since $6.76 \times 10^{-5} \ll 0.010$, then $0.010 - 6.76 \times 10^{-5} \approx 0.010$

$$\therefore K_a = \frac{(6.76 \times 10^{-5}\ \text{mol dm}^{-3})^2}{(0.010\ \text{mol dm}^{-3})} = \boxed{4.57 \times 10^{-7}\ \text{mol dm}^{-3}}$$

($4.60 \times 10^{-7}\ \text{mol dm}^{-3}$ without approximation.)

Exercise 253

First, calculate [H⁺(aq)] from pH

$pH = -\log[H^+(aq)] = 3.28$

$\log[H^+(aq)] = -3.28$

$\therefore [H^+(aq)] = \text{antilog}(-3.28) = \text{antilog}(-4 + 0.72) = 5.25 \times 10^{-4}\ \text{mol dm}^{-3}$

At equilibrium, the concentration of hydrogen ion is equal to the concentration of benzoate ion, i.e. $[C_6H_5CO_2^-(aq)] = 5.25 \times 10^{-4}$ mol dm^{-3}

	$C_6H_5CO_2H(aq)$	⇌	$H^+(aq)$	+ $C_6H_5CO_2^-(aq)$
Initial concn/mol dm^{-3}	0.0050		0	0
Equilibrium concn/mol dm^{-3}	$(0.0050 - 5.25 \times 10^{-4})$		5.25×10^{-4}	5.25×10^{-4}

$$K_a = \frac{[H^+(aq)][C_6H_5CO_2^-(aq)]}{[C_6H_5CO_2H(aq)]}$$

$$= \frac{(5.25 \times 10^{-4}\ \text{mol dm}^{-3})(5.25 \times 10^{-4}\ \text{mol dm}^{-3})}{(0.004475\ \text{mol dm}^{-3})} = \boxed{6.16 \times 10^{-5}\ \text{mol dm}^{-3}}$$

Exercise 254

$pH = -\log[H^+(aq)]$

$\log[H^+(aq)] = -pH = -3.40$

$[H^+(aq)] = \text{antilog}(-4 + 0.60) = 3.98 \times 10^{-4}\ \text{mol dm}^{-3}$

$$K_a = \frac{[CH_3CO_2^-(aq)][H^+(aq)]}{[CH_3CO_2H(aq)]} = \frac{(3.98 \times 10^{-4}\ \text{mol dm}^{-3})^2}{(0.010 - 3.98 \times 10^{-4}\ \text{mol dm}^{-3}}$$

$$= \frac{1.58 \times 10^{-7}\ \text{mol}^2\ \text{dm}^{-6}}{0.0096\ \text{mol dm}^{-3}} = \boxed{1.65 \times 10^{-5}\ \text{mol dm}^{-3}}$$

Exercise 255

$pH = -\log[H^+(aq)]$

$\log[H^+(aq)] = -pH = -2.40$

$[H^+(aq)] = \text{antilog}(-3 + 0.60) = 3.98 \times 10^{-3}$

$$K_a = \frac{[HCO_2^-(aq)][H^+(aq)]}{[HCO_2H(aq)]} = \frac{(3.98 \times 10^{-3}\ \text{mol dm}^{-3})^2}{(0.10 - 3.98 \times 10^{-3})\ \text{mol dm}^{-3}}$$

$$= \frac{1.58 \times 10^{-5}\ \text{mol}^2\ \text{dm}^{-6}}{0.096\ \text{mol dm}^{-3}} = \boxed{1.65 \times 10^{-4}\ \text{mol dm}^{-3}}$$

Exercise 256

(a) $$[H^+(aq)] = \frac{K_w}{[OH^-(aq)]} = \frac{1.00 \times 10^{-14}\ \text{mol}^2\ \text{dm}^{-6}}{0.0500\ \text{mol dm}^{-3}} = 2.00 \times 10^{-13}\ \text{mol dm}^{-3}$$

$pH = -\log[H^+(aq)] = -\log(2.00 \times 10^{-13}) = -(0.3010 - 13.0000) = \boxed{12.7}$

(b)

	$CH_3CO_2H(aq)$	⇌	$H^+(aq)$	+ $CH_3CO_2^-(aq)$
Initial concn/mol dm^{-3}	0.100		0	0
Equilibrium concn/mol dm^{-3}	$0.100 - x$		x	x

$$K_a = \frac{[H^+(aq)][CH_3CO_2^-(aq)]}{[CH_3CO_2H(aq)]}$$

$$1.69 \times 10^{-5}\ \text{mol dm}^{-3} = \frac{(x\ \text{mol dm}^{-3})(x\ \text{mol dm}^{-3})}{(0.100 - x)\ \text{mol dm}^{-3}}$$

Assuming that $x \ll 0.100$, $\left(\frac{c_0}{K_a} = \frac{0.100}{1.69 \times 10^{-5}} = > 1000\right)$

$(0.100 - x) \simeq 0.100$

$$\therefore 1.69 \times 10^{-5} = \frac{x^2}{0.100}$$

$x = \sqrt{1.69 \times 10^{-6}} = 1.30 \times 10^{-3}$

$\therefore [H^+(aq)] = 1.30 \times 10^{-3}\ \text{mol dm}^{-3}$

$pH = -\log[H^+(aq)] = -\log(1.30 \times 10^{-3}) = -(0.1139 - 3.0000) = \boxed{2.9}$

Exercise 257

	$HIO_3(aq)$	⇌	$H^+(ac)$	+ $IO_3^-(aq)$
Initial concn/mol dm^{-3}	0.100		0	0
Equilibrium concn/mol dm^{-3}	$(0.100 - x)$		x	x

$$K_a = \frac{[H^+(aq)][IO_3^-(aq)]}{[HIO_3(aq)]}$$

$$0.17\ \text{mol dm}^{-3} = \frac{x^2\ \text{mol}^2\ \text{dm}^{-6}}{(0.100 - x)\ \text{mol dm}^{-3}}$$

In this case, $\frac{c_0}{K_a} = \frac{0.100}{0.17} = 0.588 < 1000$

Therefore x cannot be ignored in the bottom line and a quadratic equation must be solved. The expression becomes:

$$0.17 = \frac{x^2}{(0.100 - x)} \quad \text{or} \quad x^2 + 0.17x - 0.017 = 0$$

$$x = \frac{-0.17 \pm \sqrt{(-0.17)^2 - 4(1 \times -0.017)}}{2}$$

$$= \frac{-0.17 \pm 0.311}{2} = 0.0705 \text{ or } -0.241 \text{ (absurd root!)}$$

$\therefore [H^+(aq)] = 7.05 \times 10^{-2}\ \text{mol dm}^{-3}$

$pH = -\log[H^+(aq)]$

$= -\log(7.05 \times 10^{-2}) = -(-2 + 0.848) = \boxed{1.15}$

(If you had wrongly ignored x on the bottom line you would have obtained a value of pH = 0.886.)

Exercise 258

(a)

Initial concn/mol dm^{-3}	0.25	const.	0	0
	$NH_3(aq)$ +	$H_2O(l)$ ⇌	$NH_4^+(aq)$ +	$OH^-(aq)$
Equilibrium concn/mol dm^{-3}	$0.25-x$	const.	x	x

$$K_b = \frac{[NH_4^+(aq)][OH^-(aq)]}{[NH_3(aq)]} = \frac{x^2 \text{ mol}^2 \text{ dm}^{-6}}{(0.25-x) \text{ mol dm}^{-3}} = 1.8 \times 10^{-5} \text{ mol dm}^{-3}$$

$$\frac{c_o}{K_b} = \frac{0.25}{1.8 \times 10^{-5}} = 13889 > 1000 \quad \therefore \quad 0.25-x \approx x$$

i.e. $\frac{x^2}{0.25} = 1.8 \times 10^{-5}$

$\therefore \quad x = \sqrt{0.25 \times 1.8 \times 10^{-5}} = \sqrt{4.5 \times 10^{-6}} = 2.1 \times 10^{-3}$

i.e. $[OH^-(aq)] = 2.1 \times 10^{-3}$ mol dm^{-3}

$pOH = -\log(2.1 \times 10^{-3}) = 2.7$

$\therefore \quad pH = 14 - 2.7 = \boxed{11.3}$

Alternatively, $[H^+(aq)] = K_w/[OH^-(aq)]$

$$= \frac{1.0 \times 10^{-14} \text{ mol}^2 \text{ dm}^{-6}}{2.1 \times 10^{-3} \text{ mol dm}^{-3}} = 4.76 \times 10^{-12} \text{ mol dm}^{-3}$$

$\therefore \quad pH = -\log[H^+(aq)] = -\log(4.76 \times 10^{-12}) = \boxed{11.3}$

(b)

Initial concn/mol dm^{-3}	0.15	const.	0	0
	$C_4H_9NH_2(aq)$ +	$H_2O(l)$ ⇌	$C_4H_9NH_3^+(aq)$ +	$OH^-(aq)$
Equilibrium concn/mol dm^{-3}	$0.15-x$	const.	x	x

$$K_b = \frac{K_w}{K_a} = \frac{[C_4H_9NH_3^+(aq)][OH^-(aq)]}{[C_4H_9NH_2(aq)]}$$

i.e. $$\frac{1.0 \times 10^{-14} \text{ mol}^2 \text{ dm}^{-6}}{1.7 \times 10^{-11} \text{ mol dm}^{-3}} = \frac{x^2 \text{ mol}^2 \text{ dm}^{-6}}{(0.15-x) \text{ mol dm}^{-3}}$$

i.e. $5.88 \times 10^{-4} = \frac{x^2}{0.15-x}$

$$\frac{c_o}{K_b} = \frac{0.15}{5.88 \times 10^{-4}} = 255 < 1000 \quad \therefore \text{ a quadratic equation must be solved}$$

$x^2 + (5.88 \times 10^{-4})x - (5.88 \times 0.15 \times 10^{-4}) = 0$

$$\therefore \quad x = \frac{-5.88 \times 10^{-4} \pm \sqrt{3.46 \times 10^{-7} + 3.53 \times 10^{-4}}}{2}$$

$$= \frac{-5.88 \times 10^{-4} \pm 1.88 \times 10^{-2}}{2}$$

x must be positive, $\therefore \quad x = \frac{1.82 \times 10^{-2}}{2} = 9.1 \times 10^{-3}$

$\therefore \quad [OH^-(aq)] = 9.1 \times 10^{-3}$ mol dm^{-3}

$pOH = -\log(9.1 \times 10^{-3}) = 2.0$

$pH = 14 - pOH = \boxed{12.0}$

Exercise 259

The method is the same as in the Worked Example. The pH values, and intermediate answers, are shown below in tabular form.

Volume of alkali/cm^3	Amount of alkali/mol	Amount of acid/mol	Excess /mol	$[OH^-(aq)]$ /mol dm^{-3}	$[H^+(aq)]$ /mol dm^{-3}	pH
0.0	-	0.00250	0.00250	-	0.100	1.00
5.0	0.00050	0.00250	0.00200	-	0.0667	1.18
10.0	0.00100	0.00250	0.00150	-	0.0429	1.37
20.0	0.00200	0.00250	0.00050	-	0.0111	1.95
24.0	0.00240	0.00250	0.00010	-	0.00204	2.69
25.0	0.00250	0.00250	-	1.00×10^{-7}	1.00×10^{-7}	7.00
26.0	0.00260	0.00250	0.00010	0.00196	5.10×10^{-12}	11.3
30.0	0.00300	0.00250	0.00050	0.00909	1.10×10^{-12}	12.0

Exercise 260

Before the equivalence point, there is an excess of weak acid, and the first step is to calculate its concentration, c_o, from the amount of the excess and the total volume. The pH is calculated from this, as shown in the Worked Example on page 129 and the subsequent exercises. Note that the simplifying assumption can be made in this case so the following expression applies:

$$[H^+(aq)] = \sqrt{K_a \times c_o}$$

The answers to the intermediate steps in the calculation are shown in tabular form.

Volume of alkali cm^3	Amount of alkali/mol	Excess of acid/mol	c_o(acid) /mol dm^{-3}	$[H^+(aq)]$ /mol dm^{-3}	pH
5.0	0.00100	0.00400	0.133	1.50×10^{-3}	2.8
10.0	0.00200	0.00300	0.0857	1.21×10^{-3}	2.9
20.0	0.00400	0.00100	0.0222	6.15×10^{-4}	3.2
24.0	0.00480	0.00020	0.00408	2.63×10^{-4}	3.6
24.5	0.00490	0.00010	0.00202	1.85×10^{-4}	3.7

After the equivalence point, there is an excess of strong alkali and the calculation is simpler. The pH is determined directly from the amount of excess OH^- ions, as in the second Worked Example on page 132 and Exercise 259.

Volume of alkali/cm^3	Amount of alkali/mol	Excess of alkali/mol	$[OH^-(aq)]$ /mol dm^{-3}	$[H^+(aq)]$ /mol dm^{-3}	pH
25.5	0.00510	0.00010	1.98×10^{-3}	5.05×10^{-12}	11.3
26.0	0.00520	0.00020	3.92×10^{-3}	2.55×10^{-12}	11.6
30.0	0.00600	0.00100	1.82×10^{-2}	5.49×10^{-13}	12.3

Exercise 261

Before the equivalence point, there is an excess of strong acid and the calculation is simply that of determining the excess of H^+ ions, as in the first Worked Example on page 132 and Exercise 259.

Volume of alkali/cm^3	Amount of alkali/mol	Excess of acid/mol	$[H^+(aq)]$ /mol/dm^{-3}	pH
5.0	0.00075	0.00300	0.100	1.0
10.0	0.00150	0.00225	0.0643	1.2
20.0	0.00300	0.00075	0.0167	1.8
24.0	0.00360	0.00015	0.00306	2.5
24.5	0.00367	0.00008	0.00162	2.8

After the equivalence point, there is an excess of weak alkali and the first step is to calculate its concentration, c_0, from the amount of excess and the total volume. The pH is calculated from this, as shown in Exercise 258. The key step is to use the expression

$$[OH^-(aq)] = \sqrt{K_b \times c_0}$$

Volume of alkali/cm^3	Amount of alkali/mol	Excess of alkali/mol	c_0(alkali) /mol dm^{-3}	$[OH^-(aq)]$ /mol dm^{-3}	pH
25.5	0.00382	0.000075	0.00148	1.63×10^{-4}	10.2
26.0	0.00390	0.00015	0.00294	2.30×10^{-4}	10.4
30.0	0.00450	0.00075	0.0136	4.95×10^{-4}	10.7

Exercise 262

(a) $K_h = \frac{K_w}{K_a}$ $\quad K_a(HCN) = 4.9 \times 10^{-10}$ mol dm^{-3}

$$K_h = \frac{1.0 \times 10^{-14}\ mol^2\ dm^{-6}}{4.9 \times 10^{-10}\ mol\ dm^{-3}} = 2.0 \times 10^{-5}\ mol\ dm^{-3}$$

Initial concn/mol dm^{-3}	0.010		0	0
	$CN^-(aq)$ + $H_2O(l)$	$\rightleftharpoons$	$HCN(aq)$ +	$OH^-(aq)$
Equilibrium concn/mol dm^{-3}	$0.010-x$		x	x

$$K_h = \frac{[HCN(aq)]\ [OH^-(aq)]}{[CN^-(aq)]}$$

$$2.0 \times 10^{-5}\ mol\ dm^{-3} = \frac{(x\ mol\ dm^{-3})\ (x\ mol\ dm^{-3})}{(0.010-x)\ mol\ dm^{-3}}$$

Assume $(0.010-x) \simeq 0.010$

$$\therefore 2.0 \times 10^{-5} = \frac{x^2}{0.010}$$

$$x^2 = 0.010 \times 2.0 \times 10^{-5} = 2.0 \times 10^{-7}$$

$$x = \sqrt{2.0 \times 10^{-7}} = 4.5 \times 10^{-4}$$

$\therefore [OH^-(aq)] = 4.5 \times 10^{-4}$ mol dm^{-3} .

$$[H^+(aq)] = \frac{K_w}{[OH^-(aq)]} = \frac{1.0 \times 10^{-14}\ mol^2\ dm^{-6}}{4.5 \times 10^{-4}\ mol\ dm^{-3}} = 2.2 \times 10^{-11}\ mol\ dm^{-3}$$

$$pH = -\log [H^+(aq)] = -\log (2.2 \times 10^{-11}) = \boxed{10.7}$$

(b) $K_h = \frac{K_w}{K_b}$ $\quad K_b(NH_3) = 1.8 \times 10^{-5}$ mol dm^{-3}

$$K_h = \frac{1.0 \times 10^{-14}\ mol^2\ dm^{-6}}{1.8 \times 10^{-5}\ mol\ dm^{-3}} = 5.6 \times 10^{-10}\ mol\ dm^{-3}$$

Initial concn/mol dm^{-3}	1.0		0	0
	$NH_4^+(aq)$	$\rightleftharpoons$	$NH_3(aq)$ +	$H^+(aq)$
Equilibrium concn/mol dm^{-3}	$1.0-x$		x	x

$$K_h = \frac{[NH_3(aq)]\ [H^+(aq)]}{[NH_4^+(aq)]}$$

$$5.6 \times 10^{-10}\ mol\ dm^{-3} = \frac{(x\ mol\ dm^{-3})\ (x\ mol\ dm^{-3})}{(1.0-x)\ mol\ dm^{-3}}$$

Assume $(1.0-x) \simeq 1.0$

$$\therefore 5.6 \times 10^{-10} = \frac{x^2}{1.0}$$

$$x = \sqrt{5.6 \times 10^{-10}} = 2.4 \times 10^{-5}$$

$\therefore [H^+(aq)] = 2.4 \times 10^{-5}$ mol dm^{-3}

$$pH = -\log [H^+(aq)] = -\log (2.4 \times 10^{-5}) = \boxed{4.6}$$

(c) $K_h = \frac{K_w}{K_a}$ $\quad K_a(HF) = 5.6 \times 10^{-4}$ mol dm^{-3}

$$K_h = \frac{1.0 \times 10^{-14}\ mol^2\ dm^{-6}}{5.6 \times 10^{-4}\ mol\ dm^{-3}} = 1.8 \times 10^{-11}\ mol\ dm^{-3}$$

Initial concn/mol dm^{-3}	0.020		0	0
	$F^-(aq)$ + $H_2O(l)$	+	$HF(aq)$ +	$OH^-(aq)$
Equilibrium concn/mol dm^{-3}	$0.020-x$		x	x

$$K_h = \frac{[HF(aq)]\ [OH^-(aq)]}{[F^-(aq)]}$$

$$1.8 \times 10^{-11}\ mol\ dm^{-3} = \frac{(x\ mol\ dm^{-3})\ (x\ mol\ dm^{-3})}{(0.020-x)\ mol\ dm^{-3}}$$

$(0.020-x) \simeq 0.020$

$$\therefore 1.8 \times 10^{-11} = \frac{x^2}{0.020}$$

$$x = \sqrt{3.6 \times 10^{-13}} = 6.0 \times 10^{-7}$$

$\therefore [OH^-(aq)] = 6.0 \times 10^{-7}$ mol dm^{-3}

$$[H^+(aq)] = \frac{K_w}{[OH^-(aq)]} = \frac{1.0 \times 10^{-14}\ mol^2\ dm^{-6}}{6.0 \times 10^{-7}\ mol\ dm^{-3}} = 1.7 \times 10^{-8}\ mol\ dm^{-3}$$

$$pH = -\log [H^+(aq)] = -\log (1.7 \times 10^{-8}) = \boxed{7.8}$$

Exercise 263

(a)

Initial concn/mol dm^{-3}	0.100	0	0
	$CH_3CH_2CO_2H(aq)$ ⇌	$CH_3CH_2CO_2^-(aq)$ +	$H^+(aq)$
Equilibrium concn/mol dm^{-3}	$(0.100 - x)$	x	x

$$K_a = \frac{[CH_3CH_2CO_2^-(aq)]\,[H^+(aq)]}{[CH_3CH_2CO_2H(aq)]}$$

$$1.3 \times 10^{-5} = \frac{x^2}{(0.100 - x)} \quad \text{(cancelling units)}$$

$$\frac{c_0}{K_a} = \frac{0.100}{1.3 \times 10^{-5}} > 1000 \qquad \therefore\ 0.100 - x \simeq 0.100$$

$$x^2 = 0.100 \times 1.3 \times 10^{-5}$$

$$x = \sqrt{1.3 \times 10^{-6}} = 1.1 \times 10^{-3}$$

$$[H^+(aq)] = \boxed{1.1 \times 10^{-3}\ \text{mol dm}^{-3}}$$

(b)

Initial concn/mol dm^{-3}	0.100	0.050	0
	$CH_3CH_2CO_2H(aq)$ ⇌	$CH_3CH_2CO_2^-(aq)$ +	$H^+(aq)$
Equilibrium concn/mol dm^{-3}	$(0.100 - x)$	$(0.050 + x)$	x

$$K_a = \frac{[CH_3CH_2CO_2^-(aq)]\,[H^+(aq)]}{[CH_3CH_2CO_2H(aq)]}$$

$$1.3 \times 10^{-5} = \frac{(0.050 + x) \times x}{(0.100 - x)} \quad \text{(cancelling units)}$$

$$\frac{c_0}{K_a} = \frac{0.050}{1.3 \times 10^{-5}} > 1000$$

$\therefore$ $(0.050 + x) \simeq 0.050$ and $(0.100 - x) \simeq 0.100$

$$\therefore\ 1.3 \times 10^{-5} = \frac{0.050\,x}{0.100}$$

$$x = \frac{1.3 \times 10^{-5} \times 0.100}{0.050} = 2.6 \times 10^{-5}$$

$$\therefore\ [H^+(aq)] = \boxed{2.6 \times 10^{-5}\ \text{mol dm}^{-3}}$$

(c) Propenoic acid solution has a greater hydrogen ion concentration than the buffer solution containing the same concentration of propanoic acid. In other words, the ionization of the weak acid is suppressed in the buffer solution. This should not surprise you, if you applied Le Chatelier's principle to the equilibrium:

$$CH_3CH_2CO_2H(aq) \rightleftharpoons CH_3CH_2CO_2^-(aq) + H^+(aq)$$

An increase in the concentration of propanoate ions will cause the equilibrium to shift to the left, increasing the concentration of un-ionized acid and decreasing the concentration of hydrogen ions.

Exercise 264

Amount of acid in solution = cV = 0.090 mol dm^{-3} x 0.010 dm^3 = 9.0 x 10^{-4} mol

Amount of salt in solution = cV = 0.15 mol dm^{-3} x 0.010 dm^3 = 3.0 x 10^{-3} mol

$$\text{Concn of acid in buffer} = \frac{n}{V} = \frac{9.0 \times 10^{-4}\ \text{mol}}{0.030\ \text{dm}^3} = 0.030\ \text{mol dm}^{-3}$$

$$\text{Concn of salt in buffer} = \frac{n}{V} = \frac{3.0 \times 10^{-3}\ \text{mol}}{0.030\ \text{dm}^3} = 0.10\ \text{mol dm}^{-3}$$

pH = $-\log [H^+(aq)]$ = 5.85

$\therefore$ $[H^+(aq)]$ = antilog (-5.85) = 1.41 x 10^{-6} mol dm^{-3}

Initial concn/mol dm^{-3}	0.030	0	0.10
	$HA(aq)$ ⇌	$H^+(aq)$ +	$A^-(aq)$
Equilibrium concn/mol dm^{-3}	0.030 - 1.41 x 10^{-6}	1.41 x 10^{-6}	0.10 + 1.41 x 10^{-6}

$$K_a = \frac{[H^+(aq)]\,[A^-(aq)]}{[HA(aq)]}$$

$$= \frac{1.41 \times 10^{-6}\ \text{mol dm}^{-3} \times (0.10 + 1.41 \times 10^{-6})\ \text{mol dm}^{-3}}{(0.030 - 1.41 \times 10^{-6})\ \text{mol dm}^{-3}}$$

$$\simeq \frac{1.41 \times 10^{-6}\ \text{mol dm}^{-3} \times 0.10\ \text{mol dm}^{-3}}{0.030\ \text{mol dm}^{-3}} = \boxed{4.7 \times 10^{-6}\ \text{mol dm}^{-3}}$$

Exercise 265

(a) Initial amount of CH_3CO_2H = concentration x volume

= 0.10 mol dm^{-3} x 0.025 dm^3 = 2.5 x 10^{-3} mol

Amount of CH_3CO_2H at half neutralization point = 1.25 x 10^{-3} mol

$$\text{Volume of NaOH required for half neutralization} = \frac{\text{amount}}{\text{concentration}}$$

$$= \frac{1.25 \times 10^{-3}\ \text{mol}}{0.10\ \text{mol dm}^{-3}} = 1.25 \times 10^{-2}\ \text{dm}^3$$

Total volume at half neutralization point

= volume of CH_3CO_2H solution + volume of NaOH solution added = 0.0375 dm^3

$$[CH_3CO_2H(aq)] = \frac{\text{amount}}{\text{volume}} = \frac{1.25 \times 10^{-3}\ \text{mol}}{0.0375\ \text{dm}^3} = \boxed{0.033\ \text{mol dm}^{-3}}$$

Since half the original acid has been neutralized and converted to ethanoate ions,

$$[CH_3CO_2^-(aq)] = [CH_3CO_2H(aq)] = \boxed{0.033\ \text{mol dm}^{-3}}$$

(b)
$$K_a = \frac{[H^+(aq)]\,[CH_3CO_2^-(aq)]}{[CH_3CO_2H(aq)]}$$

$$\therefore\ [H^+(aq)] = K_a \frac{[CH_3CO_2H(aq)]}{[CH_3CO_2^-(aq)]} = K_a \times 1 = 1.7 \times 10^{-5}\ \text{mol dm}^{-3}$$

$$\text{pH} = -\log [H^+(aq)] = -\log (1.7 \times 10^{-5}) = \boxed{4.8}$$

Exercise 266

(a) $C_3H_7CO_2H(aq) \rightleftharpoons C_3H_7CO_2^-(aq) + H^+(aq)$

$$K_a = \frac{[C_3H_7CO_2^-(aq)]\,[H^+(aq)]}{[C_3H_7CO_2H(aq)]}$$

$$\therefore [H^+(aq)] = K_a \frac{[C_3H_7CO_2H(aq)]}{[C_3H_7CO_2^-(aq)]}$$

$$= 1.5 \times 10^{-5}\ \text{mol dm}^{-3} \times \frac{1.0\ \text{mol dm}^{-3}}{1.0\ \text{mol dm}^{-3}} = 1.5 \times 10^{-5}\ \text{mol dm}^{-3}$$

$$\text{pH} = -\log [H^+(aq)] = -\log (1.5 \times 10^{-5}) = -(0.18 - 5) = \boxed{4.8}$$

(b) $$[H^+(aq)] = 1.5 \times 10^{-5}\ \text{mol dm}^{-3} \times \frac{2.0\ \text{mol dm}^{-3}}{1.0\ \text{mol dm}^{-3}} = 3.0 \times 10^{-5}\ \text{mol dm}^{-3}$$

$$\text{pH} = -\log [H^+(aq)] = -\log (3.0 \times 10^{-5}) = -(0.48 - 5) = \boxed{4.5}$$

(c) $$[H^+(aq)] = 1.5 \times 10^{-5}\ \text{mol dm}^{-3} \times \frac{2.0\ \text{mol dm}^{-3}}{0.50\ \text{mol dm}^{-3}} = 6.0 \times 10^{-5}\ \text{mol dm}^{-3}$$

$$\text{pH} = -\log [H^+(aq)] = -\log (6.0 \times 10^{-5}) = -(0.78 - 5) = \boxed{4.2}$$

(d) $$[H^+(aq)] = 1.5 \times 10^{-5}\ \text{mol dm}^{-3} \times \frac{0.50\ \text{mol dm}^{-3}}{2.0\ \text{mol dm}^{-3}} = 3.75 \times 10^{-6}\ \text{mol dm}^{-3}$$

$$\text{pH} = -\log [H^+(aq)] = -\log (3.75 \times 10^{-6}) = -(0.57 - 6) = \boxed{5.4}$$

Exercise 267

$$K_a = \frac{[H^+(aq)]\,[HCO_2^-(aq)]}{[HCO_2H(aq)]} \qquad \therefore [H^+(aq)] = K_a \times \frac{[HCO_2H(aq)]}{[HCO_2^-(aq)]}$$

Making the usual assumption that in a buffer solution, the equilibrium concentrations are equal to the concentrations on mixing:

(a) $$[H^+(aq)] = 1.6 \times 10^{-4}\ \text{mol dm}^{-3} \times \frac{0.10\ \text{mol dm}^{-3}}{1.0\ \text{mol dm}^{-3}} = 1.6 \times 10^{-5}\ \text{mol dm}^{-3}$$

$$\text{pH} = -\log [H^+(aq)] = -\log (1.6 \times 10^{-5}) = -(-5 + 0.20) = \boxed{4.8}$$

(b) $$[H^+(aq)] = 1.6 \times 10^{-4}\ \text{mol dm}^{-3} \times \frac{2.0\ \text{mol dm}^{-3}}{0.20\ \text{mol dm}^{-3}} = 1.6 \times 10^{-3}\ \text{mol dm}^{-3}$$

$$\text{pH} = -\log [H^+(aq)] = -\log (1.6 \times 10^{-3}) = -(-3 + 0.20) = \boxed{2.8}$$

(c) $$[H^+(aq)] = 1.6 \times 10^{-4}\ \text{mol dm}^{-3} \times \frac{0.50\ \text{mol dm}^{-3}}{2.5\ \text{mol dm}^{-3}} = 3.2 \times 10^{-5}\ \text{mol dm}^{-3}$$

$$\text{pH} = -\log [H^+(aq)] = -\log (3.2 \times 10^{-5}) = -(-5 + 0.51) = \boxed{4.5}$$

Exercise 268

$$K_a = \frac{[HPO_4^{2-}(aq)]\,[H^+(aq)]}{[H_2PO_4^-(aq)]} \qquad \therefore [H^+(aq)] = K_a \times \frac{[H_2PO_4^-(aq)]}{[HPO_4^{2-}(aq)]}$$

Amount of NaH_2PO_4 = 2.0 mol dm^{-3} × 0.500 dm^3 = 1.0 mol

Amount of Na_2HPO_4 = 0.50 mol dm^{-3} × 1.00 dm^3 = 0.50 mol

Again, we assume that the equilibrium amounts in a buffer solution are equal to the added amounts.

$$\therefore [H^+(aq)] = 6.2 \times 10^{-8}\ \text{mol dm}^{-3} \times \frac{1.0\ \text{mol}/1.5\ \text{dm}^3}{0.50\ \text{mol}/1.5\ \text{dm}^3} = 1.2 \times 10^{-7}\ \text{mol dm}^{-3}$$

$$\text{pH} = -\log [H^+(aq)] = -\log (1.2 \times 10^{-7}) = -(-7 + 0.08) = \boxed{6.9}$$

Exercise 269

$$K_a = \frac{[CH_3CO_2^-(aq)][H^+(aq)]}{[CH_3CO_2H(aq)]}$$

$$\therefore [H^+(aq)] = K_a \times \frac{[CH_3CO_2H(aq)]}{[CH_3CO_2^-(aq)]}$$

Initially, $[H^+(aq)] = K_a = 1.70 \times 10^{-5}$ mol dm^{-3}

$$\therefore \text{initial pH} = -\log [H^+(aq)] = -(0.23 - 5) = \boxed{4.77}$$

(a) Amount of HCl added = cV = 1.0 mol dm^{-3} × 0.0010 dm^3 = 0.0010 mol

Assuming that all the added HCl reacts with $CH_3CO_2^-$ ions to form CH_3CO_2H, the new equilibrium amounts are:

Amount of CH_3CO_2H = (0.50 mol dm^{-3} × 0.10 dm^3) + 0.0010 mol = 0.051 mol

Amount of $CH_3CO_2^-$ = (0.50 mol dm^{-3} × 0.10 dm^3) − 0.0010 mol = 0.049 mol

$$[H^+(aq)] = K_a \frac{[CH_3CO_2H(aq)]}{[CH_3CO_2^-(aq)]} = 1.70 \times 10^{-5}\ \text{mol dm}^{-3} \times \frac{0.051\ \text{mol}/0.101\ \text{dm}^3}{0.049\ \text{mol}/0.101\ \text{dm}^3}$$

$$= 1.77 \times 10^{-5}\ \text{mol dm}^{-3}$$

$$\therefore \text{new pH} = -\log [H^+(aq)] = -(0.25 - 5) = \boxed{4.75} \quad \boxed{0.02\ \text{decrease}}$$

(b) Amount of KOH added = cV = 1.5 mol dm^{-3} × 0.0020 dm^3 = 0.0030 mol

Assuming that all the added KOH reacts with CH_3CO_2H to form $CH_3CO_2^-$ ions, the new equilibrium amounts are:

Amount of CH_3CO_2H = (0.50 mol dm^{-3} × 0.10 dm^3) − 0.030 mol = 0.047 mol

Amount of $CH_3CO_2^-$ = (0.50 mol dm^{-3} × 0.10 dm^3) + 0.030 mol = 0.053 mol

$$[H^+(aq)] = K_a \frac{[CH_3CO_2H(aq)]}{[CH_3CO_2^-(aq)]} = 1.70 \times 10^{-5}\ \text{mol dm}^{-3} \times \frac{0.047\ \text{mol}/0.102\ \text{dm}^3}{0.053\ \text{mol}/0.102\ \text{dm}^3}$$

$$= 1.51 \times 10^{-5}\ \text{mol dm}^{-3}$$

$$\therefore \text{new pH} = -\log [H^+(aq)] = -(0.18 - 5) = \boxed{4.82} \quad \boxed{0.05\ \text{increase}}$$

Exercise 270

(a) $K_a = \frac{[C_3H_7CO_2^-(aq)]\,[H^+(aq)]}{[C_3H_7CO_2H(aq)]}$

$[H^+(aq)] = K_a \frac{[C_3H_7CO_2H(aq)]}{[C_3H_7CO_2^-(aq)]} = 1.5 \times 10^{-5}\ mol\ dm^{-3} \times \frac{2.0\ mol\ dm^{-3}}{0.50\ mol\ dm^{-3}}$

$= 6.0 \times 10^{-5}\ mol\ dm^{-3}$

Initial pH = $-\log [H^+(aq)] = -(0.78 - 5) =$ **4.22**

(b) Amount of NaOH added = cV = $2.0\ mol\ dm^{-3} \times 0.020\ dm^3 = 0.040\ mol$

Assuming that all the added NaOH reacts with $C_3H_7CO_2H$ to form $C_3H_7CO_2^-$, the new equilibrium amounts are:

Amount of $C_3H_7CO_2H$ = $(2.00\ mol\ dm^{-3} \times 1.0\ dm^3) - 0.040\ mol = 1.96\ mol$

Amount of $C_3H_7CO_2^-$ = $(0.50\ mol\ dm^{-3} \times 1.0\ dm^3) + 0.040\ mol = 0.54\ mol$

$[H^+(aq)] = K_a \frac{[C_3H_7CO_2H(aq)]}{[C_3H_7CO_2^-(aq)]} = 1.5 \times 10^{-5}\ mol\ dm^{-3} \times \frac{1.96\ mol/1.02\ dm^3}{0.54\ mol/1.02\ dm^3}$

$= 5.4 \times 10^{-5}\ mol\ dm^{-3}$

New pH = $-\log [H^+(aq)] = -(0.73 - 5) =$ **4.27** **0.05 increase**

(c) Amount of HCl added = cV = $0.50\ mol\ dm^{-3} \times 0.100\ dm^3 = 0.050\ mol$

Assuming that all the added HCl reacts with $C_3H_7CO_2^-$ ions to form $C_3H_7CO_2H$, the new equilibrium amounts are:

Amount of $C_3H_7CO_2H$ = $(2.00\ mol\ dm^{-3} \times 1.0\ dm^3) + 0.05\ mol = 2.05\ mol$

Amount of $C_3H_3CO_2^-$ = $(0.50\ mol\ dm^{-3} \times 1.0\ dm^3) - 0.05\ mol = 0.45\ mol$

$[H^+(aq)] = K_a \frac{[C_3H_7CO_2H(aq)]}{[C_3H_7CO_2^-(aq)]} = 1.5 \times 10^{-5}\ mol\ dm^{-3} \times \frac{2.05\ mol/1.10\ dm^3}{0.45\ mol/1.10\ dm^3}$

$= 6.8 \times 10^{-5}\ mol\ dm^{-3}$

New pH = $-\log [H^+(aq)] = -(0.83 - 5) =$ **4.17** **0.05 decrease**

Exercise 271

$K_a = \frac{[CH_3CO_2^-(aq)]\,[H^+(aq)]}{[CH_3CO_2H(aq)]}$ $\therefore$ $[H^+(aq)] = K_a \times \frac{[CH_3CO_2H(aq)]}{[CH_3CO_2^-(aq)]}$

Initial amount of CH_3CO_2H = $1.5\ mol\ dm^{-3} \times 0.50\ dm^3 = 0.75\ mol$

Initial amount of $CH_3CO_2^-$ = $0.50\ mol\ dm^{-3} \times 1.0\ dm^3 = 0.50\ mol$

$[H^+(aq)] = 1.7 \times 10^{-5}\ mol\ dm^{-3} \times \frac{0.75\ mol/1.5\ dm^3}{0.50\ mol/1.5\ dm^3} = 2.55 \times 10^{-5}\ mol\ dm^{-3}$

pH = $-\log [H^+(aq)] = -\log (2.55 \times 10^{-5}) = -(-5 + 0.41) =$ **4.6**

The amount of added NaOH = $0.10\ mol\ dm^{-3} \times 0.050\ dm^3 = 0.0050\ mol$

Assuming that virtually all the added NaOH reacts with CH_3CO_2H to form $CH_3CO_2^-$ ions,

amount of CH_3CO_2H = initial amount - amount reacted

$= (0.75\ mol \times \frac{150\ cm^3}{1500\ cm^3}) - 0.0050\ mol = 0.070\ mol$

amount of $CH_3CO_2^-$ = $(0.50\ mol \times \frac{150\ cm^3}{1500\ cm^3}) + 0.0050\ mol = 0.055\ mol$

$[H^+(aq)] = 1.7 \times 10^{-5}\ mol\ dm^{-3} \times \frac{0.070\ mol/0.20\ dm^3}{0.055\ mol/0.20\ dm^3} = 2.16 \times 10^{-5}\ mol\ dm^{-3}$

pH = $-\log [H^+(aq)] = -\log (2.16 \times 10^{-5}) = -(-5 + 0.33) =$ **4.7**

Thus, the pH is **increased by 0.1 units.**

Exercise 272

(a) Amount of hydrogen ion added = concentration x volume

$= 1.0\ mol\ dm^{-3} \times 0.0010\ dm^3$

$= 1.0 \times 10^{-3}\ mol$

Total volume of solution after contamination = $101\ cm^3 = 0.101\ dm^3$

$[H^+(aq)] = \frac{amount}{volume} = \frac{1.0 \times 10^{-3}\ mol}{0.101\ dm^3} = 9.9 \times 10^{-3}\ mol\ dm^{-3}$

pH = $-\log[H^+(aq)] = -\log (9.9 \times 10^{-3}) = -(0.996 - 3) =$ **2.0**

The pH of the water has decreased from 7.0 to 2.0, i.e. by 5.0 units

(b) Amount of hydroxide ion added = concentration x volume

$= 1.5\ mol\ dm^{-3} \times 0.0020\ dm^3$

$= 3.0 \times 10^{-3}\ mol$

Total volume of solution after contamination = $102\ cm^3 = 0.102\ dm^3$

$[OH^-(aq)] = \frac{amount}{volume} = \frac{3.0 \times 10^{-3}\ mol}{0.102\ dm^3} = 0.029\ mol\ dm^{-3}$

$K_a = [H^+(aq)]\,[OH^-(aq)]$

$[H^+(aq)] = \frac{K_w}{[OH^-(aq)]} = \frac{1.0 \times 10^{-14}\ mol^2\ dm^{-6}}{0.029\ mol\ dm^{-3}} = 3.4 \times 10^{-13}\ mol\ dm^{-3}$

pH = $-\log [H^+(aq)] = -\log 3.4 \times 10^{-13} = -(0.53 - 13) =$ **12.5**

The pH of the water has increased from 7.0 to 12.5, i.e. by 5.5 units

Exercise 273

$NH_4^+(aq) \rightleftharpoons NH_3(aq) + H^+(aq)$

$$K_a = \frac{[NH_3(aq)]\,[H^+(aq)]}{[NH_4^+(aq)]}$$

Amount of NH_4^+ added $= cV = 0.20\ \text{mol dm}^{-3} \times 0.75\ \text{dm}^3 = 0.15\ \text{mol}$

Amount of NH_3 added $= cV = 0.10\ \text{mol dm}^{-3} \times 0.75\ \text{dm}^3 = 0.075\ \text{mol}$

$$[H^+(aq)] = K_a \frac{[NH_4^+(aq)]}{[NH_3(aq)]} = 6.00 \times 10^{-10}\ \text{mol dm}^{-3} \times \frac{0.15\ \text{mol}/1.5\ \text{dm}^3}{0.075\ \text{mol}/1.5\ \text{dm}^3}$$
$$= 1.2 \times 10^{-9}\ \text{mol dm}^{-3}$$

$pH = -\log [H^+(aq)] = -\log (1.2 \times 10^{-9}) = -(-9 + 0.08) = \boxed{8.9}$

Exercise 274

(a) $pH = pK_a - \log \frac{[\text{acid}]}{[\text{conjugate base}]}$

$\therefore \log \frac{[CH_3CO_2H(aq)]}{[CH_3CO_2^-(aq)]} = pK_a - pH = 4.74 - 4.70 = 0.04$

$\therefore \frac{[CH_3CO_2H(aq)]}{[CH_3CO_2^-(aq)]} = \text{antilog } 0.04 = \boxed{1.1}$

i.e. the solutions should be mixed in the ratio 1.1 volumes of ethanoic acid to one volume of sodium ethanoate.

(b) $\log \frac{[CH_3CO_2H(aq)]}{[CH_3CO_2^-(aq)]} = pK_a - pH = 4.74 - 4.40 = 0.34$

$\therefore \frac{[CH_3CO_2H(aq)]}{[CH_3CO_2^-(aq)]} = \text{antilog } 0.34 = \boxed{2.2}$

i.e. the solutions should be mixed in the ratio 2.2 volumes of ethanoic acid to one volume of sodium ethanoate.

Note that we again assume that, due to the presence of ethanoate ion, the ionization of ethanoic acid is suppressed. Therefore, the concentration of hydrogen ion is small compared to the concentration of either acid or base, so that we can take initial concentrations as being equal to equilibrium concentrations.

Exercise 275

(a) $pK_a = -\log K_a = -\log (1.58 \times 10^{-4}) = 3.8$

$pH = pK_a - \log \frac{[HCO_2H(aq)]}{[HCO_2^-(aq)]}$

Since $pH = pK_a$, $\log \frac{[HCO_2H(aq)]}{[HCO_2^-(aq)]} = 0$

i.e. $[HCO_2^-(aq)] = [HCO_2H(aq)] = \boxed{0.50\ \text{mol dm}^{-3}}$

(b) $pH = pK_a - \log \frac{[HCO_2H(aq)]}{[HCO_2^-(aq)]}$

$4.1 = 3.8 - \log \frac{[HCO_2H(aq)]}{[HCO_2^-(aq)]}$

$\therefore \frac{[HCO_2H(aq)]}{[HCO_2^-(aq)]} = \text{antilog } (3.8 - 4.1) = \text{antilog } -0.3 = \text{antilog } (-1 + 0.7)$
$= 5 \times 10^{-1}$ or 0.5

$\therefore [HCO_2^-(aq)] = 2[HCO_2H(aq)] = \boxed{1.0\ \text{mol dm}^{-3}}$

Exercise 276

$pH = pK_a - \log \frac{[HA(aq)]}{[A^-(aq)]}$

$\therefore \log \frac{[C_2H_5CO_2H(aq)]}{[C_2H_5CO_2^-(aq)]} = pK_a - pH = 4.9 - 4.0 = 0.9$

$\therefore \frac{[C_2H_5CO_2H(aq)]}{[C_2H_5CO_2^-(aq)]} = \text{antilog } 0.9 = 7.9$

Assuming that, in a buffer solution, the equilibrium amounts are equal to the amounts dissolved, the solution should be mixed in the ratio 7.9 parts of propanoic acid to 1 part of sodium propanoate.

Volume of 1.0 M $C_2H_5CO_2H$ $= 100\ \text{cm}^3 \times \frac{7.9}{8.9} = \boxed{89\ \text{cm}^3}$

Volume of 1.0 M $C_2H_5CO_2Na$ $= 100\ \text{cm}^3 \times \frac{1.0}{8.9} = \boxed{11\ \text{cm}^3}$

Exercise 277

$pH = pK_a - \log \frac{[HA(aq)]}{[A^-(aq)]}$

$\therefore \log \frac{[HPO_4^{2-}(aq)]}{[PO_4^{3-}(aq)]} = pK_a - pH = 12.4 - 11.8 = 0.6$

$\therefore \frac{[HPO_4^{2-}(aq)]}{[PO_4^{3-}(aq)]} = \text{antilog } 0.6 = 4.0$

Assuming that, in a buffer solution the equilibrium amounts are equal to the amounts dissolved, the solution should be mixed in the ratio 4 parts of Na_2HPO_4 to 1 part of Na_3PO_4.

Volume of 0.50 M Na_2HPO_4 $= 200\ \text{cm}^3 \times \frac{4}{5} = \boxed{160\ \text{cm}^3}$

Volume of 0.50 M Na_3PO_4 $= 200\ \text{cm}^3 \times \frac{1}{5} = \boxed{40\ \text{cm}^3}$

Exercise 278

$pH = pK_a - \log \frac{[\text{acid}]}{[\text{conjugate base}]}$

$\therefore \log \frac{[NH_4^+(aq)]}{[NH_3(aq)]} = pK_a - pH = 9.3 - 9.5 = -0.2$

$\therefore \frac{[NH_4^+(aq)]}{[NH_3(aq)]} = \text{antilog } (-0.2) = \text{antilog } (-1 + 0.8) = 0.63$

$[NH_4^+(aq)] = 0.63 \times [NH_3(aq)] = 0.63 \times 0.40\ \text{mol dm}^{-3} = \boxed{0.25\ \text{mol dm}^{-3}}$

Exercise 279

First, notice that the expression for pH given in the question is in a slightly different form from the one we have used in the chapter. However, the two are equivalent because

$$+\log \frac{[\text{base}]}{[\text{acid}]} = -\log \frac{[\text{acid}]}{[\text{base}]}$$

(a) The substance most likely to change the pH of the blood is carbon dioxide on its way to the lungs. CO_2 dissolves in blood to give carbonic(IV) acid, which is normally dissociated in solution.

$$CO_2(g) + H_2O(l) \rightleftharpoons H_2CO_3(aq)$$

$$H_2CO_3(aq) \rightleftharpoons HCO_3^-(aq) + H^+(aq)$$

As the concentration of free hydrogen ions increases, the equilibrium

$$H_2CO_3(aq) \rightleftharpoons HCO_3^-(aq) + H^+(aq)$$

shifts to the left, removing them from solution and preventing a drop in pH.

(We have assumed throughout this answer that the concentration of hydrogen ion produced by the dissociation of carbonic(IV) acid is small compared with the concentrations of the carbonic(IV) acid itself and the hydrogencarbonate anion. Therefore, we take the concentrations given in the question as equilibrium concentrations.)

(b) $[H^+(aq)] = 7.9 \times 10^{-7}$ mol dm^{-3}

$$pH = -\log [H^+(aq)] = -\log (7.9 \times 10^{-7}) = -(0.90 - 7) = 6.10$$

$$pH = -\log K_a + \log \frac{[\text{base}]}{[\text{acid}]}$$

$6.10 = -\log K_a + 0$ (since [base] = [acid])

$\therefore \log K_a = \boxed{-6.10}$

(c) $pH = -\log K_a + \log \frac{[\text{base}]}{[\text{acid}]}$

$$7.4 = -(-6.10) + \log \frac{[\text{base}]}{[\text{acid}]}$$

$$\therefore \log \frac{[\text{base}]}{[\text{acid}]} = 7.4 - 6.10 = 1.3$$

$$\therefore \frac{[\text{base}]}{[\text{acid}]} = \text{antilog } 1.3 = \boxed{20}$$

(d) Since the ratio of concentrations of base to acid is 20:1,

$$[\text{base}] = \frac{20}{21} \times 2.52 \times 10^{-2} \text{ mol dm}^{-3} = \boxed{2.4 \times 10^{-2} \text{ mol dm}^{-3}}$$

$$[\text{acid}] = \frac{1}{21} \times 2.52 \times 10^{-2} \text{ mol dm}^{-3} = \boxed{1.2 \times 10^{-3} \text{ mol dm}^{-3}}$$

Exercise 280

(a) (i) $pH = pK_a - \log \frac{[HIn(aq)]}{[In^-(aq)]}$

(ii) Assuming that acid is added to alkali in the titration, the indicator would initially be predominantly in its base form, In^-, which is blue. As more acid is added, the concentration of HIn increases and the first sign of colour change would appear when $[In^-(aq)] \approx 10[HIn(aq)]$

Then $pH = pK_a - \log\frac{1}{10} = 4.7 + 1.0 = \boxed{5.7}$

The intermediate colour, green, would appear when $[In^-(aq)] \approx [HIn(aq)]$.

Then $pH = pK_a - \log 1 = 4.7 - 0 = \boxed{4.7}$

The colour would appear to change completely to yellow by the time $[HIn(aq)] \approx 10[In^-(aq)]$.

Then $pH = pK_a - \log 10 = 4.7 - 1.0 = \boxed{3.7}$

(b) $K_b(CH_3CO_2^-) = \frac{K_w}{K_a(CH_3CO_2H)} = \frac{1.0 \times 10^{-14} \text{ mol}^2 \text{ dm}^{-6}}{1.7 \times 10^{-5} \text{ mol dm}^{-3}}$

$= 5.9 \times 10^{-10}$ mol dm^{-3}

$[CH_3CO_2^-(aq)] = 0.10$ mol dm^{-3} (since the volume is doubled)

Then $[OH^-(aq)] = \sqrt{K_b \times c_0} = \sqrt{5.9 \times 10^{-10} \text{ mol dm}^{-3} \times 0.10 \text{ mol dm}^{-3}}$

$= \sqrt{5.9 \times 10^{-11} \text{ mol}^2 \text{ dm}^{-6}} = 7.7 \times 10^{-6}$ mol dm^{-3}

pH = 14 - pOH = 14 - 5.1 = 8.9

A suitable indicator must have $pK_a \approx$ pH at equivalence, e.g. phenolphthalein (pK_a = 9.3) or thymol blue (pK_a = 8.9).

Exercise 281 (a) (i) 3.3 (ii) 10.7 (b) 2.9

Exercise 282 1.76×10^{-4} mol dm^{-3}

Exercise 283 4.70

Exercise 284 (a) 0.201 mol dm^{-3} (b) 2.07

Exercise 285

(a) (i) 2.95 (ii) 8.93×10^{-12} mol dm^{-3} (iii) 0.125 mol (iv) red

(b) 1.6×10^{-5} mol dm^{-3} (c) 1.33 mol dm^{-3}

Exercise 286

(b) (i) 0.125 mol dm^{-3} (ii) 4.9   (iii) 1.3×10^{-5} mol dm^{-3}

(iv) Up to the equivalence point, the curve should be of the same general shape but higher, i.e. the pH is greater for the same volume of added alkali. After the equivalence point, the curves should be identical.

Exercise 287   3.2 mol dm^{-3}

Exercise 288   pH 4.5 → pH 4.3

Exercise 289

(a) $Pt(s),H_2(g) \mid H^+(aq,1.0\ M) \vdots Zn^{2+}(aq,1.0\ M) \mid Zn(s)$ $\Delta E^\ominus = -0.76$ V

$Pt(s),H_2(g) \mid H^+(aq,1.0\ M) \vdots Cu^{2+}(aq,1.0\ M) \mid Cu(s)$ $\Delta E^\ominus = +0.35$ V

$Pt(s),H_2(g) \mid H^+(aq,1.0\ M) \vdots Ag^+(aq,1.0\ M) \mid Ag(s)$ $\Delta E^\ominus = +0.81$ V

(b) $Zn^{2+}(aq) + 2e^- \rightleftharpoons Zn(s)$ $E^\ominus = -0.76$ V

$Cu^{2+}(aq) + 2e^- \rightleftharpoons Cu(s)$ $E^\ominus = +0.35$ V

$Ag^+(aq) + e^- \rightleftharpoons Ag(s)$ $E^\ominus = +0.81$ V

Exercise 290

(a) Method 1: $2\overset{+5}{N}O_3^- \rightarrow \overset{+4}{N}_2O_4$ 1 e per N atom, 2e's in all

Method 2: Total charge on left = 2(−1) + 4(+1) = +2

Total charge on right = 0

∴ 2 e's must be added.

$2NO_3^-(aq) + 4H^+(aq) + 2e^- \rightarrow N_2O_4(g) + 2H_2O(l)$

(b) Method 1: $2H_2\overset{+4}{S}O_3 \rightarrow \overset{+2}{S}_2O_3^{2-}$ 2 e's per S atom, 4 e's in all

Method 2: Total charge on left = 2(+1) = +2

Total charge on right = −2

∴ 4 e's must be added.

$2H_2SO_3(aq) + 2H^+(aq) + 4e^- \rightarrow S_2O_3^{2-}(aq) + 3H_2O(l)$

(c) Method 1: $O\overset{+1}{I}^- \rightarrow \overset{-1}{I}^-$ 2 e's added

Method 2: Total charge on left = −1

Total charge on right = −1 + 2(−1) = −3

∴ 2 e's must be added.

$OI^-(aq) + H_2O(l) + 2e^- \rightarrow I^-(aq) + 2OH^-(aq)$

Exercise 291

(a) $Pt(s),H_2(g) \mid H^+(aq) \vdots Ag^+(aq) \mid Ag(s)$

Substituting into the expression:

$$\Delta E^\ominus = E^\ominus_R - E^\ominus_L$$

$$\Delta E^\ominus = +0.80\ V - 0.00\ V = +0.80\ V$$

The positive e.m.f. means the cell reaction would take place from left to right as shown in the cell diagram:

$Pt(s),H_2(g) \mid H^+(aq) \vdots Ag^+(aq) \mid Ag(s)$

$H_2(g) \rightarrow 2H^+(aq) + 2e^-$ $Ag^+(aq) + e^- \rightarrow Ag(s)$

Adding the first half-equation to twice the second (in order to transfer the same number of electrons) gives the overall reaction:

$H_2(g) + 2Ag^+(aq) \rightarrow 2H^+(aq) + Ag(s)$

(b) $Ni(s) \mid Ni^{2+}(aq) \vdots Zn^{2+}(aq) \mid Zn(s)$

Substituting into the expression:

$$\Delta E^\ominus = E^\ominus_R - E^\ominus_L$$

$$\Delta E^\ominus = -0.76\ V - (-0.25\ V) = -0.51\ V$$

The negative sign on the e.m.f. means the cell reaction would take place from right to left as shown in the cell diagram:

$Ni(s) \mid Ni^{2+}(aq) \vdots Zn^{2+}(aq) \mid Zn(s)$

$Ni^{2+}(aq) + 2e^- \rightarrow Ni(s)$ $Zn(s) \rightarrow Zn^{2+}(aq) + 2e^-$

Adding these half-equations gives the overall reaction:

$Zn(s) + Ni^{2+}(aq) \rightarrow Zn^{2+}(aq) + Ni(s)$

(c) $Ni(s) \mid Ni^{2+}(aq) \vdots [NO_3^-(aq) + 3H^+(aq)],[HNO_2(aq) + H_2O(l)] \mid Pt$

Substituting into the expression:

$$\Delta E^\ominus = E^\ominus_R - E^\ominus_L$$

$$\Delta E^\ominus = +0.94\ V - (-0.25\ V) = +1.19\ V$$

The positive e.m.f. means the cell reaction would take place from left to right as shown in the cell diagram:

$Ni(s) \mid Ni^{2+}(aq) \vdots [NO_3^-(aq) + 3H^+(aq)],[HNO_2(aq) + H_2O(l)] \mid Pt$

$Ni(s) \rightarrow Ni^{2+}(aq) + 2e^-$ $NO_3^-(aq) + 3H^+(aq) + xe^- \rightarrow HNO_2(aq) + H_2O(l)$

To work out x in the right-hand half-cell:

$\overset{+5}{N}O_3^- \rightarrow H\overset{+3}{N}O_2$ 2 e's must be added, i.e. $x = 2$

Adding the two half-equations:

$NO_3^-(aq) + 3H^+(aq) + 2e^- \rightarrow HNO_2(aq) + H_2O(l)$

$Ni(s) \rightarrow Ni^{2+}(aq) + 2e^-$

gives the overall reaction:

$Ni(s) + NO_3^-(aq) + 3H^+(aq) \rightarrow Ni^{2+}(aq) + HNO_2(aq) + H_2O(l)$

Exercise 291 (continued)

(d) $Pt \mid 2H_2SO_3(aq),[4H^+(aq) + S_2O_6^{2-}(aq)] \vdots Cr^{3+}(aq) \mid Cr(s)$

Substituting into the expression:

$$\Delta E^\ominus = E^\ominus_R - E^\ominus_L$$

$$\Delta E^\ominus = -0.74\ V - (+0.57\ V) = -1.31\ V$$

The negative e.m.f. means the cell reaction would take place from right to left as shown in the cell diagram:

$Pt \mid 2H_2SO_3(aq),[4H^+(aq) + S_2O_6^{2-}(aq)] \vdots Cr^{3+}(aq) \mid Cr(s)$

$S_2O_6^{2-}(aq) + 4H^+(aq) + xe^- \rightarrow 2H_2SO_3(aq) \quad Cr(s) \rightarrow Cr^{3+}(aq) + 3e^-$

To work out x in the left-hand cell:

$$\overset{+5}{S_2}O_6^{2-} \rightarrow 2H_2\overset{+4}{S}O_3 \qquad 1\ e\ \text{per S atom, i.e. } x = 2$$

Adding the two half-equations:

$3 \times [S_2O_6^{2-}(aq) + 4H^+(aq) + 2e^- \rightarrow 2H_2SO_3(aq)]$

$2 \times [Cr(s) \rightarrow Cr^{3+}(aq) + 3e^-]$

gives the overall equation:

$3S_2O_6^{2-}(aq) + 12H^+(aq) + 2Cr(s) \rightarrow 6H_2SO_3(aq) + 2Cr^{3+}(aq)$

Exercise 292

(a) The two half-equations are:

$Br_2 + 2e^- \rightleftharpoons 2Br^-(aq)$

$2I^-(aq) \rightleftharpoons I_2(aq) + 2e^-$

which combine to make the cell:

$Pt \mid I^-(aq),I_2(aq) \vdots Br_2(aq),Br^-(aq) \mid Pt$

Substituting into the expression:

$$\Delta E^\ominus = E^\ominus_R - E^\ominus_L$$

$$\Delta E^\ominus = +1.07\ V - (+0.54\ V) = +0.53\ V$$

Since $\Delta E^\ominus$ is positive, the reaction would be expected to 'go'.

(b) The two half-equations are:

$Br_2(aq) + 2e^- \rightleftharpoons 2Br^-(aq)$

$2Cl^-(aq) \rightleftharpoons Cl_2(aq)$

which combine to make the cell:

$Pt \mid Cl^-(aq),Cl_2(aq) \vdots Br_2(aq),Br^-(aq) \mid Pt$

Substituting into the expression:

$$\Delta E^\ominus = E^\ominus_R - E^\ominus_L$$

$$\Delta E^\ominus = +1.07\ V - (+1.36\ V) = -0.29\ V$$

Since $\Delta E^\ominus$ is negative, the reaction would not 'go'.

(c) The two half-equations are:

$Zn(s) \rightleftharpoons Zn^{2+}(aq) + 2e^-$

$2Fe^{3+}(aq) + 2e^- \rightleftharpoons 2Fe^{2+}(aq)$

which combine to make the cell:

$Zn(s) \mid Zn^{2+}(aq) \vdots Fe^{3+}(aq),Fe^{2+}(aq) \mid Pt$

Substituting into the expression:

$$\Delta E^\ominus = E^\ominus_R - E^\ominus_L$$

$$\Delta E^\ominus = +0.77\ V - (-0.76\ V) = +1.53\ V$$

Since $\Delta E^\ominus$ is positive, the reaction would be expected to 'go'.

(d) The two half-equations are:

$2MnO_4^-(aq) + 16H^+(aq) + 10e^- \rightleftharpoons 8H_2O(l) + 2Mn^{2+}(aq)$

$10Br^-(aq) \rightleftharpoons 5Br_2(aq) + 10e^-$

which combine to make the cell:

$Pt \mid 2Br^-(aq),Br_2(aq) \vdots [MnO_4^-(aq) + 8H^+(aq)],[4H_2O(l) + Mn^{2+}(aq)] \mid Pt$

(Note that the stoichiometric coefficients are reduced to the smallest whole numbers which give the correct ratio.)

Substituting into the expression:

$$\Delta E^\ominus = E^\ominus_R - E^\ominus_L$$

$$\Delta E^\ominus = +1.51\ V - (+1.09\ V) = +0.42\ V$$

Since $\Delta E^\ominus$ is positive, the reaction would be expected to 'go'.

(e) The two half-equations are:

$2MnO_4^-(aq) + 16H^+(aq) + 10e^- \rightleftharpoons 8H_2O(l) + 2Mn^{2+}(aq)$

$5Cu(s) \rightleftharpoons 5Cu^{2+}(aq) + 10e^-$

which combine to make the cell:

$Cu(s) \mid Cu^{2+}(aq) \vdots [MnO_4^-(aq) + 8H^+(aq)],[4H_2O(l) + Mn^{2+}(aq)] \mid Pt$

Substituting into the expression:

$$\Delta E^\ominus = E^\ominus_R - E^\ominus_L$$

$$\Delta E^\ominus = +1.51\ V - (+0.34\ V) = +1.17\ V$$

Since $\Delta E^\ominus$ is positive the reaction would be expected to 'go'.

(f) The two half-equations are:

$3S_2O_8^{2-}(aq) + 6e^- \rightleftharpoons 6SO_4^{2-}(aq)$

$2Cr^{3+}(aq) + 7H_2O \rightleftharpoons Cr_2O_7^{2-}(aq) + 14H^+(aq) + 6e^-$

which combine to make the cell:

$Pt \mid [2Cr^{3+}(aq) + 7H_2O(l)],[Cr_2O_7^{2-}(aq) + 14H^+(aq)] \vdots$
$S_2O_8^{2-}(aq),2SO_4^{2-}(aq) \mid Pt$

Substituting into the expression:

$$\Delta E^\ominus = E^\ominus_R - E^\ominus_L$$

$$\Delta E^\ominus = +2.01\ V - (+1.33\ V) = +0.68\ V$$

Since $\Delta E^\ominus$ is positive, the reaction would be expected to 'go'.

Exercise 293

(a) $Mn^{2+}(aq) + 2e^- \rightleftharpoons Mn(s)$; $E^\ominus = -1.19$ V

$Pb^{2+}(aq) + 2e^- \rightleftharpoons Pb(s)$; $E^\ominus = -0.13$ V

$Mn(s) + Pb^{2+}(aq) \rightarrow Mn^{2+}(aq) + Pb(s)$

(b)

$S(s) + 2e^- \rightleftharpoons S^{2-}(aq)$; $E^\ominus = -0.48$ V

$I_2(aq) + 2e^- \rightleftharpoons 2I^-(aq)$; $E^\ominus = +0.54$ V

(The equations were not tabulated in the correct order in the question.)

$S^{2-}(aq) + I_2(aq) \rightarrow S(s) + 2I^-(aq)$

(c)

$2H^+(aq) + 2e^- \rightleftharpoons H_2(g)$; $E^\ominus = 0.00$ V

$Ag^+(aq) + e^- \rightleftharpoons Ag(s)$; $E^\ominus = +0.80$ V

$H_2(g) + 2Ag^+(aq) \rightarrow 2H^+(aq) + 2Ag(s)$

Exercise 294

(a) The half-equations are listed in order of increasingly positive values of $E^\ominus$. Therefore, for the Fe^{3+},Fe^{2+} reaction to proceed from right to left, it must be combined with either of the half-reactions listed below it, and these will proceed from left to right. The substances which should be able to oxidize $Fe^{2+}(aq)$ to $Fe^{3+}(aq)$ are acidified hydrogen peroxide solution and cobalt(III) ions.

(b)

$Fe^{3+} + e^- \rightleftharpoons Fe^{2+}(aq)$

$H_2O_2(l) + 2H^+(aq) + 2e^- \rightleftharpoons 2H_2O(l)$

$2Fe^{2+}(aq) + H_2O_2(l) + 2H^+(aq) \rightarrow 2Fe^{3+}(aq) + 2H_2O(l)$

$Fe^{3+}(aq) + e^- \rightleftharpoons Fe^{2+}(aq)$

$Co^{3+}(aq) + e^- \rightleftharpoons Co^{2+}(aq)$

$Fe^{2+}(aq) + Co^{3+}(aq) \rightarrow Fe^{3+}(aq) + Co^{2+}(aq)$

Exercise 295

$[Ag^+(aq)]$/mol dm^{-3}	$\log([Ag^+(aq)]$/mol $dm^{-3})$	ΔE/V	$E(Ag^+\|Ag)$/V
0.00010	-4.0	+0.25	+0.59
0.00033	-3.5	+0.28	+0.62
0.0010	-3.0	+0.31	+0.65
0.0033	-2.5	+0.34	+0.68
0.010	-2.0	+0.38	+0.72
0.10	-1.0	+0.43	+0.77

Exercise 295 (continued)

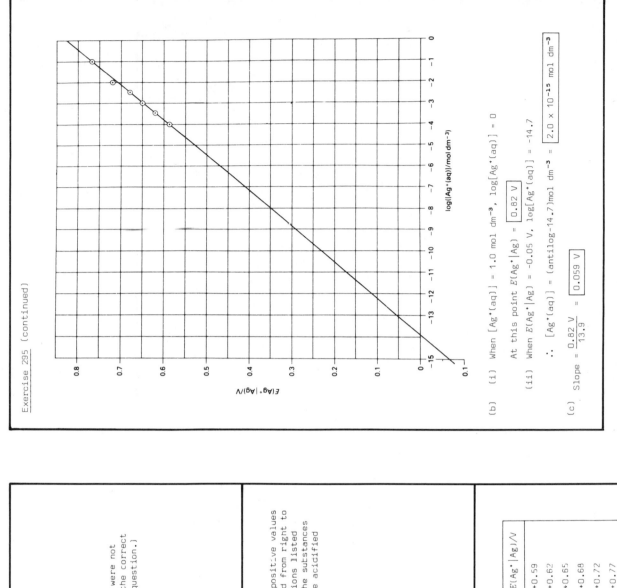

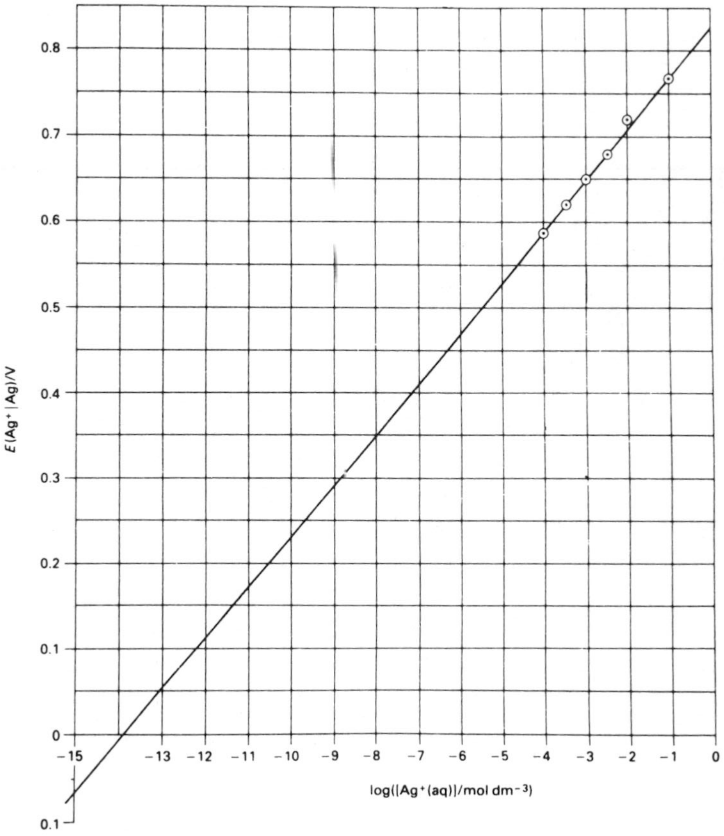

(b) (i) When $[Ag^+(aq)] = 1.0$ mol dm^{-3}, $\log[Ag^+(aq)] = 0$

At this point $E(Ag^+|Ag)$ = **0.82 V**

(ii) When $E(Ag^+|Ag) = -0.05$ V, $\log[Ag^+(aq)] = -14.7$

$\therefore$ $[Ag^+(aq)]$ = (antilog-14.7)mol dm^{-3} = **2.0×10^{-15} mol dm^{-3}**

(c) Slope = $\frac{0.82 \text{ V}}{13.9}$ = **0.059 V**

Exercise 296

(a)

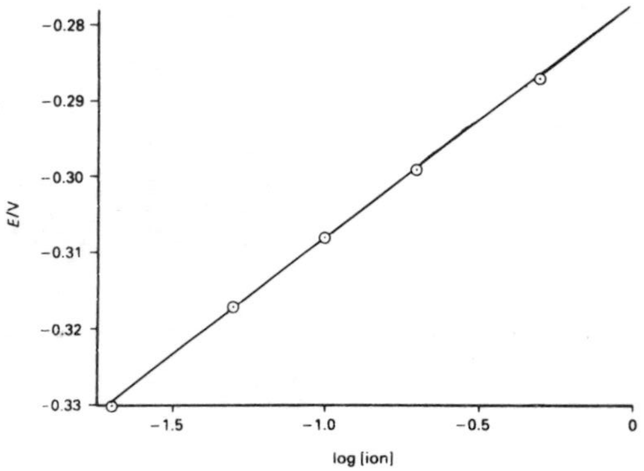

(b) Extrapolating the graph to log[ion] = 0 (i.e. [ion] = 1), gives the standard electrode potential, -0.277 V.

(c) The slope of the graph = $\dfrac{-0.277\ \text{V} - (-0.330\ \text{V})}{-1.72}$ = 0.031 V

Since slope = $\dfrac{0.060}{n}$ where n = charge on metal ion,

0.031 V = $\dfrac{0.060\ \text{V}}{n}$ $\quad\therefore\quad n = \dfrac{0.060\ \text{V}}{0.031\ \text{V}} = \boxed{2}$

Exercise 297

(a) Slope of the graph = 0.030 V

Since slope = $\dfrac{0.06\ \text{V}}{n}$, 0.030 V = $\dfrac{0.060\ \text{V}}{n}$

$\therefore\ n = \dfrac{0.060}{0.030} = \boxed{2}$

(b) $E^{\ominus}$ is the value of E when $[X^{n+}(aq)]$ = 1.0 mol dm^{-3} and $\log[X^{n+}(aq)] = 0$.

$\therefore\ E^{\ominus}(X^{n+},X) = \boxed{-1.19\ \text{V}}$

(c) $X(s)\ |\ X^{n+}(aq)\ \vdots\ Zn^{2+}(aq)\ |\ Zn(s)$

$\Delta E^{\ominus} = E^{\ominus}_R - E^{\ominus}_L$ = -0.76 V -(-1.19 V) = $\boxed{+0.43\ \text{V}}$

Exercise 298

$[X^{2+}(aq)] : [X^{n+}(aq)]$	E/V	$\log([X^{n+}(aq)]/[X^{2+}(aq)])$
8 : 1	0.113	-0.90
3 : 1	0.125	-0.48
1 : 2	0.149	+0.30
1 : 8	0.167	+0.90

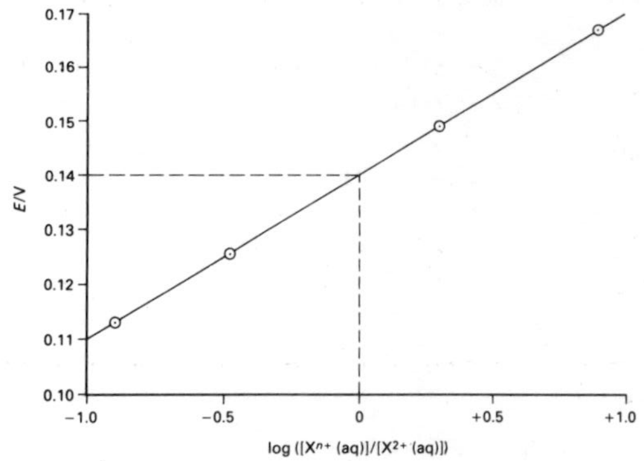

(a) $E^{\ominus}$ is the value of E when $\log\dfrac{[X^{n+}(aq)]}{[X^{2+}(aq)]} = 0$

i.e. $E^{\ominus} = \boxed{+0.14\ \text{V}}$

(b) The slope of the graph = 0.030 V = $\dfrac{0.060\ \text{V}}{n}$

where n refers to the number of electrons transferred.

$\therefore\ n = \dfrac{0.06\ \text{V}}{0.03\ \text{V}} = 2$

i.e. there are two electrons transferred in the process:

$X^{n+}(aq) + 2e^- \rightarrow X^{2+}(aq)$ $\quad\therefore\ n = \boxed{4}$

Exercise 299

(a) $E^{\ominus}(Ag^+|Ag)$ = +0.80 V (from data book)

Substituting into the Nernst equation:

$$E = E^{\ominus} + \frac{0.059\ V}{n}\log[\text{ion}]$$

$$0.50\ V = 0.80\ V + (0.059\ V \times \log[Ag^+(aq)])$$

$$\therefore \log[Ag^+(aq)] = \frac{0.50\ V - 0.80\ V}{0.059\ V} = \frac{-0.30\ V}{0.059\ V} = -5.1$$

$$\therefore [Ag^+(aq)] = \text{antilog}(-5.1) = \boxed{7.9 \times 10^{-6}\ \text{mol dm}^{-3}}$$

(b) On the graph from Experiment 4,

E = 0.50 V corresponds to $\log[Ag^+(aq)] = -5.5$

$$\therefore [Ag^+(aq)] = \text{antilog}(-5.5) = \boxed{3.2 \times 10^{-6}\ \text{mol dm}^{-3}}$$

(c) $Ag^+(aq) + Cl^-(aq) \rightarrow AgCl(s)$

$\therefore n(Ag^+) = n(Cl^-)$

Since $n = cV$

1.0×10^{-5} mol dm^{-3} x 50 cm^3 = 0.010 mol dm^{-3} x V (NaCl)

$$\therefore V = 50\ \text{cm}^3 \times \frac{1.0 \times 10^{-5}}{0.010} = 0.05\ \text{cm}^3$$

(d) 0.05 cm^3 is roughly the volume of one drop from a burette. It is impossible to titrate with any accuracy using such a small volume. If very dilute NaCl(aq) were used in order to give a larger volume, the concentrations would be so small that it would be impossible to determine the end-point.

Exercise 300

System	E/V $(Ag^+|Ag)$	$\log[Ag^+(aq)]$	$[Ag^+(aq)]$ /mol dm^{-3}	$[X^-(aq)]$ /mol dm^{-3}	K_s/mol^2 dm^{-6}
1. AgCl	+0.340	-8.20	6.3×10^{-9}	3.3×10^{-2}	2.1×10^{-10}
2. AgBr	+0.205	-10.40	4.0×10^{-11}	3.3×10^{-2}	1.3×10^{-12}
3. AgI	-0.020	-14.00	1.0×10^{-14}	3.3×10^{-2}	3.3×10^{-16}
4. $AgIO_3$	+0.480	-5.80	1.6×10^{-6}	3.3×10^{-2}	5.3×10^{-8}

Calculations

1. Values of $\log[Ag^+(aq)]$ corresponding to the values of $E(Ag^+|Ag)$ in column 2 are shown in column 3. In each case, the antilogarithm gives a value for $[Ag^+(aq)]$ shown in column 4.
2. The concentration of X^- ions at equilibrium is given by assuming a complete reaction:

   $$Ag^+(aq) + X^-(aq) \rightarrow AgX(s)$$

   10 cm^3 of 0.10 M $AgNO_3$ react with 10 cm^3 of 0.10 M KX leaving an excess of 10 cm^3 of 0.10 M KX in a total volume of 30 cm^3 of mixture.

   $\therefore [X^-(aq)]$ = 0.10 mol dm^{-3} x 10 cm^3/30 cm^3 = 3.3×10^{-2} mol dm^{-3}
3. $K_s = [Ag^+(aq)] \times [X^-(aq)]$

   Values of K_s in column 6 are, therefore, obtained by multiplying the values in columns 4 and 5.

Exercise 301

The graph from Exercise 295 gives the concentrations of silver ions corresponding to the measured electrode potentials as follows:

AgCl; E = +0.34 V $\log[Ag^+(aq)]$ = -8.2 $[Ag^+(aq)]$ = 6.3×10^{-9} mol dm^{-3}

AgBr; E = 0.16 V $\log[Ag^+(aq)]$ = -11.2 $[Ag^+(aq)]$ = 6.3×10^{-12} mol dm^{-3}

AgI; E = -0.06 V $\log[Ag^+(aq)]$ = -14.8 $[Ag^+(aq)]$ = 1.6×10^{-15} mol dm^{-3}

In each case, $[X^-(aq)]$ is found by assuming that all the halide ions come from the excess KX(aq) after the complete reaction:

$$Ag^+(aq) + X^-(aq) \rightarrow AgX(s)$$

10 cm^3 of 0.10 M $AgNO_3$ react with 10 cm^3 of 0.10 M KX, leaving an excess of 5 cm^3 of 0.10 M KX in a total volume of 25 cm^3 of mixture.

$\therefore [X^-(aq)]$ = 0.10 mol dm^{-3} x 5 cm^3/25 cm^3 = 2.0×10^{-2} mol dm^{-3}

$K_s(AgCl) = [Ag^+(aq)][Cl^-(aq)]$

$= 6.3 \times 10^{-9}$ mol dm^{-3} x 2.0×10^{-2} mol dm^{-3} = $\boxed{1.3 \times 10^{-10}\ \text{mol}^2\ \text{dm}^{-6}}$

$K_s(AgBr) = [Ag^+(aq)][Br^-(aq)]$

$= 6.3 \times 10^{-12}$ mol dm^{-3} x 2.0×10^{-2} mol dm^{-3} = $\boxed{1.3 \times 10^{-12}\ \text{mol}^2\ \text{dm}^{-6}}$

$K_s(AgI) = [Ag^+(aq)][I^-(aq)]$

$= 1.6 \times 10^{-15}$ mol dm^{-3} x 2.0×10^{-2} mol dm^{-3} = $\boxed{3.2 \times 10^{-17}\ \text{mol}^2\ \text{dm}^{-6}}$

Exercise 302

(a) $E = E^{\ominus} + \frac{0.059\ V}{n} \log[\text{ion}]$

$+0.48\ V = +0.80\ V + 0.059\ V \log[Ag^+(aq)]$

$\log[Ag^+(aq)] = \frac{0.48\ V - 0.80\ V}{0.059\ V} = \frac{-0.32\ V}{0.059\ V} = -5.4$

$\therefore\ [Ag^+(aq)] = \text{antilog}(-5.4) = \boxed{4.0 \times 10^{-6}\ mol\ dm^{-3}}$

(b) $[CrO_4^{2-}(aq)]$ is found by assuming that complete precipitation occurs according to the equation:

$$2Ag^+(aq) + CrO_4^{2-}(aq) \rightarrow Ag_2CrO_4(s)$$

10 cm^3 of 1.0 M $AgNO_3$ react with 5 cm^3 of 1.0 M K_2CrO_4, leaving an excess of 5 cm^3 of the latter in a total volume of 20 cm^3 of mixture.

$\therefore\ [CrO_4^{2-}(aq)] = 1.0\ mol\ dm^{-3} \times 5\ cm^3/20\ cm^3 = 0.25\ mol\ dm^{-3}$

(c) $K_s(Ag_2CrO_4) = [Ag^+(aq)]^2[CrO_4^{2-}(aq)]$

$= (4.0 \times 10^{-6}\ mol\ dm^{-3})^2 \times 0.25\ mol\ dm^{-3}$

$= \boxed{4.0 \times 10^{-12}\ mol^3\ dm^{-9}}$

Exercise 303

(a) Assuming the Fe^{3+},Fe^{2+} half-cell is standard, the cell diagram is:

Pt | Fe^{2+}(aq, 1.0 M), Fe^{3+}(aq, 1.0 M) ¦ Ag^+(aq, x M) | Ag(s)

At equilibrium, $\Delta E = 0$,

$\therefore\ E(Ag^+|Ag) = E^{\ominus}(Fe^{3+},Fe^{2+}) = +0.77\ V$

From the graph (Exercise 295), $\log[Ag^+(aq)] = -1.0$

$\therefore\ [Ag^+(aq)] = \text{antilog}(-1.0) = 0.10\ mol\ dm^{-3}$

From the chemical equation, $K_c = \frac{[Fe^{3+}(aq)]}{[Fe^{2+}(aq)][Ag^+(aq)]}$

$\therefore\ K_c = \frac{1.0\ mol\ dm^{-3}}{1.0\ mol\ dm^{-3} \times 0.10\ mol\ dm^{-3}} = \boxed{10\ mol^{-1}\ dm^3}$

(b) Assuming the Cu^{2+},Cu^+ half-cell is standard,* the cell diagram is:

Pt | Cu^{2+}(aq), 1.0 M), Cu^+(aq, 1.0 M) ¦ Ag^+(aq, x M) | Ag(s)

At equilibrium, $\Delta E = 0$

$\therefore\ E(Ag^+\ Ag) = E^{\ominus}(Cu^{2+},Cu^+) = +0.15\ V$

From the graph (Exercise 295), $\log[Ag^+(aq)] = -11.3$

$\therefore\ [Ag^+(aq)] = \text{antilog}(-11.3) = 5.0 \times 10^{-12}\ mol\ dm^{-3}$

From the chemical equation, $K_c = \frac{[Cu^{2+}(aq)]}{[Cu^+(aq)][Ag^+(aq)]}$

$\therefore\ K_c = \frac{1.0\ mol\ dm^{-3}}{1.0\ mol\ dm^{-3} \times 5.0 \times 10^{-12}\ mol\ dm^{-3}} = \boxed{2.0 \times 10^{11}\ mol^{-1}\ dm^3}$

*It is valid to make this assumption for the purposes of calculation even though such a half-cell cannot be constructed in practice due to the disproportionation of Cu^+ to Cu^{2+} and Cu.

Exercise 303 (continued)

(c) Assuming the $Co^{2+}|Co$ half-cell is standard (it is equally acceptable to choose the $Pb^{2+}|Pb$ half-cell) the cell diagram is:

Co(s) | Co^{2+}(aq) ¦ Pb^{2+}(aq) | Pb(s)

At equilibrium, $\Delta E = 0$

$\therefore\ E(Pb^{2+}|Pb) = E^{\ominus}(Co^{2+}|Co) = -0.28\ V$

Using the Nernst equation, since no graph is available:

$E(Pb^{2+}|Pb) = E^{\ominus}(Pb^{2+}|Pb) + \frac{0.059\ V}{n} \log[Pb^{2+}(aq)]$

i.e. $-0.28\ V = -0.13\ V + 0.0295\ V \log[Pb^{2+}(aq)]$

$\therefore\ \log[Pb^{2+}(aq)] = \frac{-0.28\ V + 0.13\ V}{0.0295\ V} = \frac{-0.15\ V}{0.0295\ V} = -5.1$

$\therefore\ [Pb^{2+}(aq)] = \text{antilog}(-5.1) = 7.9 \times 10^{-6}\ mol\ dm^{-3}$

From the chemical equation, $K_c = \frac{[Co^{2+}(aq)]}{[Pb^{2+}(aq)]}$

$\therefore\ K_c = \frac{1.0\ mol\ dm^{-3}}{7.9 \times 10^{-6}\ mol\ dm^{-3}} = \boxed{1.3 \times 10^5}$

Exercise 304

(a) $E = E^{\ominus} + 0.06\ V \log\frac{[Fe^{3+}(aq)]}{[Fe^{2+}(aq)]}$

$= 0.77\ V + \left(0.06\ V \times \log\frac{1.0}{0.6}\right) = 0.77\ V + 0.01\ V = \boxed{0.78\ V}$

(b) $E = E^{\ominus} + 0.06\ V \log[Ag^+(aq)]$

$\therefore\ \log[Ag^+(aq)] = \frac{E - E^{\ominus}}{0.06\ V} = \frac{0.78 - 0.80}{0.06} = -0.33$

$\therefore\ [Ag^+(aq)] = (\text{antilog} - 0.33)\ mol\ dm^{-3} = \boxed{0.46\ mol\ dm^{-3}}$

(c) Nothing would happen. The tendency for electrons to be transferred between Fe^{2+} and Fe^{3+} ions (via the silver) is exactly balanced by the tendency for electrons to be transferred between silver metal and silver ions, i.e. an equilibrium exists:

$$Fe^{3+}(aq) + Ag(s) \rightleftharpoons Fe^{2+}(aq) + Ag^+(aq)$$

(d) $K_c = \frac{[Fe^{2+}(aq)]}{[Fe^{3+}(aq)]} \times [Ag^+(aq)]$

$= \frac{0.6\ mol\ dm^{-3}}{1.0\ mol\ dm^{-3}} \times 0.46\ mol\ dm^{-3} = \boxed{0.28\ mol\ dm^{-3}}$

Exercise 305

(a) $Cu(s) \mid Cu^{2+}(aq) \,\vdots\, Br_2(aq),Br^-(aq) \mid Pt$

$\Delta E^{\ominus} = E^{\ominus}(Br^-, Br_2) - E^{\ominus}(Cu^{2+}|Cu)$

$= +1.09\ V - 0.34\ V = \boxed{+0.75\ V}$

Since $\Delta E^{\ominus}$ is positive, the reaction proceeds from left to right as indicated in the cell diagram:

$Cu(s) + Br_2(aq) \rightarrow Cu^{2+}(aq) + 2Br^-(aq); \quad \Delta E^{\ominus} = +0.75\ V$

(b) $nF\Delta E^{\ominus} = 2.3\ RT \log K_c$

$\therefore \log K_c = \dfrac{nF\Delta E^{\ominus}}{2.3\ RT} = \dfrac{2 \times 9.65 \times 10^4\ C\ mol^{-1} \times 0.75\ V}{2.3 \times 8.31\ J\ K^{-1}\ mol^{-1} \times 298\ K}$

$= 25.4\ C\ V\ J^{-1} = 25.4$

$\therefore K_c = \text{antilog } 25.4 = \boxed{2.5 \times 10^{25}\ mol^2\ dm^{-6}}$

This very large value of K_c indicates that the reaction goes virtually to completion. (Note that the unit is not given by the calculation but by reference to the equation and the expression for K_c.)

Exercise 306

(a) $Fe^{2+}(aq) \rightleftharpoons Fe^{3+}(aq) + e^-$

$Cr_2O_7^{2-}(aq) + 14H^+(aq) + 6e^- \rightleftharpoons 2Cr^{3+}(aq) + 7H_2O(l)$

(b) $Pt \mid Fe^{2+}(aq),Fe^{3+}(aq) \,\vdots\, [Cr_2O_7^{2-}(aq) + H^+(aq)],[Cr^{3+}(aq) + H_2O(l)] \mid Pt$

(c) $\Delta E^{\ominus} = E^{\ominus}_R - E^{\ominus}_L = 1.33\ V - 0.77\ V = \boxed{0.56\ V}$

(d) $nF\Delta E^{\ominus} = 2.3\ RT \log K_c$

$\therefore \log K_c = \dfrac{nF\Delta E^{\ominus}}{2.3\ RT} = \dfrac{6 \times 9.65 \times 10^4\ C\ mol^{-1} \times 0.56\ V}{2.3 \times 8.31\ J\ K^{-1}\ mol^{-1} \times 298\ K} = 57$

$\therefore K_c = \text{antilog } 57 = \boxed{1.0 \times 10^{57}\ mol^{-13}\ dm^{39}}$

The unit is not given by the calculation, but by reference to the expression for K_c:

$$K_c = \frac{[Fe^{3+}(aq)]^6[Cr^{3+}(aq)]^2}{[Fe^{2+}(aq)]^6[Cr_2O_7^{2-}(aq)][H^+(aq)]^{14}}$$

Exercise 307

(a) $Pt \mid Fe^{2+}(aq),Fe^{3+}(aq) \mid Ag^+(aq) \mid Ag(s)$

$\Delta E^{\ominus} = E^{\ominus}_R - E^{\ominus}_L = 0.80\ V - 0.77\ V = +0.03\ V$

$nF\Delta E^{\ominus} = 2.3\ RT \log K_c$

$\therefore \log K_c = \dfrac{nF\Delta E^{\ominus}}{2.3\ RT} = \dfrac{1 \times 9.65 \times 10^4\ C\ mol^{-1} \times 0.03\ V}{2.3 \times 8.31\ J\ K^{-1}\ mol^{-1} \times 298\ K} = 0.51$

$\therefore K_c = \text{antilog } 0.51 = \boxed{3.2\ mol^{-1}\ dm^3}$

(b) $Pt \mid [Cr^{3+}(aq),H_2O(l)],[Cr_2O_7^{2-}(aq),14H^+(aq)] \,\vdots\, Cl_2(aq),Cl^-(aq) \mid Pt$

$\Delta E^{\ominus} = E^{\ominus}_R - E^{\ominus}_L = 1.36\ V - 1.33\ V = +0.03\ V$

$nF\Delta E^{\ominus} = 2.3\ RT \log K_c$

$\therefore \log K_c = \dfrac{nF\Delta E^{\ominus}}{2.3\ RT} = \dfrac{6 \times 9.65 \times 10^4\ C\ mol^{-1} \times 0.03\ V}{2.3 \times 8.31\ J\ K^{-1}\ mol^{-1} \times 298\ K} = 3.0$

$\therefore K_c = \text{antilog } 3.0 = \boxed{1.0 \times 10^3\ mol^{16}\ dm^{-48}}$

Exercise 308

(a)

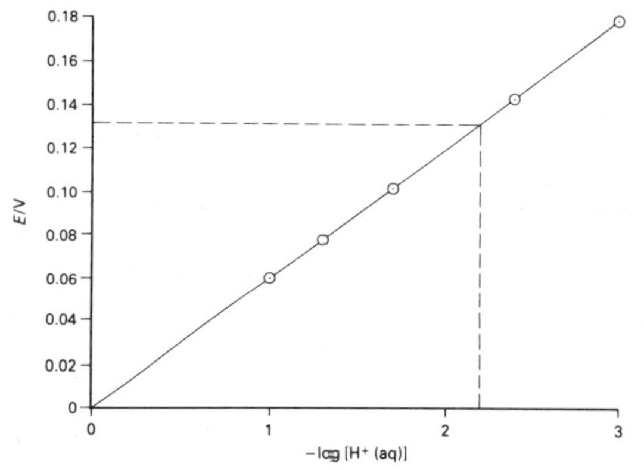

(b) From the graph, when $\Delta E = +0.132\ V$,

$-\log[H^+(aq)] = 2.4 \qquad \therefore\ pH = \boxed{2.4}$

Exercise 309

(b) (i) $Ag^+(aq)$ (ii) $Cu(s)$, $Cu^+(aq)$, $Zn(s)$ and $Li(s)$

(c) (ii) +1.10 V (Cu electrode positive)

(iii) $Zn(s) + Cu^{2+}(aq) \rightarrow Zn^{2+}(aq) + Cu(s)$

(iv) From Zn to Cu.

Exercise 310 (a), (b) and (d)

Exercise 311

(a) $Cr^{3+}(aq) + e^- \rightleftharpoons Cr^{2+}(aq)$

$Ce^{4+}(aq) + e^- \rightleftharpoons Ce^{3+}(aq)$

$Cr_2O_7^{2-}(aq) + 14H^+(aq) + 6e^- \rightleftharpoons 2Cr^{3+}(aq) + 7H_2O(l)$

(b) (i) $\Delta E^\ominus = +1.70\ V - (-0.41\ V) = +2.11\ V$

(ii) $\Delta E^\ominus = +1.70\ V - (+1.33\ V) = +0.37\ V$

(c) (i) $Cr^{2+}(aq) + Ce^{4+}(aq) \rightarrow Cr^{3+}(aq) + Ce^{3+}(aq)$

(ii) $2Cr^{3+}(aq) + 7H_2O(l) + 6Ce^{4+}(aq) \rightarrow Cr_2O_7^{2-}(aq) + 14H^+(aq) + 6Ce^{3+}(aq)$

Exercise 312 (e) (i) 4.7×10^{-6} mol dm^{-3} (ii) 2.2×10^{-11} mol^2 dm^{-6}

Exercise 313 $E^\ominus = +0.34$ V

Exercise 314 (a) (ii) +0.32 V (b) 2.1×10^{11}

Exercise 315

(a) In each case, $p = p^oX$

X_{hexane}	0.00	0.200	0.400	0.600	0.800	1.00
p_{hexane}/mmHg	0.00	74.0	149	223	298	372
$X_{heptane}$	0.00	0.200	0.400	0.600	0.800	1.00
$p_{heptane}$/mmHg	0.00	25.4	50.8	76.2	102	127

(b) & (c)

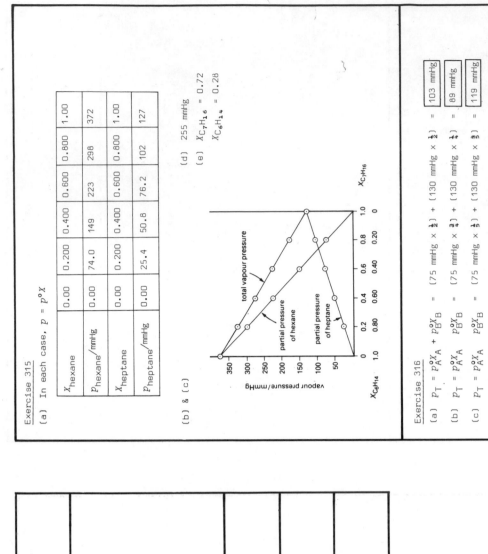

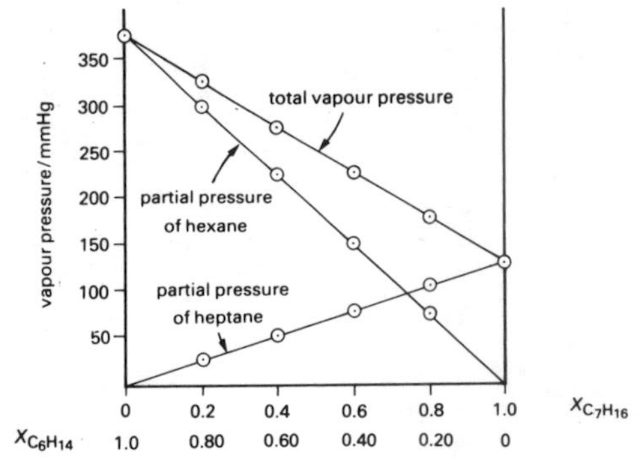

(d) 255 mmHg

(e) $X_{C_7H_{16}} = 0.72$

$X_{C_6H_{14}} = 0.28$

Exercise 316

(a) $p_T = p^o_AX_A + p^o_BX_B = (75\ mmHg \times \frac{1}{2}) + (130\ mmHg \times \frac{1}{2}) =$ **103 mmHg**

(b) $p_T = p^o_AX_A \quad p^o_BX_B = (75\ mmHg \times \frac{3}{4}) + (130\ mmHg \times \frac{1}{4}) =$ **89 mmHg**

(c) $p_T = p^o_AX_A \quad p^o_BX_B = (75\ mmHg \times \frac{1}{5}) + (130\ mmHg \times \frac{4}{5}) =$ **119 mmHg**

Exercise 317

(a) In an equimolar liquid mixture, $X_{hexane} = X_{heptane} = 0.5$

The total vapour pressure is given by the general formula:

$$p = X_A p_A^o + X_B p_B^o$$

$$= (0.5 \times 56\ 000\ N\ m^{-2}) + (0.5 \times 24\ 000\ N\ m^{-2})$$

$$= \boxed{40\ 000\ N\ m^{-2}}$$

(b) The partial pressure of heptane in the vapour $= 0.5 \times 56\ 000\ N\ m^{-2}$

$= 28\ 000\ N\ m^{-2}$

In the vapour, $p_{heptane} = X_{heptane}\ p_{total}$

$$\therefore\ X_{heptane} = \frac{p_{heptane}}{p_{total}} = \frac{28\ 000\ N\ m^{-2}}{40\ 000\ N\ m^{-2}} = \boxed{0.7}$$

Exercise 318

(a) and (c) See graph.

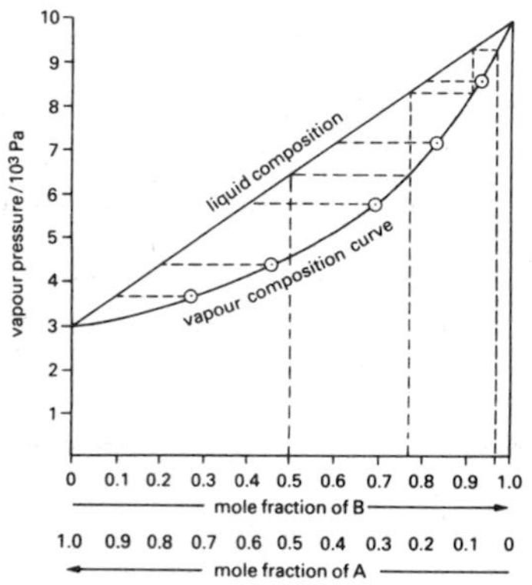

(b)

X_B (liquid)	p_T /kPa	X_B (vapour)
0.1	3.6	0.27
0.2	4.3	0.45
0.4	5.7	0.70
0.6	7.1	0.83
0.8	8.5	0.93

(d) $X_B = 0.77$

(e) $X_B = 0.97$

Exercise 319

(a) The total vapour pressure exerted by the mixture is equal to the sum of their separate saturation vapour pressures.

(b) The boiling-point of the mixture is 98 °C since the total vapour pressure exerted at this temperature equals atmospheric pressure.

(c) The boiling-point of the mixture is lower than either of the boiling-points of the pure liquids. Extrapolation of the vapour pressure curves for water and nitrobenzene show that they will cross the dotted line of atmospheric pressure at temperatures higher than 98 °C.

(d) The mixture will steam distil at 98 °C.

Exercise 320

$$\frac{\text{mass of bromobenzene}}{\text{mass of water}} = \frac{\text{r.m.m. of bromobenzene}}{\text{r.m.m. of water}} \times \frac{p_{bromobenzene}}{p_{water}}$$

$$= \frac{157 \times (101 - 85.8)\text{kPa}}{18 \times 85.8\ \text{kPa}} = \frac{1.54}{1}$$

$$\therefore\ \%\text{ by mass of bromobenzene} = \frac{1.54}{1 + 1.54} \times 100 = \boxed{60.6\%}$$

Exercise 321

(a) If steam is blown through a mixture of two immiscible liquids such as phenylamine and water, the mixture will boil below the boiling-point of either of the pure liquids. This is because the total vapour pressure is the sum of the vapour pressures of the pure liquids alone. The distillate, collected in a receiver flask, contains both liquids but has a far higher proportion of phenylamine than the original mixture.

Steam distillation is most successful when the liquid or the solid to be separated

(i) is immiscible with or insoluble in water,

(ii) has a high relative molecular mass, and

(iii) exerts a high vapour pressure at about 100 °C.

Any impurities present must be non-volatile under the conditions used. Steam distillation is particularly useful for the purification of substances that decompose at temperatures near their boiling-points.

Thus, steam distillation is successful for the phenylamine/water mixture because the two liquids are immiscible, while phenylamine has a high relative molecular mass and is readily oxidized on direct heating.

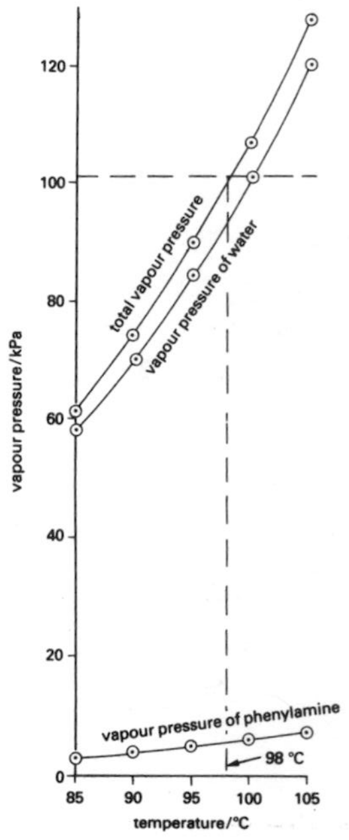

(b) (i) The phenylamine-water mixture will steam-distil at about 38 °C.

(ii) $\dfrac{\text{mass of phenylamine}}{\text{mass of water}}$

$$= \frac{M(\text{phenylamine})}{M(\text{water})} \times \frac{p(\text{phenylamine})}{p(\text{water})}$$

$$= \frac{93}{18} \times \frac{6}{94} = \frac{0.33}{1}$$

$\therefore$ % by mass of phenylamine

$$= \frac{0.33}{1.33} \times 100 = \boxed{24.8\%}$$

Exercise 322

X_{water}	X_{sugar}	p/mmHg	$\frac{p^o - p}{p^o}$
1.000	0.000	17.5	0.000
0.950	0.050	16.6	0.051
0.900	0.100	15.8	0.097
0.850	0.150	14.9	0.150

The values in the fourth column are almost the same as those in the second column.

(b) (i) The relative lowering of the vapour pressure is equal to the mole fraction of the solute in the solution. (Note that this is only applicable to dilute solutions.)

(ii) $\frac{p^o - p}{p^o} = X_{solute}$

Exercise 323

(a)

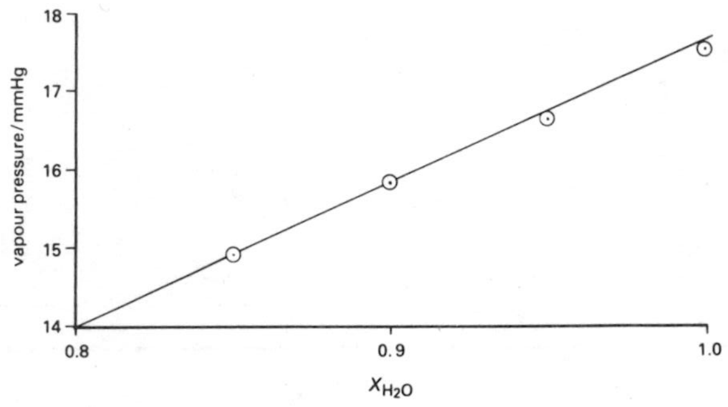

(b) The vapour pressure of the solution is directly proportional to the mole fraction of water. (Strictly speaking, you should show that your straight line passes through the origin but, since the sugar is non-volatile, you can assume that $p = 0$ when $X_{water} = 0$.)

Exercise 324

(a) Slope $= k = \frac{(17.5 - 14.0)\text{ mmHg}}{0.200} = \boxed{17.5\text{ mmHg}}$

(b) $k = p^o$

(c) $p_{H_2O} = p^o X_{H_2O}$

(d) $p_T = p^o_A X_A + p^o_B X_B$

i.e. $p_{H_2O} + p_{sugar} = p^o_{H_2O} X_{H_2O} + p^o_{sugar} X_{sugar}$

But $p_{sugar} = 0$ and $p^o_{sugar} = 0$

$\therefore\ p_{H_2O} = p^o_{H_2O} X_{H_2O}$

(e) $\frac{p^o - p}{p^o} = X_{solute}$

or $1 - \frac{p}{p^o} = X_{solute}$

But $X_{solute} = 1 - X_{H_2O}$ and $p^o = p^o_{H_2O}$

$\therefore\ 1 - \frac{p}{p^o_{H_2O}} = 1 - X_{H_2O}$

$\therefore\ \frac{p}{p^o_{H_2O}} = X_{H_2O}$ or $p = p^o_{H_2O} X_{H_2O}$

Exercise 325

$\frac{p^o - p}{p^o} = X_{solute}$ i.e. $p^o - p = X_{solute} p^o$

$\therefore\ p = p^o(1 - X_{solute})$

(a) Amount of solute $= 9.0\text{ g}/180\text{ g mol}^{-1} = 0.050\text{ mol}$

Amount of water $= 50\text{ g}/18\text{ g mol}^{-1} = 2.78\text{ mol}$

$\therefore\ X_{glucose} = \frac{0.050}{2.78 + 0.05} = 0.018$

$\therefore\ p = p^o(1 - X_{glucose}) = 31.2\text{ kPa} \times 0.982 = \boxed{30.6\text{ kPa}}$

(b) Amount of ions in solute = 2.0 mol

Amount of water $= 1000\text{ g}/18\text{ g mol}^{-1} = 55.6\text{ mol}$

$\therefore\ X_{Na^+Cl^-} = \frac{2.0}{55.6 + 2.0} = 0.035$

$\therefore\ p = p^o(1 - X_{Na^+Cl^-}) = 31.2\text{ kPa} \times 0.965 = \boxed{30.1\text{ kPa}}$

(c) Amount of ions in solute = 3.0 mol

$\therefore\ X_{Mg^{2+}2Cl^-} = \frac{3.0}{55.6 + 3.0} = 0.051$

$\therefore\ p = p^o(1 - X_{Mg^{2+}2Cl^-}) = 31.2\text{ kPa} \times 0.949 = \boxed{29.6\text{ kPa}}$

Exercise 326

(a) The relative lowering of the vapour pressure of a dilute solution of a non-volatile solute is equal to the mole fraction of the solute, i.e.

$$\frac{p^o - p}{p^o} = X_{solute}$$

(b)

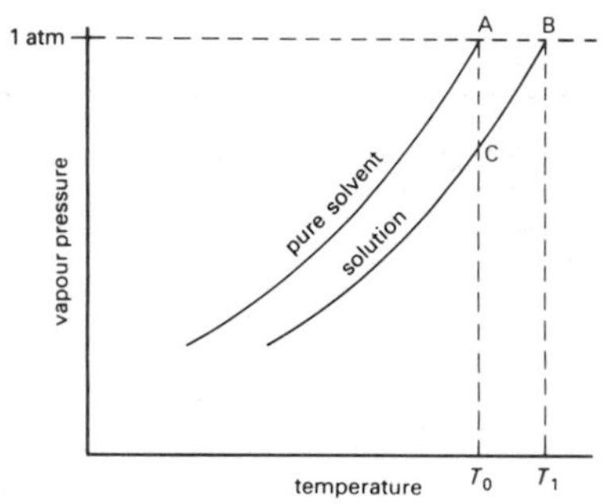

The vapour pressure of the solution is lower than that of the pure solvent because the solute is non-volatile. The solute reduces the concentration (and therefore the vapour pressure) of the solvent, but does not itself contribute to the total vapour pressure.

The reduction in vapour pressure (AC) means that a higher temperature is required to bring the vapour pressure up to atmospheric pressure, i.e. the boiling-point is increased from T_0 to T_1 (AB).

(c) The vapour pressure rises because, at higher temperatures, more molecules have sufficient energy to escape from the liquid surface against the attraction of neighbouring molecules.

(d) (i) Amount of X $= \frac{8.4 \text{ g}}{168 \text{ g mol}^{-1}} = 0.050$ mol

Amount of ethanol $= \frac{69 \text{ g}}{46 \text{ g mol}^{-1}} = 1.5$ mol

Mole fraction of X $= \frac{0.050 \text{ mol}}{(0.050 + 1.5)\text{mol}} = 0.033$

(ii) $\frac{p^o - p}{p^o} = X$, i.e. $\frac{26.6 \text{ kPa} - p}{26.6 \text{ kPa}} = 0.033$

$\therefore$ 26.6 kPa $- p$ = 0.033 x 26.6 kPa

$\therefore$ p = 26.6 kPa (1 -0.033) = **25.7 kPa**

(iii) 69 g of ethanol dissolves 8.4 g of X

$\therefore$ 1000 g of ethanol dissolves $\frac{8.4}{69}$ x 1000 g = 122 g of X

$\therefore \Delta T_b = K_b \frac{m}{M} = 1.15 \text{ K kg mol}^{-1} \times \frac{122 \text{ g kg}^{-1}}{168 \text{ g mol}^{-1}} = 0.84$ K

The boiling-point of the mixture is 78 °C + 0.84 °C = 78.84 °C

Exercise 327

Triangles ABC and ADE are similar (i.e. have exactly the same shape).

$\therefore \frac{AC}{AB} = \frac{AE}{AD}$ = constant, k

$\therefore$ AC = kAB and AE = kAD

But AC and AE are values of the lowering of vapour pressure, $p^o - p$, and AB and AD are the corresponding elevations of the boiling-point, ΔT.

$\therefore p^o - p = k\Delta T$

Raoult's law states that $\frac{p^o - p}{p^o} = X_{solute}$ or $p^o - p = p^oX_{solute}$

$\therefore p^oX_{solute} = k\Delta T$ or $\Delta T = k'p^oX_{solute}$

For a dilute solution, $X_{solute} \propto$ concentration $\propto \frac{m}{M}$

$\therefore \Delta T = K_b \times \frac{m}{M}$

Exercise 328

Rearranging $\Delta T = K_b \times \frac{m}{M}$ gives $M = \frac{K_b m}{\Delta T}$

56 g of water has 4.6 g of solute

$\therefore$ 1000 g of water has $\frac{4.6}{56}$ x 1000 g = 82.1 g of solute

Substituting m = 82.1 g kg^{-1} in the expression above gives:

$M = \frac{K_b m}{\Delta T} = \frac{0.52 \text{ K kg mol}^{-1} \times 82.1 \text{ g kg}^{-1}}{0.71 \text{ K}}$ = **60 g mol^{-1}**

Exercise 329

In each case $m = \frac{\text{mass of solute}}{\text{mass of solvent}} \times 1000 \text{ g kg}^{-1}$

$\Delta T = K_b \times \frac{m}{M}$ $\therefore M = \frac{K_b m}{\Delta T}$

(a) $m = \frac{2.11 \text{ g}}{100 \text{ g}} \times 1000 \text{ g kg}^{-1} = 21.1 \text{ g kg}^{-1}$

$M = \frac{0.52 \text{ K kg mol}^{-1} \times 21.1 \text{ g kg}^{-1}}{0.15 \text{ K}}$ = **73 g mol^{-1}**

(b) $m = \frac{1.82 \text{ g}}{58.7 \text{ g}} \times 1000 \text{ g kg}^{-1} = 31.0 \text{ g kg}^{-1}$

$M = \frac{1.15 \text{ K kg mol}^{-1} \times 31.0 \text{ g kg}^{-1}}{0.55 \text{ K}}$ = **65 g mol^{-1}**

(c) $m = \frac{1.07 \text{ g}}{49.1 \text{ g}} \times 1000 \text{ g kg}^{-1} = 21.8 \text{ g kg}^{-1}$

$M = \frac{2.53 \text{ K kg mol}^{-1} \times 21.8 \text{ g kg}^{-1}}{0.67 \text{ K}}$ = **82 g mol^{-1}**

(d) $m = \frac{4.13 \text{ g}}{71.6 \text{ g}} \times 1000 \text{ g kg}^{-1} = 57.7 \text{ g kg}^{-1}$

$M = \frac{1.71 \text{ K kg mol}^{-1} \times 57.7 \text{ g kg}^{-1}}{1.07 \text{ K}}$ = **92 g mol^{-1}**

Exercise 330

(a) Sodium chloride solution contains two moles of ions for each mole of solid dissolved. Since the lowering of vapour pressure (and hence the elevation of boiling-point) depends on the number of solute particles present, 1 mol of NaCl would have twice the effect of 1 mol of a non-ionic solute. Conversely, the same effect is produced by half the amount of NaCl; consequently, the molar mass appears to be half its true value.

(b) One mole of $FeCl_3$ produces four moles of ions in aqueous solution (ignoring the small degree of hydrolysis). Therefore, the apparent molar mass is $\frac{1}{4}(56 + 106.5)\ g\ mol^{-1} = \boxed{40.6\ g\ mol^{-1}}$

(c) K_b for water is much smaller than for other solvents. The elevation of boiling-point is therefore much smaller than would be observed for the same amount of solute in another solvent, and would be difficult to measure.

(d) Inspection of the expression, $\Delta T = K_b \times \frac{m}{M}$

shows that as M increases, ΔT decreases. The smaller the value of ΔT, the more difficult it is to measure accurately.

Exercise 331

(a) T_0 = freezing-point of solvent
T_1 = freezing-point of dilute solution
T_2 = freezing-point of less dilute solution

(b) The more concentrated (less dilute) solution has the lowest freezing-point.

(c) The line AB represents the vapour pressure of the solid solvent at temperatures below its freezing-point.

Exercise 332

(a) One mole of solute particles (ions or molecules) dissolved in 1 kg of sulphuric acid will depress the freezing-point of the acid by 6.1 K.

(b) $2H_2SO_4 + HNO_3 \rightarrow 2HSO_4^- + H_3O^+ + NO_2^+$

One mole of HNO_3 results in the formation of four moles of ions.

$\therefore\ \Delta T_f = 4.0 \times 6.1\ K = \boxed{24.4\ K}$

Exercise 333

(a) $\Delta T = K_f \times \frac{m}{M}$

$\therefore\ K_f = M\Delta T \times \frac{1}{m} = 128\ g\ mol^{-1} \times 2.2\ K \times \frac{100\ g}{5.63\ g} \times \frac{1}{1000\ g\ kg^{-1}}$

$= \boxed{5.00\ K\ kg\ mol^{-1}}$

(b) $\Delta T = K_f \times \frac{m}{M}$

$\therefore\ M = \frac{K_f}{\Delta T} \times m$

$= \frac{5.00\ K\ kg\ mol^{-1}}{1.1\ K} \times \frac{5.83\ g}{100\ g} \times 1000\ g\ kg^{-1} = \boxed{265\ g\ mol^{-1}}$

This shows that aluminium chloride is dimerized in benzene.
$M_r(Al_2Cl_6) = (2 \times 27) + (6 \times 35.5) = 267.$

Exercise 334

$\Delta T = K_f \times \frac{m}{M}$

$\therefore\ M = \frac{K_f}{\Delta T} \times m$

$= \frac{7.10\ K\ kg\ mol^{-1}}{4.8\ K} \times \frac{1.54\ g}{17.84\ g} \times 1000\ g\ kg^{-1} = \boxed{128\ g\ mol^{-1}}$

Exercise 335

(a) $\Pi V = nRT$

$\therefore\ \Pi = n\frac{RT}{V} = \frac{0.100\ mol \times 0.0821\ atm\ dm^3\ K^{-1}\ mol^{-1} \times 298\ K}{1.00\ dm^3} = \boxed{2.45\ atm}$

(b) $\Pi V = nRT = \frac{m}{M}RT$

$\therefore\ \Pi = \frac{mRT}{MV} = \frac{36.0\ g \times 0.0821\ atm\ dm^3\ K^{-1}\ mol^{-1} \times 293\ K}{180\ g\ mol^{-1} \times 0.250\ dm^3} = \boxed{19.2\ atm}$

(c) $\Pi = \frac{mRT}{MV} = \frac{7.80\ g \times 0.0821\ atm\ dm^3\ K^{-1}\ mol^{-1} \times 308\ K}{59.0\ g\ mol^{-1} \times 0.100\ dm^3} = \boxed{33.4\ atm}$

(d) $\Pi = \frac{mRT}{MV} = \frac{2.15\ g \times 0.0821\ atm\ dm^3\ K^{-1}\ mol^{-1} \times 293\ K}{60.0\ g\ mol^{-1} \times 1.50\ dm^3} = \boxed{0.575\ atm}$

Exercise 336

(a) $\frac{\Pi V}{T} = nR = \text{constant} \quad \therefore \quad \frac{\Pi_1 V_1}{T_1} = \frac{\Pi_2 V_2}{T_2}$ (compare with the combined gas law)

$\therefore \quad \Pi_2 = \Pi_1 \times \frac{V_1}{V_2} \times \frac{T_2}{T_1} = 1.12 \text{ atm} \times \frac{1}{2} \times \frac{298 \text{ K}}{288 \text{ K}} = \boxed{0.579 \text{ atm}}$

(b) $\Pi V = nRT = \frac{m}{M}RT$

$\therefore \quad M = \frac{mRT}{\Pi V} = \frac{42.1 \text{ g} \times 0.0821 \text{ atm dm}^3 \text{ K}^{-1} \text{ mol}^{-1} \times 308 \text{ K}}{3.45 \text{ atm} \times 1.00 \text{ dm}^3} = \boxed{309 \text{ g mol}^{-1}}$

(c) $M = \frac{mRT}{\Pi V} = \frac{72.0 \text{ g} \times 0.0821 \text{ atm dm}^3 \text{ K}^{-1} \text{ mol}^{-1} \times 303 \text{ K}}{2.89 \text{ atm} \times 0.525 \text{ dm}^3} = \boxed{1180 \text{ g mol}^{-1}}$

Exercise 337

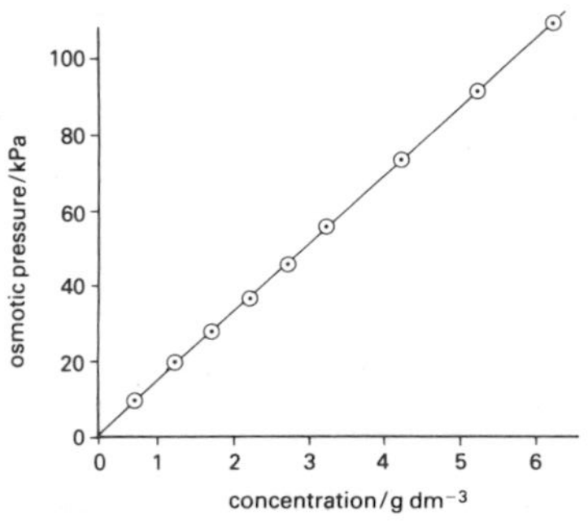

The slope of the graph = Π/c

$= \frac{110 \text{ kPa}}{6.00 \text{ g dm}^{-3}} = \frac{110 \text{ kPa}}{6.00 \text{ kg m}^{-3}}$

$= 18.3 \text{ kPa m}^3 \text{ kg}^{-1}$

Substituting kPa

$= 10^3 \text{ kg m}^{-1} \text{ s}^{-2}$

$\Pi/c =$

$18.3 \times 10^3 \text{ kg m}^{-1} \text{ s}^{-2} \text{ m}^3 \text{ kg}^{-1}$

$= 18.3 \times 10^3 \text{ m}^2 \text{ s}^{-2}$

But $\frac{\Pi}{c} = \frac{RT}{M}$

$\therefore \quad M = \frac{RT}{\Pi/c} =$

$\frac{8.31 \text{ kg m}^2 \text{ s}^{-2} \text{ mol}^{-1} \text{ K}^{-1} \times 298 \text{ K}}{18.3 \times 10^3 \text{ m}^2 \text{ s}^{-2}}$

$= 0.135 \text{ kg mol}^{-1}$

$= \boxed{135 \text{ g mol}^{-1}}$

Exercise 338

(a) $\Pi V = nRT$

$\therefore \quad \Pi = \frac{nRT}{V} = \frac{0.0010 \text{ mol} \times 0.0821 \text{ atm dm}^3 \text{ K}^{-1} \text{ mol}^{-1} \times 298 \text{ K}}{1.0 \text{ dm}^3} = \boxed{0.024 \text{ atm}}$

(b) $NaCl(aq) \rightarrow Na^+(aq) + Cl^-(aq)$

$\therefore$ there are twice as many ions in (b) as there are molecules in (a)

$\therefore \quad \Pi = 0.024 \text{ atm} \times 2 = \boxed{0.048 \text{ atm}}$

(c) $Fe_2(SO_4)_3(aq) \rightarrow 2Fe^{3+}(aq) + 3SO_4^{2-}(aq)$

$\therefore$ there are five times as many ions in (c) as there are molecules in (a)

$\therefore \quad \Pi = 0.024 \times 5 = \boxed{0.120 \text{ atm}}$

Exercise 339

Π/c is quoted in units incompatible with those given for R.

Π should be in N m^{-2} (N = kg m s^{-2}) and c in g dm^{-3}

Consider the mass, m, of a column of butanone, 110 cm tall with a cross-section of 1.00 cm^2.

$m = 110 \text{ cm} \times 1.00 \text{ cm}^2 \times 0.80 \text{ g cm}^{-3} = 88 \text{ g}$

This column exerts a force over an area of 1 cm^2, given by

$m \times g = 88 \text{ g} \times 981 \text{ cm s}^{-2} = 8.633 \times 10^4 \text{ g cm s}^{-2}$

i.e. the pressure exerted is $8.633 \times 10^4 \text{ g cm s}^{-2} \text{ cm}^{-2}$

$= 8.633 \times 10^4 \text{ g cm}^{-1} \text{ s}^{-2}$

Substituting g = 10^{-3} kg and cm^{-1} = 10^2 m^{-1} gives

$\Pi = 8.633 \times 10^3 \text{ kg m}^{-1} \text{ s}^{-2}$

$\therefore \quad \frac{\Pi}{c} = 110 \text{ cm}^4 \text{ g}^{-1} = \frac{110 \text{ cm butanone}}{1.00 \text{ g cm}^{-3}} = \frac{8.633 \times 10^3 \text{ kg m}^{-1} \text{ s}^{-2}}{1.00 \text{ g cm}^{-3}}$

But $1.00 \text{ g cm}^{-3} = 1.00 \text{ kg dm}^{-3} = 1.00 \times 10^3 \text{ kg m}^{-3}$

$\therefore \quad \frac{\Pi}{c} = \frac{8.633 \times 10^3 \text{ kg m}^{-1} \text{ s}^{-2}}{1.00 \times 10^3 \text{ kg m}^{-3}} = 8.633 \text{ m}^2 \text{ s}^{-2}$

From the equation $\frac{\Pi}{c} = \frac{RT}{M}$

$M = \frac{RT}{\Pi/c} = \frac{8.31 \text{ kg m}^2 \text{ s}^{-2} \text{ mol}^{-1} \text{ K}^{-1} \times 300 \text{ K}}{8.633 \text{ m}^2 \text{ s}^{-2}} = 289 \text{ kg mol}^{-1}$

$= \boxed{289{,}000 \text{ g mol}^{-1}}$

Exercise 340 (b) 16 mmHg

Exercise 341 (a) (i) 7.34 kPa (ii) 7.29 kPa

(b) (i) −0.52 °C (ii) −1.26 °C

100.14 °C 100.35 °C

Exercise 342 Mercury(II) nitrate: 109 (dissociated)

Mercury(II) chloride: 269 (not dissociated)

Exercise 343 P_4

Exercise 344 (b) 100.10 °C (c) (ii) 0.030

Exercise 345 (a) 3.12 atm (b) 0.0401 atm

Exercise 346 1.76×10^5 g mol^{-1}

Exercise 347

(a)

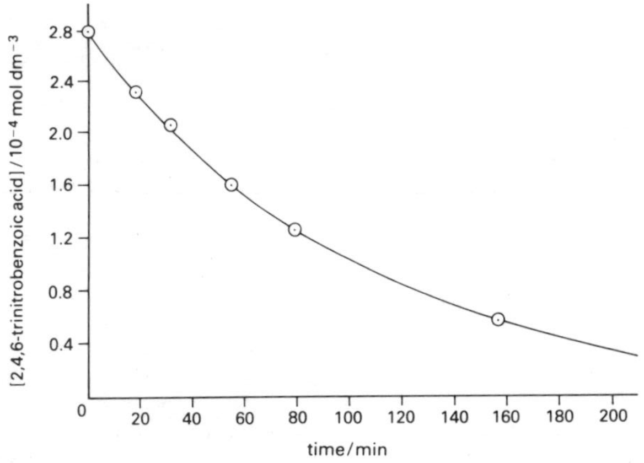

(b)

Time /min	Concentration /mol dm^{-3}	Slope /mol dm^{-3} min^{-1}	Rate /mol dm^{-3} min^{-1}
10	2.50×10^{-4}	$-\frac{2.74 \times 10^{-4}}{112.5}$	2.44×10^{-6}
50	1.66×10^{-4}	$-\frac{2.49 \times 10^{-4}}{151.5}$	1.64×10^{-6}
100	1.04×10^{-4}	$-\frac{2.04 \times 10^{-4}}{199.5}$	1.02×10^{-6}
150	0.63×10^{-4}	$-\frac{1.58 \times 10^{-4}}{249}$	0.635×10^{-6}

Exercise 347 (continued)

(c)

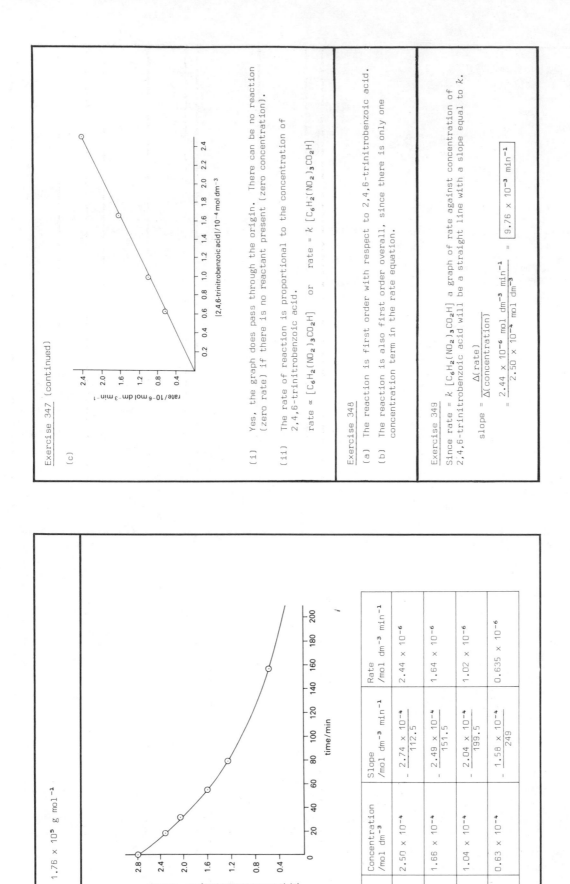

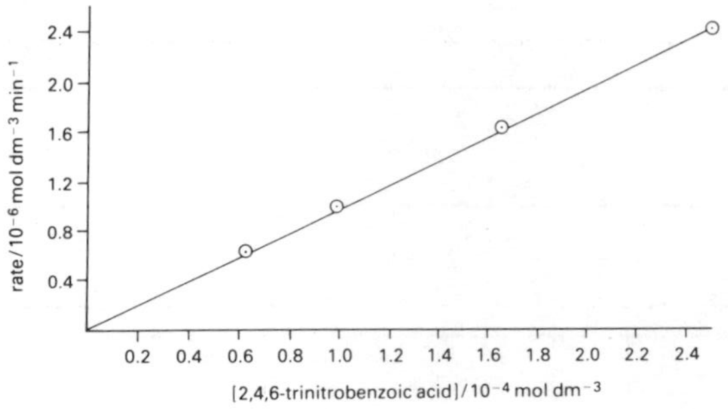

(i) Yes, the graph does pass through the origin. There can be no reaction (zero rate) if there is no reactant present (zero concentration).

(ii) The rate of reaction is proportional to the concentration of 2,4,6-trinitrobenzoic acid.

rate $\propto$ [$C_6H_2(NO_2)_3CO_2H$] or rate = k [$C_6H_2(NO_2)_3CO_2H$]

Exercise 348

(a) The reaction is first order with respect to 2,4,6-trinitrobenzoic acid.

(b) The reaction is also first order overall, since there is only one concentration term in the rate equation.

Exercise 349

Since rate = k [$C_6H_2(NO_2)_3CO_2H$] a graph of rate against concentration of 2,4,6-trinitrobenzoic acid will be a straight line with a slope equal to k.

$$\text{slope} = \frac{\Delta(\text{rate})}{\Delta(\text{concentration})}$$

$$= \frac{2.44 \times 10^{-6}\ \text{mol dm}^{-3}\ \text{min}^{-1}}{2.50 \times 10^{-4}\ \text{mol dm}^{-3}} = \boxed{9.76 \times 10^{-3}\ \text{min}^{-1}}$$

Exercise 350

(a) From the graph in Exercise 347, the slope of a tangent to the curve at time = 0 is:

$$\text{slope} = \frac{2.77 \times 10^{-4}\ \text{mol dm}^{-3}}{102.5\ \text{min}} = \boxed{2.70 \times 10^{-6}\ \text{mol dm}^{-3}\ \text{min}^{-1}}$$

This is the initial rate of reaction (at time = 0).

(b) To calculate the rate constant for each of the times listed simply substitute values for rate and concentration in the rate equation. The table below gives a summary of the values:

Time /min	Concentration /mol dm^{-3}	Rate /mol dm^{-3} min^{-1}	k/min^{-1}
0	2.77×10^{-4}	2.70×10^{-6}	9.75×10^{-3}
10	2.50×10^{-4}	2.44×10^{-6}	9.76×10^{-3}
50	1.66×10^{-4}	1.64×10^{-6}	9.88×10^{-3}
100	1.04×10^{-4}	1.02×10^{-6}	9.81×10^{-3}
150	0.63×10^{-4}	0.635×10^{-6}	10.1×10^{-3}

(c) Average value for $k = \boxed{9.86 \times 10^{-3}\ \text{min}^{-1}}$

(d) The value from the graph is likely to be more accurate. When you drew the best straight line between the points on the rate concentration graph in Exercise 347, you were smoothing out some of the inaccuracies due to experimental error - effectively averaging over the whole range. In this exercise, parts (b) and (c), you averaged values from only five rates and concentrations.

Exercise 351

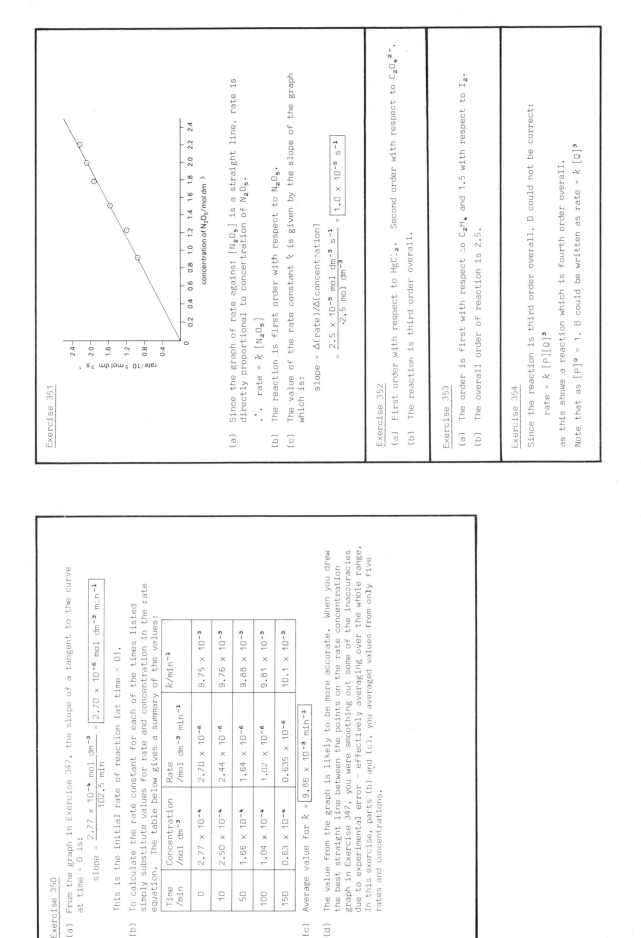

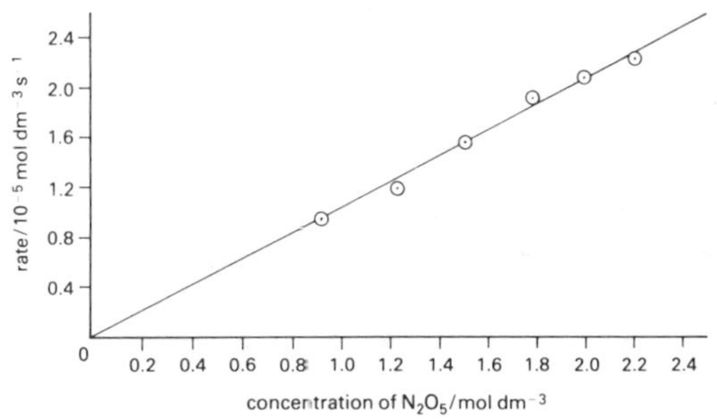

(a) Since the graph of rate against $[N_2O_5]$ is a straight line, rate is directly proportional to concentration of N_2O_5.

$\therefore$ rate = $k\,[N_2O_5]$

(b) The reaction is first order with respect to N_2O_5.

(c) The value of the rate constant k is given by the slope of the graph which is:

$$\text{slope} = \Delta(\text{rate})/\Delta(\text{concentration})$$

$$= \frac{2.5 \times 10^{-5}\ \text{mol dm}^{-3}\ \text{s}^{-1}}{2.5\ \text{mol dm}^{-3}} = \boxed{1.0 \times 10^{-5}\ \text{s}^{-1}}$$

Exercise 352

(a) First order with respect to $HgCl_2$. Second order with respect to $C_2O_4^{2-}$.

(b) The reaction is third order overall.

Exercise 353

(a) The order is first with respect to C_2H_4 and 1.5 with respect to I_2.

(b) The overall order of reaction is 2.5.

Exercise 354

Since the reaction is third order overall, D could not be correct:

rate = $k\,[P][Q]^3$

as this shows a reaction which is fourth order overall.

Note that as $[P]^0 = 1$, B could be written as rate = $k\,[Q]^3$

Exercise 355

rate = $k\,[S_2O_8^{2-}(aq)][I^-(aq)]$ $\therefore$ $k = \dfrac{\text{rate}}{[S_2O_8^{2-}(aq)][I^-(aq)]}$

Let rate = x mol dm^{-3} s^{-1}, $[S_2O_8^{2-}(aq)]$ = y mol dm^{-3} and $[I^-(aq)]$ = z mol dm^{-3}

$\therefore\ k = \dfrac{x \text{ mol dm}^{-3}\text{ s}^{-1}}{(y \text{ mol dm}^{-3})(z \text{ mol dm}^{-3})} = \left[\dfrac{x}{yz}\right]\dfrac{\text{mol dm}^{-3}\text{ s}^{-1}}{\text{mol}^2\text{ dm}^{-6}} = \left[\dfrac{x}{yz}\right]\text{ dm}^3\text{ mol}^{-1}\text{ s}^{-1}$

i.e. the unit is $\boxed{\text{dm}^3\text{ mol}^{-1}\text{ s}^{-1}}$

Exercise 356

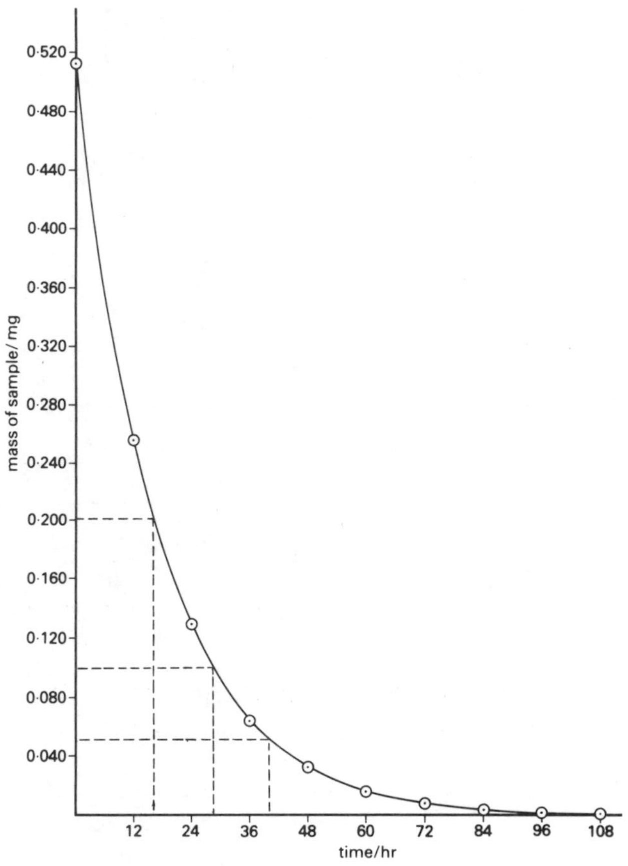

Exercise 356 (continued)

(a) (i) 28 hr - 16 hr = 12 hr

(ii) 40 hr - 28 hr = 12 hr

(b) Theoretically, the amount would never reach zero, as this is an exponential process.

(c) There are two ways of answering this question.

(i) The graph shows the mass of the sample decreasing by half every twelve hours, i.e. the half-life is 12 hours and 10 half-lives = 120 hours. Since activity is proportional to mass, the percentage of activity remaining after 10 half-lives

$$= \frac{\text{mass after 120 hours}}{\text{original mass}} \times 100$$

$$= \frac{0.0005 \text{ g}}{0.512 \text{ g}} \times 100 = \boxed{0.1\%}$$

(ii) Without reference to the graph, an answer is obtained more precisely: fraction left after 10 half-lives

$$= \tfrac{1}{2} \times \tfrac{1}{2} \times \tfrac{1}{2} \times \tfrac{1}{2} \times \tfrac{1}{2} \times \tfrac{1}{2} \times \tfrac{1}{2} \times \tfrac{1}{2} \times \tfrac{1}{2} \times \tfrac{1}{2}$$

$$= (\tfrac{1}{2})^{10} = 9.77 \times 10^{-4}$$

$$\% = \text{fraction} \times 100 = \boxed{0.0977\%}$$

Exercise 357

(a) The half-life of a reactant in a chemical reaction (or in radioactive decay) is the time taken for the concentration of a substance (or the amount, if solid) to fall to half its initial value.

(b) (i) The half-life of X is 10 mins.

(ii) First order.

(c)

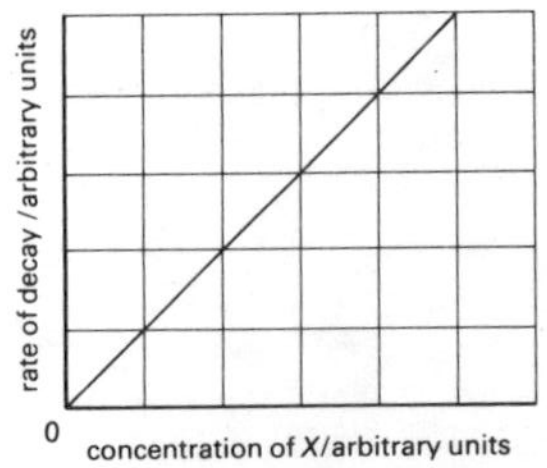

Exercise 358

(a) $^{14}_{7}N + ^{1}_{0}n \rightarrow ^{14}_{6}C + ^{1}_{1}H$

(b) The relative activity of the sample compared with the activity of new wood

$$= \frac{\text{activity of sample}}{\text{activity of new wood}} = \frac{7.5 \text{ counts min}^{-1}\text{ g}^{-1}}{15.0 \text{ counts min}^{-1}\text{ g}^{-1}} = 0.50$$

Since the sample has an activity of half that of new wood, one half-life has elapsed.

∴ year in which tomb was built = 1982 AD - 5730 = **3748 BC**

The method is not as accurate as this answer implies. It would be safer to say the tomb was built between 3700 BC and 3800 BC.

(c) The amount of carbon-14 in the atmosphere is assumed to be constant so that the % of carbon-14 in freshly-grown wood is also constant.

Exercise 359

Since the reaction is first order, the half-life is constant.

Half the benzenediazonium chloride decomposes in 40 mins giving 40 cm^3 gas

A further half decomposes in a further 40 mins giving 20 cm^3 gas

A further half decomposes in a further 40 mins giving 10 cm^3 gas

Adding these figures shows that 70 cm^3 of gas is evolved in **120 mins**

Exercise 360

(a)

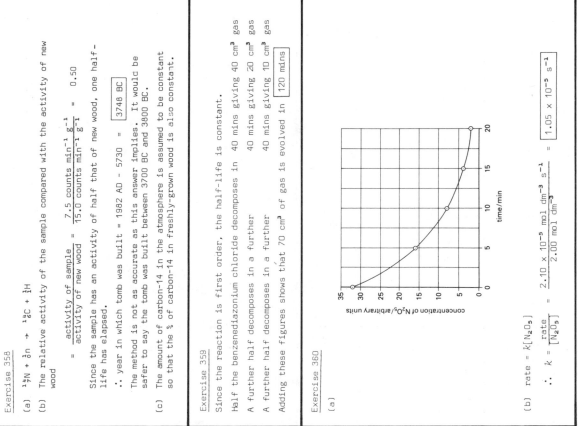

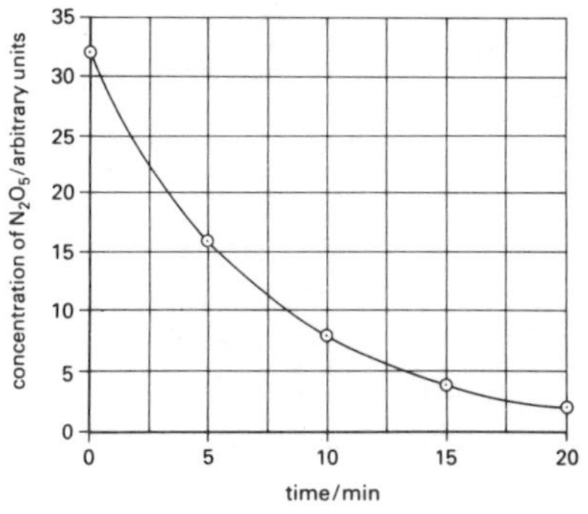

(b) rate = $k[N_2O_5]$

$$\therefore\ k = \frac{\text{rate}}{[N_2O_5]} = \frac{2.10 \times 10^{-5} \text{ mol dm}^{-3}\text{ s}^{-1}}{2.00 \text{ mol dm}^{-3}} = \boxed{1.05 \times 10^{-5}\text{ s}^{-1}}$$

Exercise 361

(a) (i) Consider Experiments 1, 4 and 5, in which initial [B] is constant.

Doubling [A] doubles the rate, and tripling [A] triples the rate.

∴ with respect to A, reaction is **first order**

(ii) Consider Experiments 1, 2 and 3, in which initial [A] is constant.

Doubling [B] multiplies rate by 4 (2^2)

Tripling [B] multiplies rate by 9 (3^2)

∴ with respect to B, reaction is **second order**

(b) rate = $k[A][B]^2$

(c) Rearranging the expression in (b) gives $k = \frac{\text{rate}}{[A][B]^2}$

$$\therefore\ k = \frac{2.0 \times 10^{-3} \text{ mol dm}^{-3}\text{ min}^{-1}}{(0.100 \text{ mol dm}^{-3})^3} = \boxed{2.00 \text{ mol}^{-2}\text{ dm}^6\text{ min}^{-1}}$$

Exercise 362

(a) Consider Experiments 1, 2 and 3, in which initial [NO] is constant.

Doubling [H_2] doubles the rate, and tripling [H_2] triples the rate.

∴ with respect to [H_2], the reaction is **first order** i.e. m = 1

(b) Consider Experiments 4, 5 and 6, in which initial [H_2] is constant.

Doubling [NO] multiplies rate by 4 (2^2)

Tripling [NO] multiplies rate by 9 (3^2)

∴ with respect to [NO], the reaction is **second order** i.e. n = 2

Exercise 363

(a) The hydrogencarbonate reacts with the acid catalyst and effectively stops the reaction to enable the titration to be done.

(b) Since the iodine concentration is proportional to the volume of thiosulphate solution used, the rate of change of iodine concentration is proportional to the slope of the graph. This is constant over the period illustrated.

(c) Rate $= \frac{\Delta V}{\Delta t} = \frac{(20 - 17)\ cm^3}{30\ min} = 0.10\ cm^3\ min^{-1}$

(d) The concentrations of propanone and acid are much greater than the initial concentration of iodine and can therefore be assumed to remain effectively constant. As the concentration of iodine falls, the rate of reaction remains constant; the reaction is therefore zero order with respect to iodine.

(e) Doubling the concentration should double the rate, i.e. $0.20\ cm^3\ min^{-1}$.

(f) & (g) There are two possible answers here. Either show the volume remaining constant over a period of 30 minutes, because the reaction, in the absence of a catalyst, proceeds too slowly to be measured; or show the volume decreasing slowly at first and then more rapidly over a longer period of time because there are a few hydrogen ions in any aqueous solution to catalyse the reaction and more are produced as the reaction proceeds.

The second alternative is theoretically sound but, in practice, the first corresponds to observation.

Exercise 364

(a) In each case the rate is obtained by substituting in the equation:

rate $= k\,[H_2(g)][I_2(g)]$

A. Rate $= 8.58 \times 10^{-5}\ mol^{-1}\ dm^3\ s^{-1} \times 0.010\ mol\ dm^{-3} \times 0.050\ mol\ dm^{-3}$

$= \boxed{4.29 \times 10^{-8}\ mol\ dm^{-3}\ s^{-1}}$

B. Rate $= 8.58 \times 10^{-5}\ mol^{-1}\ dm^3\ s^{-1} \times 0.020\ mol\ dm^{-3} \times 0.050\ mol\ dm^{-3}$

$= \boxed{8.58 \times 10^{-8}\ mol\ dm^{-3}\ s^{-1}}$

C. Rate $= 8.58 \times 10^{-5}\ mol^{-1}\ dm^3\ s^{-1} \times 0.020\ mol\ dm^{-3} \times 0.10\ mol\ dm^{-3}$

$= \boxed{1.72 \times 10^{-7}\ mol\ dm^{-3}\ s^{-1}}$

(b) Considering Experiments A and B, doubling the initial concentration of hydrogen at constant initial concentration of iodine doubles the rate of reaction.

Considering Experiments B and C, doubling the initial concentration of iodine at constant initial concentration of hydrogen doubles the rate of reaction.

Exercise 365

(a)

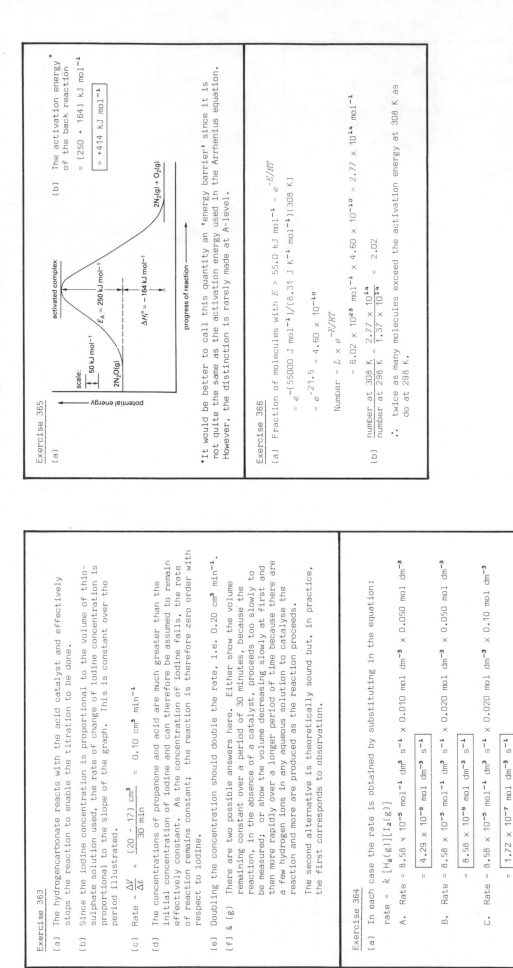

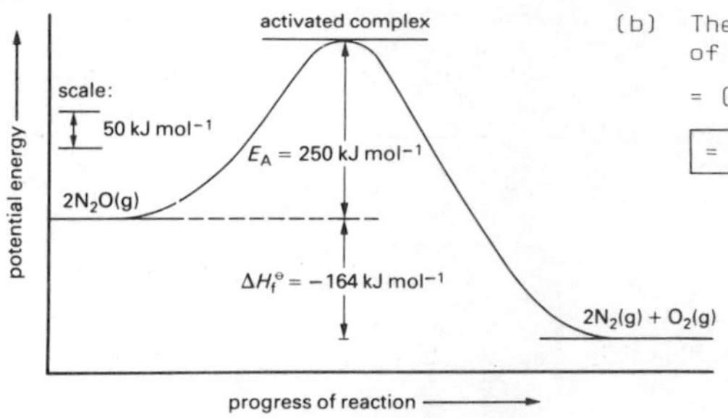

(b) The activation energy* of the back reaction

$= (250 + 164)\ kJ\ mol^{-1}$

$\boxed{= +414\ kJ\ mol^{-1}}$

*It would be better to call this quantity an 'energy barrier' since it is not quite the same as the activation energy used in the Arrhenius equation. However, the distinction is rarely made at A-level.

Exercise 366

(a) Fraction of molecules with $E > 55.0\ kJ\ mol^{-1} = e^{-E/RT}$

$= e^{-(55000\ J\ mol^{-1})/(8.31\ J\ K^{-1}\ mol^{-1})(308\ K)}$

$= e^{-21.5} = 4.60 \times 10^{-10}$

Number $= L \times e^{-E/RT}$

$= 6.02 \times 10^{23}\ mol^{-1} \times 4.60 \times 10^{-10} = 2.77 \times 10^{14}\ mol^{-1}$

(b) $\frac{\text{number at 308 K}}{\text{number at 298 K}} = \frac{2.77 \times 10^{14}}{1.37 \times 10^{14}} = 2.02$

$\therefore$ twice as many molecules exceed the activation energy at 308 K as do at 298 K.

Exercise 367

(a) The variables in the equation

$$\log_{10}k = \log_{10}A - \frac{E_a}{2.3\ RT}$$

are k and T.

(b) $\log_{10}k$ is analogous to y, m to $E_a/2.3R$, x to $1/T$ and $\log_{10}A$ to c.

(c) The slope, $-E_a/2.3R$, is always negative and $1/T$ is always positive. $\log_{10}k$ may be positive or negative according to the magnitude of k (it is usually negative, corresponding to small values of k). The plot may take either of the two forms shown, with (2) the more probable.

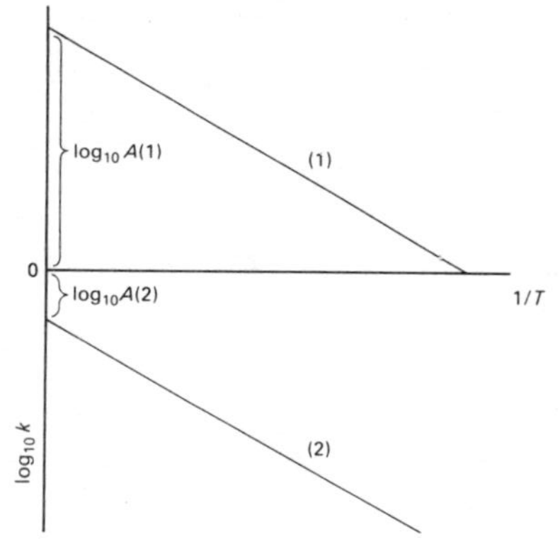

(d) The intercept on the vertical axis is equal to $\log_{10}A$. If the value of $\log_{10}A$ is obtained by reading the graph then A can be calculated. The slope of the straight line is equal to $-E_a/2.3\ R$. If the slope of the line is calculated from the graph then $E_a/2.3\ R$ can also be calculated, since R is a known constant.

Exercise 368

(a)

T/K	$1/T$	k/dm³ mol⁻¹ s⁻¹	log$k$
633	1.58×10^{-3}	1.78×10^{-5}	-4.75
666	1.50×10^{-3}	1.07×10^{-4}	-3.97
697	1.43×10^{-3}	5.01×10^{-4}	-3.30
715	1.40×10^{-3}	1.05×10^{-3}	-2.98
781	1.28×10^{-3}	1.51×10^{-2}	-1.82

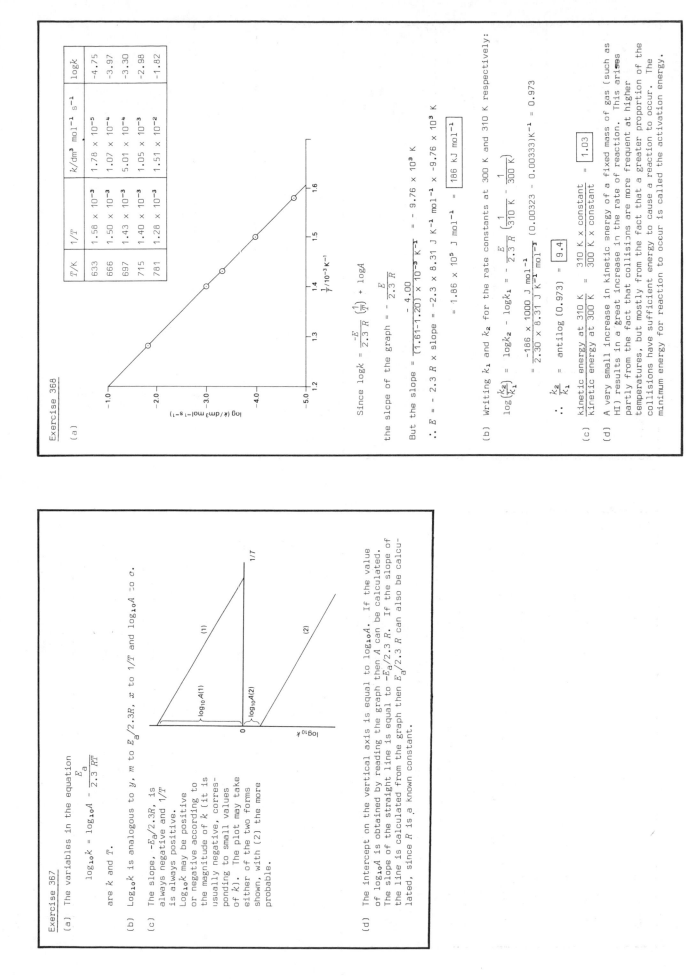

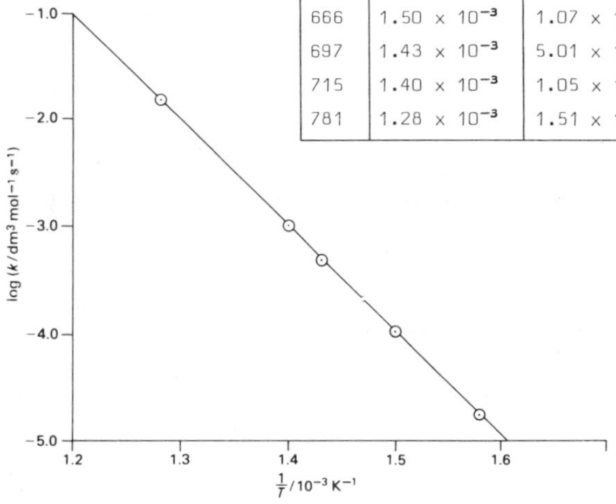

Since $\log k = \frac{-E}{2.3\ R}\left(\frac{1}{T}\right) + \log A$

the slope of the graph $= -\frac{E}{2.3\ R}$

But the slope $= \frac{-4.00}{(1.61-1.20) \times 10^{-3}\ K^{-1}} = -9.76 \times 10^3$ K

$\therefore E = -2.3\ R \times \text{slope} = -2.3 \times 8.31\ J\ K^{-1}\ mol^{-1} \times -9.76 \times 10^3\ K$

$= 1.86 \times 10^5\ J\ mol^{-1} = \boxed{186\ kJ\ mol^{-1}}$

(b) Writing k_1 and k_2 for the rate constants at 300 K and 310 K respectively:

$$\log\left(\frac{k_2}{k_1}\right) = \log k_2 - \log k_1 = -\frac{E}{2.3\ R}\left(\frac{1}{310\ K} - \frac{1}{300\ K}\right)$$

$$= \frac{-186 \times 1000\ J\ mol^{-1}}{2.30 \times 8.31\ J\ K^{-1}\ mol^{-1}}(0.00323 - 0.00333)K^{-1} = 0.973$$

$$\therefore \frac{k_2}{k_1} = \text{antilog}\ (0.973) = \boxed{9.4}$$

(c) $\frac{\text{kinetic energy at 310 K}}{\text{kinetic energy at 300 K}} = \frac{310\ K \times \text{constant}}{300\ K \times \text{constant}} = \boxed{1.03}$

(d) A very small increase in kinetic energy of a fixed mass of gas (such as HI) results in a great increase in the rate of reaction. This arises partly from the fact that collisions are more frequent at higher temperatures, but mostly from the fact that a greater proportion of the collisions have sufficient energy to cause a reaction to occur. The minimum energy for reaction to occur is called the activation energy.

Exercise 369

Rate constants (k) are obtained by plotting loss in mass against time. Activation energy is obtained by plotting $\log k$ against $1/T$.

Experiment 1 at 0.5 °C			Experiment 2 at 13.5 °C			Experiment 3 at 25 °C		
Time /min	Mass /mg	Loss /mg	Time /min	Mass /mg	Loss /mg	Time /min	Mass /mg	Loss /mg
0	130	0	0	130	0	0	130	0
20	120	10	10	115	15	4	116	14
60	98	32	15	106	24	8	103	27
80	86	44	30	86	44	12	87	43

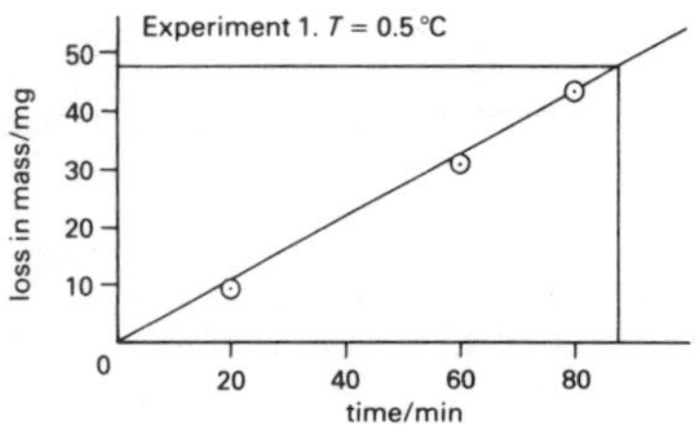

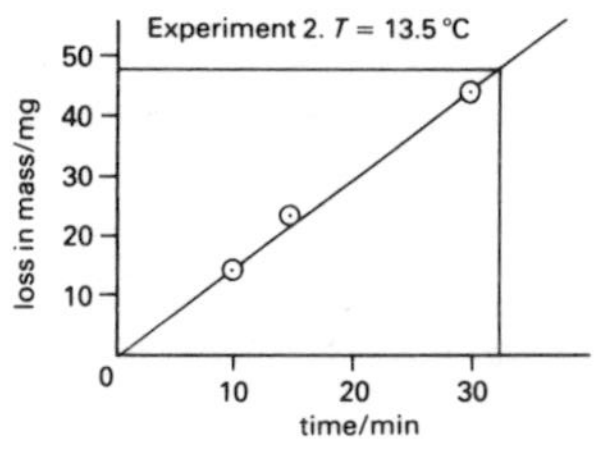

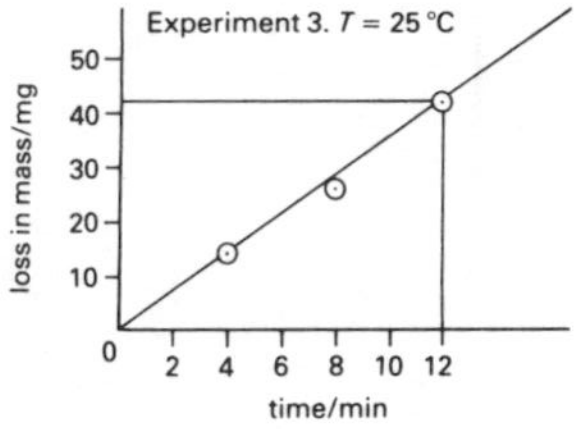

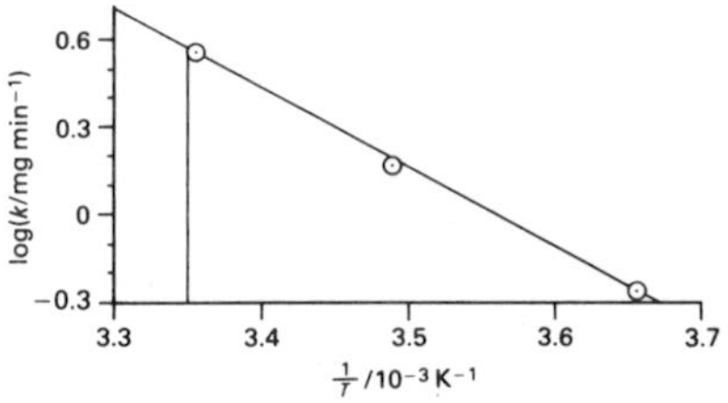

In each experiment, the straight line graph shows that the rate of loss of mass is constant. The rate constant, k, is the slope of the line.

1. $k = \dfrac{48.0 \text{ mg}}{88.0 \text{ min}} = \boxed{0.55 \text{ mg min}^{-1}}$

2. $k = \dfrac{48.0 \text{ mg}}{32.5 \text{ min}} = \boxed{1.48 \text{ mg min}^{-1}}$

3. $k = \dfrac{43.0 \text{ mg}}{12.0 \text{ min}} = \boxed{3.58 \text{ mg min}^{-1}}$

$k = B \times 10^{-E/2.3\,RT}$ (B is a constant) $\therefore \log_{10} k = \log_{10} B - \left(\dfrac{E}{2.3\,R} \times \dfrac{1}{T}\right)$

Slope of graph $= \dfrac{(-0.30 - 0.58) \text{mg min}^{-1}}{(3.67 - 3.35) \times 10^{-3} \text{ K}^{-1}} = -2.75 \times 10^3 \text{ K} = \dfrac{-E}{2.30\,R}$

$\therefore E = 2.30 \times 8.31 \text{ J K}^{-1} \text{ mol}^{-1} \times 2.75 \times 10^3 \text{ K} = \boxed{52.6 \text{ kJ mol}^{-1}}$

If the cobalt were not rotated, the rates of reaction (and hence the rate constants) would be slightly smaller, due to the time taken for $S_2O_8^{2-}$ ions to diffuse towards the metal surface. Activation energy would be unaffected.

Exercise 370

(a) E_a for the forward reaction (catalysed) is 184 kJ mol^{-1}.
E_a for the back reaction (uncatalysed) is 236 kJ mol^{-1}.

(b) E_a for the forward reaction (catalysed) is 59 kJ mol^{-1}.
E_a for the back reaction (catalysed) is 111 kJ mol^{-1}.

(c) The enthalpy change is ± 52 kJ mol^{-1}.

Exercise 371

(a) Initial rate = slope at $t = 0 \approx 6.3 \times 10^{-7}$ mol dm^{-3} s^{-1}

Rate at $t = 50$ s $\approx 4.5 \times 10^{-6}$ mol dm^{-3} s^{-1}

Thus, the rate of reaction increases 7-fold over the first 50 seconds. This increase in the rate indicates that one of the products of the reaction is acting as a catalyst. The catalyst is likely to be the transition metal ions, Mn^{2+}(aq).

(b) Two effects are operating during this reaction: the production of catalyst and the decrease in concentration of reactants. At the start of the reaction the first effect dominates; catalyst is being produced and speeds up the rate of the reaction as long as there is an adequate supply of reactant. Towards the end of the reaction, however, the second effect becomes more important. The concentration of reactants has fallen to a low level and so the rate of reaction decreases, even though there is an adequate supply of catalyst.

Exercise 372

(a) $-\dfrac{d[PH_3(g)]}{dt} = 4 \times \dfrac{d[P_4(g)]}{dt} = \dfrac{2}{3} \times \dfrac{d[H_2(g)]}{dt}$

(b) (i) $\dfrac{d[H_2(g)]}{dt} = -\dfrac{3}{2}\,\dfrac{d[PH_3(g)]}{dt} = -\dfrac{3}{2} \times (-2.4 \times 10^{-3} \text{ mol dm}^{-3} \text{ s}^{-1})$
$= \boxed{3.6 \times 10^{-3} \text{ mol dm}^{-3} \text{ s}^{-1}}$

(ii) $\dfrac{d[P_4(g)]}{dt} = -\dfrac{1}{4}\,\dfrac{d[PH_3(g)]}{dt} = -\dfrac{1}{4} \times (-2.4 \times 10^{-3} \text{ mol dm}^{-3} \text{ s}^{-1})$
$= \boxed{6.0 \times 10^{-4} \text{ mol dm}^{-3} \text{ s}^{-1}}$

Exercise 373

Let c = concentration of 2,4,6-trinitrobenzoic acid.
If the reaction is first order,

$$\log c = -kt/2.30 + \log c_0$$

t/min	0	18	31	55	79	157
$c/10^{-4}$ mol dm^{-3}	2.77	2.32	2.05	1.59	1.26	0.58
log (c/mol dm^{-3})	-3.56	-3.63	-3.69	-3.80	-3.90	-4.24

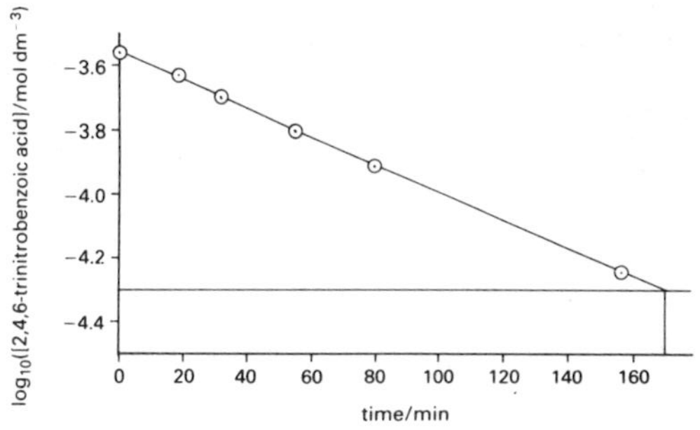

Since the graph is a straight line, the reaction is first order. The slope of the graph = $-k/2.30$

$$\therefore \quad k = -2.30 \times \frac{-4.30 - (-3.56)}{(165-0)\text{ min}} = \boxed{1.03 \times 10^{-2}\ \text{min}^{-1}}$$

Exercise 374

Let c = concentration of dinitrogen pentoxide

Then $\log c = -kt/2.30 + \log c_0$

t/s	0	250	500	750	1000	1500	2000	2500
c/mol dm^{-3}	2.33	1.95	1.68	1.42	1.25	0.95	0.70	0.50
log c	0.367	0.290	0.225	0.152	0.097	-0.022	-0.155	-0.301

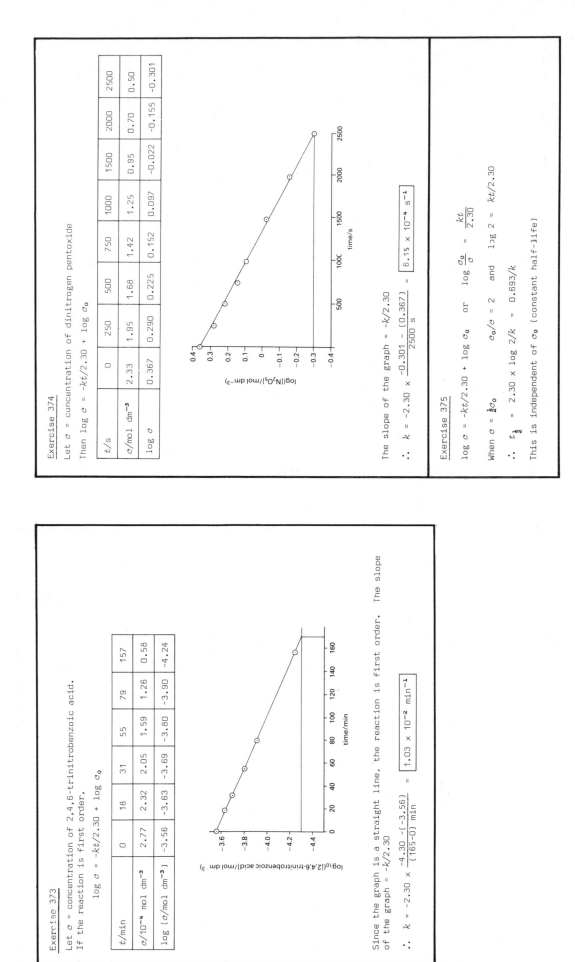

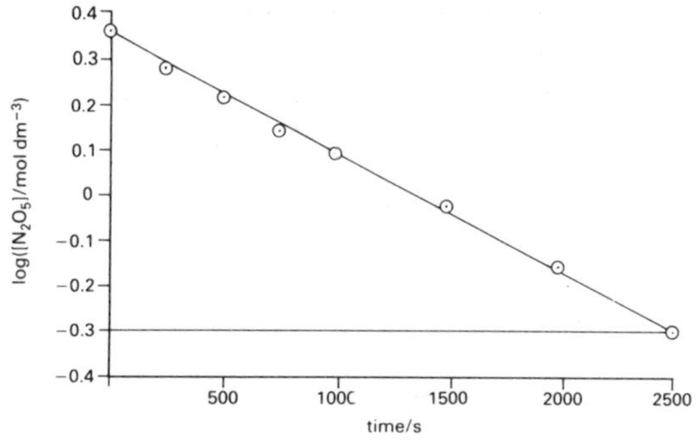

The slope of the graph = $-k/2.30$

$$\therefore \quad k = -2.30 \times \frac{-0.301 - (0.367)}{2500\text{ s}} = \boxed{6.15 \times 10^{-4}\ \text{s}^{-1}}$$

Exercise 375

$\log c = -kt/2.30 + \log c_0$ or $\log \frac{c_0}{c} = \frac{kt}{2.30}$

When $c = \frac{1}{2}c_0$ $\quad c_0/c = 2$ and $\log 2 = kt/2.30$

$\therefore \quad t_{\frac{1}{2}} = 2.30 \times \log 2/k = 0.693/k$

This is independent of c_0 (constant half-life)

Exercise 376

Let c = [OH^-(aq)]

Then $t = 1/kc - 1/c_0$

t/min	3	5	7	10	15	21	25
$c/10^{-3}$ mol dm^{-3}	7.40	6.34	5.50	4.64	3.63	2.88	2.54
$(1/c)$/dm^3 mol^{-1}	135	158	182	216	275	347	394

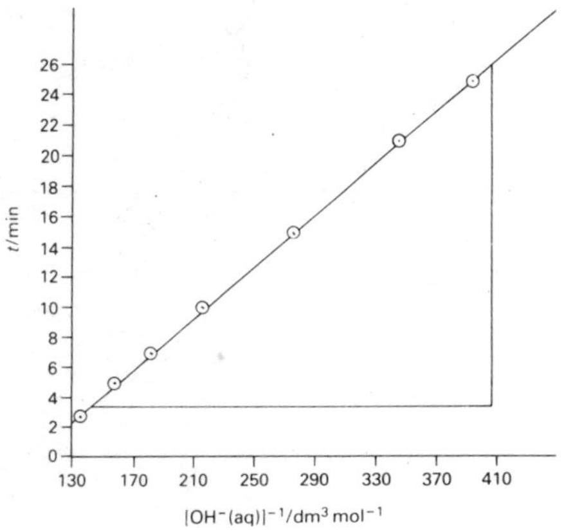

The slope of the graph = $1/k$

$$\therefore\ k = 1/\text{slope} = \frac{(406 - 140)\ \text{mol}^{-1}\ \text{dm}^3}{(26.0 - 3.4)\ \text{min}} = \boxed{11.8\ \text{mol}^{-1}\ \text{dm}^3\ \text{min}^{-1}}$$

Exercise 377 Zero order. (Rate remains constant.)

Exercise 378 dm^3 mol^{-1} s^{-1}

Exercise 379 32

Exercise 380

(a) Curve A: 1.82 unit of concentration min^{-1}

Curve B: 1.25 unit of concentration min^{-1}

Curve C: 0.625 unit of concentration min^{-1}

(b) (i) Iodide ions. (ii) First order.

(c) (i) 0.75 min. (ii) 1.5 min. First order.

Exercise 381

(b) (i) x = 1 (ii) ~ 5.0 × 10^{-4} s^{-1}

(c) 2772 s

Exercise 382 (c) 119 kJ mol^{-1}

Exercise 383 (c) (i) 1st order (constant half-life)

Exercise 384 (b) 2 (d) | 2.04 × 10^{-3} dm^3 mol^{-1} s^{-1}

Exercise 385 16 800 years

Exercise 386

(a) 3 (b) 4 (c) 2 (d) 3 (e) 3 (f) 5 (g) 4 (h) 3 or 4

(i) 3, 4, 5 or 6

Exercise 387 (a) 208 g (b) 0.649 dm^3

Exercise 388 (a) 3.4 m (b) 76 cm^3

Exercise 389 (a) 3.0 × 10^{-3} mol (b) 67.8 g

Exercise 390 (a) 0.48 mol (b) 2 g cm^{-3}